QUANTITATIVE ANALYSES
in Wildlife Science

Wildlife Management and Conservation
Paul R. Krausman, Series Editor

QUANTITATIVE ANALYSES
in Wildlife Science

EDITED BY

LEONARD A. BRENNAN, ANDREW N. TRI,
AND BRUCE G. MARCOT

Published in association with *THE WILDLIFE SOCIETY*

JOHNS HOPKINS UNIVERSITY PRESS | BALTIMORE

Johns Hopkins University Press
2715 North Charles Street
Baltimore, Maryland 21218-4363
www.press.jhu.edu

Library of Congress Cataloging-in-Publication Data

Names: Brennan, Leonard A. (Leonard Alfred), editor.
Title: Quantitative analyses in wildlife science / edited by
    Leonard A. Brennan, Andrew N. Tri, and Bruce G. Marcot.
Description: Baltimore : Johns Hopkins University Press,
    2019. | Series: Wildlife management and conservation |
    Includes bibliographical references and index.
Identifiers: LCCN 2018051292 | ISBN 9781421431079
    (hardcover : alk. paper) | ISBN 1421431076 (hardcover :
    alk. paper) | ISBN 9781421431086 (electronic) |
    ISBN 1421431084 (electronic)
Subjects: LCSH: Population biology—Mathematical models. |
    Population biology—Data processing.
Classification: LCC QH352 .Q36 2019 | DDC 577.8/8—dc23
LC record available at https://lccn.loc.gov/2018051292

A catalog record for this book is available from the British
Library.

Special discounts are available for bulk purchases of this book.
For more information, please contact Special Sales at
410-516-6936 or specialsales@press.jhu.edu.

Johns Hopkins University Press uses environmentally
friendly book materials, including recycled text paper that is
composed of at least 30 percent post-consumer waste,
whenever possible.

Contents

Contributors

William M. Block
Rocky Mountain Research Station
USDA Forest Service
Flagstaff, Arizona, USA

Leonard A. Brennan
Caesar Kleberg Wildlife Research
    Institute and
Department of Rangeland and
    Wildlife Sciences
Texas A&M University–Kingsville
Kingsville, Texas, USA

Stephen T. Buckland
Centre for Research into
    Ecological and Environmental
    Modelling
University of St. Andrews
St Andrews, Fife, UK

Christopher J. Chizinski
School of Natural Resources
University of Nebraska–Lincoln
Lincoln, Nebraska, USA

Evan G. Cooch
Department of Natural
    Resources
Cornell University
Ithaca, New York, USA

Raymond J. Davis
Northwest Forest Plan Interagency
    Monitoring Program
USDA Forest Service–Pacific
    Northwest Region
Corvallis, Oregon, USA

Stephen J. DeMaso
Gulf Coast Joint Venture
US Fish and Wildlife Service
Lafayette, Louisiana, USA

Randy W. DeYoung
Caesar Kleberg Wildlife Research
    Institute and
Department of Rangeland and
    Wildlife Sciences
Texas A&M University–Kingsville
Kingsville, Texas, USA

Jane Elith
School of BioSciences
University of Melbourne
Parkville, Victoria, Australia

Joseph J. Fontaine
U.S. Geological Survey, Nebraska
    Cooperative Fish and Wildlife
    Research Unit
School of Natural Resources

University of
    Nebraska–Lincoln
Lincoln, Nebraska, USA

R. J. Gutiérrez
Department of Fisheries,
    Wildlife and Conservation
    Biology
University of Minnesota
St. Paul, Minnesota, USA

Julie A. Heinrichs
Natural Resource Ecology
    Laboratory
Colorado State University
Fort Collins, Colorado, USA

Mevin B. Hooten
U.S. Geological Survey
Colorado Cooperative Fish and
    Wildlife Research Unit
Departments of Fish, Wildlife,
    and Conservation Biology
    and Statistics
Colorado State University
Fort Collins, Colorado, USA

Julianna M. A. Jenkins
Pacific Northwest Research
    Station

USDA Forest Service
Corvallis, Oregon, USA

Zachry S. Ladin
Department of Entomology and
    Wildlife Ecology
University of Delaware
Newark, Delaware, USA.

Damon B. Lesmeister
Pacific Northwest Research Station
USDA Forest Service
Corvallis, Oregon, USA

Daniel W. Linden
Greater Atlantic Regional Fisheries
    Office
NOAA National Marine Fisheries
    Service
Gloucester, Massachusetts, USA

Jeffrey J. Lusk
Wildlife Division
Nebraska Game and Parks
    Commission
Lincoln, Nebraska, USA

Bruce G. Marcot
Pacific Northwest Research Station
USDA Forest Service
Portland, Oregon, USA

David L. Miller
Centre for Research into
    Ecological and Environmental
    Modelling
University of St. Andrews
St Andrews, Fife, UK

Michael L. Morrison
Department of Wildlife &
    Fisheries Sciences
Texas A&M University
College Station,
    Texas, USA

Eric Rexstad
Centre for Research into
    Ecological and Environmental
    Modelling
University of St. Andrews
St Andrews, Fife, UK

Jamie S. Sanderlin
Rocky Mountain Research
    Station
USDA Forest Service
Flagstaff, Arizona, USA

Joseph P. Sands
Migratory Birds and Habitat
    Program
U.S. Fish and Wildlife Service
Portland, Oregon, USA

Erica F. Stuber
Max Planck–Yale Center for
    Biodiversity, Movement and
    Global Change
Yale University
New Haven, CT

Chris Sutherland
Department of Environmental
    Conservation
University of Massachusetts
Amherst, Massachusetts, USA

Andrew N. Tri
Caesar Kleberg Wildlife
    Research Institute and
Department of Rangeland and
    Wildlife Sciences
Texas A&M University–
    Kingsville
Kingsville, Texas, USA

David B. Wester
Caesar Kleberg Wildlife
    Research Institute and
Department of Rangeland
    and Wildlife Sciences
Texas A&M University–
    Kingsville
Kingsville, Texas, USA

Gary C. White
Department of Fish, Wildlife,
    and Conservation Biology
Colorado State University
Fort Collins, Colorado, USA

Christopher K. Williams
Department of Entomology
    and Wildlife Ecology
University of Delaware
Newark, Delaware, USA

Damon L. Williford
Texas Parks and Wildlife
    Department
Perry R. Bass Marine
    Fisheries Research Station
    and Hatchery
Palacios, Texas, USA

Foreword

R. J. Gutiérrez

Aldo Leopold wrote *Game Management* in 1933, and with it he introduced the nascent profession of wildlife management as a discipline that would be shaped by quantitative analysis of field data. He also recognized at the time that wildlife management was as much "art" as it was science. But his vision was the stuff of dreams because the tools wildlife professionals needed for routine quantitative assessments of data were decades away. Lacking such tools, the wildlife profession quickly became an amalgam of specializations wherein biologists managed habitats and individual species primarily through trial-and-error learning.

The initial specializing emphasis in wildlife management was on exploited species, mainly because the overexploitation of wildlife in the past required wildlife professionals to focus on species recovery and legal protection. Yet, the philosophical foundation of the profession has always been much broader than game species—a fact often misunderstood by many ecologists and other conservationists outside the wildlife profession. To wit, the founding policy statement published in the first issue of the *Journal of Wildlife Management* clearly showed that "wildlife management" as envisioned by the founders of our emerging profession was inclusive of all wildlife: "Wildlife management is not restricted to game management, though game management is recognized as an important branch of wildlife management. It embraces the practical ecology of *all vertebrates and*

their plant and animal associates" (*Journal of Wildlife Management* 1973, volume 1/2:1; emphasis mine). These visionary founders of our profession codified the encompassing nature of *and* responsibility for managing all wildlife resources, thereby setting our profession on a course that naturally led to increasing sophistication in quantitative analysis of wildlife problems. This disciplinary evolution occurred for many reasons, but it was clear that "art" as a means for achieving wildlife management was not reproducible. Therefore, management actions predicated on art (i.e., expert opinion, although we see in this book that it still has value) were easily criticized by the public because they were not objectively defensible. Over time public advocacy groups have coalesced around their special interests, which has increased the need for wildlife managers and scientists to be confident (i.e., have a low scientific uncertainty) that their recommendations could pass muster of both public and legal evaluation. The dilemma of defensibility was also recognized by the founders of wildlife management: "One of the glaring faults connected with wildlife administration in this country has been the propagandizing and misrepresentation of plans as if they were things already accomplished" (*Journal of Wildlife Management* 1937, volume 1/2:2). Because the underpinnings of most plans should have been scientifically based, they wanted to move wildlife management into the realm of objectivity as envisioned by Leopold and, therefore, make management strategies defensible. So the evolution of quantitative approaches in wildlife management was a consequence of its founding principles because they increased defensibility of management strategies. It was not a random or serendipitous event.

Leopold and the other founders of wildlife management not only established a philosophical challenge to require high-quality science as requisite to achieving the goals of wildlife conservation, but also provided an initial framework that challenged our profession to adopt quantitative analysis of ecological data as the basis for wildlife management. Now the conservation stakes have been raised even higher in an era when the public, courts, and politicians scrutinize almost all our work. In meeting these challenges, wildlife professionals have become world leaders in analysis of wildlife population dynamics and habitat selection in terms of both application and development of quantitative tools such as statistical methods and analytical software.

Ironically, the majority of wildlife professionals enter the field because of their interest in wildlife conservation, not their interest in statistics and mathematics. With the ever-rising dependence of the discipline on quantitative approaches, there has been tension between field biologists and statisticians and between professionals and the public, as well as among biologists about issues such as their inability to meet assumptions of statistical approaches owing to the nature of real-world data, the technical understanding gap between professionals and the public about what constitutes meaningful insights about wildlife populations, and how best to analyze data. These conflicts have resulted in a familiar return to specialization in various aspects of statistical analysis.

The professional bar set by our founders was high but prophetically correct. They recognized the problems of politicization of wildlife research and management, so they established a goal for wildlife research to be able to withstand such politicization, although they likely did not envision the sometimes withering scrutiny that wildlife research and researchers sometimes incur today—or even that science would come under broad attack. Yet, modern wildlife professionals have realized that the only way to meet a high standard of research as well as pass scrutiny by a receptive public is through the application of rigorous research designs, robust quantitative analysis of data, and proper inference of results. Concomitantly, wildlife professionals have realized they need to maintain transparency in their wildlife research—what they did, when they did it, where they did it, how they did it. But the advances in modern analysis have been so rapid and occurred on so many fronts, that field biologists (even dedicated researchers) have often fallen quickly behind in their understanding of these advances, which has made it difficult for professionals to remain accessible to the public and to each other. Thankfully, wildlife biologists have also been among the leaders in dissemination of information about quantitative analysis, which enhances accessibility. One of the ways they have done this has been through publication of synthetic books where experts have been gathered to present the "state of the art" about a subject. *Quantitative Analysis in Wildlife Science* is such a book. But it is masterful not just in its amalgamation of topics and experts who contributed chapters but in its selection of topics and the background of its authors. The authors are researchers of applied problems, and their topics are those central to the wildlife management enterprise. Even though the authors have strong analytical expertise, I felt as if they were speaking to me as a field biologist.

Any nonfiction book is an entry to knowledge—some are better than others at conveying that knowledge. As I read this book, I found myself wistfully thinking, "I wish I had this book when I was a graduate student both to aid my dissertation research and as a review guide for my doctoral exams." Reading these chapters was an energizing exercise for it allowed me to review methods with which I was familiar and those with which I was not.

Every wildlife biologist and most wildlife administrators realize that Aldo Leopold's vision to strengthen our profession through quantitative rigor was correct; we absolutely needed this to achieve defensibility of our recommendations by the public in general and the courts specifically. But most field biologists, like me, as well as most managers, are not broadly versed in quantitative methods. Even if we have some understanding of quantitative methods, we surely do not understand all of them or we have forgotten the details of those we have not used in a long time. Therefore, this book provides both an entry and a refresher to a variety of quantitative methods that are extremely valuable to all of us—wildlife students, field biologists, researchers, and managers. It does so in a way that is accessible, comprehensive, and interesting.

Preface

This is a book about some approaches to analyzing data collected in the course of conducting research in wildlife ecology and science. No book can serve as a be-all, end-all reference for such as topic, and we make no pretenses for such a thing here. Rather, our goal was to produce a volume that can give graduate students and other wildlife researchers an entree into some of the more widely used approaches to data analyses today. Space limitations forced us to make hard decisions about which topics to include and which ones to leave out. As we note in the summary chapter, the topics we excluded could clearly be the backbone of a companion volume to this book sometime in the near future. At the same time, we also included some topics such as causal modeling, machine-learning, and boosted regression trees, which are not widely used by wildlife scientists today.

At the outset, we encouraged chapter authors to inject philosophical perspectives along with their technical expertise and guidance about a particular analytical approach. Often analysis of a data set can be as much an art as a science. Meeting the assumptions of a particular analytical approach—or dealing with what happens if certain assumptions are not met—can be a particularly vexing challenge that is too often overlooked during the course of conducting quantitative analyses. Such philosophical perspectives are too often absent from textbooks on quantitative analyses of ecological data. As editors, we hope the readers of this book find both the quantitative approaches and the philosophical perspectives useful in their research endeavors.

Acknowledgments

LAB: Support from the C. C. Charlie Winn Endowed Chair in the Richard M. Kleberg Jr. Center for Quail Research provided the space and time to work on this project. I appreciate the hard work of all the authors who wrote the chapters in this book, and I appreciate their patience as we worked through the review, editorial, and production processes. Special thanks to Rocky Gutiérrez for writing the foreword and to Vince Burke for supporting this book project from the beginning. The Caesar Kleberg Wildlife Research Institute provided a generous subsidy to support the publication costs of this book.

ANT: Thank you to P. Turk, for helping instill a passion for applied quantitative ecology. I thank the authors who created wonderful and illuminating content that will provide invaluable assistance to readers of this text. I thank my colleagues, and co-editors for their insight, critique, and editorial assistance in the preparation of this text. I am thankful to my mentor, L. Brennan, for spearheading this initiative and bringing a newly minted quantitative guy along for the ride. Support from the Caesar Kleberg Wildlife Research Institute, Texas A & M University–Kingsville, and the East Foundation provided me with the time to embark on this project. I also thank J. and P. Tri for their patience; I'm sorry I didn't get you out after a few more grouse or pheasants, but your kind support was helpful to me.

BGM: Mastering quantitative analyses in wildlife science is an ongoing, life-long pursuit, as methods and tools are constantly evolving. The best learning for contributions to this volume has come from a wide array of my colleagues, project team members, mentors, and students alike, too many to list. I dedicate my contributions here to Dr. E. Charles Meslow to whom I owe the academic career that set the sails on my journey, and I thank my co-editors for their patience and support. Further support for my time comes from the U.S. Forest Service, Pacific Northwest Research Station.

QUANTITATIVE ANALYSES
in Wildlife Science

1 Introduction

LEONARD A. BRENNAN,
ANDREW N. TRI, AND
BRUCE G. MARCOT

Although we often hear that data speak for themselves, their voices can be soft and sly.
—Mosteller et al. (1983:234)

During the past half century or so, wildlife science has become increasingly—some might say extremely—quantitative. This edited volume addresses many of the most widely used contemporary approaches to the analyses of wildlife populations and habitats. By "populations" we mean various aggregations of a species or groups of species of wild vertebrates. By "habitats" we mean the requisite spaces and resources—such as food and vegetation, among other things—that are required to sustain populations of a particular species of wild vertebrate. To a wildlife scientist who has not been formally trained as a biometrician, the wide array of available quantitative techniques can be bewildering, overwhelming, or both. Thus, we sought to create a book that can serve graduate students and early-career wildlife professionals, as well as established researchers, by providing an entry-level portal to many of the analytical techniques used today in wildlife science. We hope that we have accomplished this aim.

No one person can be a master of all trades when it comes to biometric assessment of wildlife populations. Furthermore, the topic of each chapter in this book has been addressed in depth in several—and in most cases many—book-length monographs. Under the circumstances, we think that it is important to provide conceptual and quantitative overviews of modern analytical methods that can be applied in wildlife research and conservation. Although the chapters in this book emphasize analyses of empirical data, we also seek to provide substantial coverage of various modeling techniques that use both real and simulated data. We want to inspire readers to delve further into their selected topics by consulting the literature, software, and programming codes that make up the foundations of the chapters in this book. Consultation of colleagues who are experts on particular analytical topics is also, of course, highly valuable.

This book covers quantitative and analytical approaches to the analyses of data sets that are typically encountered by wildlife scientists and managers. Our editorial philosophy was to ask the chapter authors to stress the conceptual and numerical basics of the particular analytic approach covered. We also asked that they consider the structure of the data, and the structure of variation in those data, in the overall context of meeting assumptions and drawing the strongest possible inferences from the analytic tool being used. Each chapter also discusses merits and limitations of the particular analytic approach(es) covered.

This book is divided into five parts: I. General Statistical Methods, II. Estimation of Abundance and Demographic Parameters, III. Dynamic Modeling of

Processes, IV. Analyses of Spatially Based Data on Animals and Resources, and V. Numerical Methods.

General Statistical Methods

General statistical methods, especially those considered to be "frequentist methods," or classical statistical methods such as t-tests and other analytical techniques taught in introductory statistics classes, have been widely used by wildlife scientists for several generations. Misuse and abuse of frequentist methods was—and in some cases continues to be—rampant. The influential paper on the statistical insignificance of significance testing by Johnson (1999) was among the first efforts by wildlife biometricians to point out that p-values do not correspond to the strength of inferences that can be derived from a data set. Nevertheless, it is safe to say that all university wildlife programs in North America require students to master applications of frequentist statistical methods such as t-tests, ANOVAs (analysis of variance tests), and related analyses. Thus, we assume that a graduate student or early-career wildlife professional who picks up this book has been well versed in these general techniques. As discussed by Vogt et al. (2014), selecting the appropriate use of quantitative, qualitative, or mixed methods of analyses is also important.

With these considerations in mind, we begin the section on general statistical methods with coverage of some important aspects of regression, which is one of the most flexible and powerful analytical tools available. While linear regression is widely used for a variety of applications, we felt it was important to cover regression from the standpoint of both linear and nonlinear as well as parametric and nonparametric perspectives. Although few data from the natural world fit in a linear, parametric context, this is a launching point for branching into nonlinear and nonparametric approaches, which are presented in Chapter 2.

Classical multiple regression, in which there are two or more independent variables that predict the nature of a dependent variable, forms the conceptual and analytical bases of many multivariate models and analyses covered in Chapter 3. The 1980s and 1990s saw a massive burst in applications of multivariate statistical analyses in wildlife and ecology. The advent of preprogrammed statistical packages such as SAS and SPSS provided user-friendly access to complex computer coding, and the direct connection between matrices of data on vegetation composition and structure in relation to the presence-absence of animals resulted in a wave of wildlife-habitat relationships data in the peer-reviewed literature (Verner et al. 1986).

Although the widespread applications of multivariate analyses seemed to peak in the 1980s and 1990s, they remain a valuable family of analytical tools for gaining insights from combined univariate data sets. A classic application involves morphological data on metrics of various body-size variables (such as culmen, tarsus, and wing-chord lengths of birds), which can be used to provide high probabilities of correctly predicting whether a specimen is a male or female when such variables are combined in a multivariate model rather than analyzed separately in a univariate context. There are more than 100 examples of such applications in the ornithological literature alone. Comparable examples also are present in the peer-reviewed literature on wildlife-habitat relationships, which shows how multivariate techniques and models can add value and perspective to univariate analyses. Chapter 3 also covers the major statistical assumptions of multivariate modeling (e.g., multicollinearity) and how to test for or detect whether data sets meet these assumptions.

The landmark books by Burnham and Anderson (1998, 2002) have inspired a tsunami of interest in and application of information-theoretic analyses of data in wildlife science. That this occurred in the wake of skepticism about frequentist techniques is more than a coincidence. As wildlife scientists became aware of the shortcomings of frequentist statistics, information-theoretic, or IT, analyses quickly became a favored alternative that exploded in the wildlife literature (Guthery et al. 2005).

During the past two decades, IT analyses, especially those based on Akiake's Information Criterion, or AIC, have become widespread in wildlife science. There are numerous other information-theoretic criteria that are largely unknown, underappreciated, or both. Depending on their use, IT methods can provide researchers a means to compare multiple competing hypotheses and provide potentially strong inferences. One of the strengths of Chapter 4 on comparing ecological models is the comparison of AIC and the Bayesian Information Criterion, or BIC, which continues to be a topic of keen interest among biometricians and therefore should be of interest to the rest of us.

Demographic Estimation: Survival and Population Change

An understanding of vital rates and how to analyze and interpret them is essential to the management and conservation of virtually any vertebrate population. This field has fundamentally changed over past decades. From principles of exponential growth to complex survival and density estimation techniques, demographic estimation remains at the core of many wildlife studies. Chapter 5 addresses estimation approaches that are rooted in mark-recapture techniques. Although this is a topic that could certainly fill several volumes, this chapter is organized in such a way as to directly lead the reader to more in-depth materials.

Nearly four decades ago, the landmark monograph by Burnham et al. (1980) and the widespread availability of the TRANSECT program made distance-sampling techniques widely available to wildlife researchers. Since that time, the subsequent DISTANCE program has evolved to present a wide and flexible array of distance-sampling analyses that can be used across a variety of taxa and habitats. Chapter 6 covers both basic design and analytical elements of distance sampling and sets the reader up to be in a position to appreciate and understand comprehensive coverage of distance-sampling topics in the recent book by Buckland et al. (2015).

Whether we like it or not, an ever-increasing number of terrestrial vertebrate species persist at relatively low population levels that preclude data-hungry techniques such as mark-recapture or distance sampling. Sometimes, it is simpler and more cost-efficient to estimate occupancy of sites. This simple measure gives researchers an index with which to assess population persistence, community diversity, and changes in occupancy of a species over time. Chapter 7 presents an overview of occupancy techniques that provide wildlife researchers with powerful tools for sampling small, fragmented or widely isolated populations and analyzing the resulting data using statistical methods that allow for strong inferences.

Dynamic Modeling of Processes

Wildlife scientists and managers are now arguably more appreciative of the value of long-term monitoring data than they ever have been. Long-term data sets are becoming more prevalent in wildlife and ecology, and the analyses of such data can be challenging, especially in light of suboptimal designs, missing data, and other failures to meet assumptions of analytical methods. Furthermore, long-term monitoring and survey data such as the Breeding Bird Survey, the Christmas Bird Count, and the like are being used in a manner not originally conceived when such programs were implemented. Chapter 8 covers fundamental aspects of designing a monitoring project for obtaining strong long-term inferences from both wildlife habitats and populations.

Wildlife habitats and wildlife populations exist as complex and highly interrelated systems, much like an ecosystem. Unfortunately, most wildlife research is not conducted at the system level, meaning to untangle species functions and complex relationships between environmental conditions and responses by individuals and populations. Instead, researchers and managers often take an isolated or reductionist approach to an investigation, and the result, more often than not, is an extremely limited set of inferences. Systems analysis and simulation techniques give researchers a holistic set of outcomes that allow them to make strong inferences in a variety of situations, especially in relation to habitats, populations, and the interplay between them. Chapter 9 provides a simple and straightforward gateway for the reader to enter the world of applied systems analysis and simulation in wildlife science.

During the past couple of decades, individual-based (also called "agent-based") models have been applied to several different aspects of wildlife science such as analysis of harvest data, animal movements, and population dynamics. Although they are capable of linking processes to observations, these models are typically not well known outside of the community of biometricians. Chapter 10 introduces them.

No population can increase over time without encountering some kind of intrinsic mechanism that will ultimately limit its numbers. Originally defined as "inversity" by Paul Errington in the 1940s, density dependence is a phenomenon that has both fascinated and vexed vertebrate population ecologists for at least eight decades (Errington 1945). Assessing density dependence requires a suite of tools and long-term data to determine under what circumstances the phenomenon becomes influential on population demographics. Chapter 11 gives the reader an introduction to the tools and concepts required for analyzing density dependence.

Analysis of Spatially Based Data on Animals and Resources

How animals use resources in relation to their availability is a cross-cutting theme in wildlife science and management. The purpose of Chapter 12 is to provide the reader with background and concepts related to resource-selection analyses, including topics such as selection functions. These have been the status quo in resource-selection studies; however, other, equally powerful, techniques have become available in recent years. There are strong potential linkages between this chapter and Chapter 7 on oc-

cupancy modeling because of challenges related to gaining inferences from incomplete data.Identifying such linkages can help overcome such challenges. Multivariate approaches also come into play with certain resource-selection analyses.

Spatial statistics represent a new frontier in the analysis of wildlife habitat and population data. Their application can range from home ranges to entire landscapes. With the exception of perhaps Bayesian techniques, few, if any, analytical techniques have this kind of power and flexibility. Because things in nature are clustered, spatial autocorrelation exists, to some extent, in every ecological system. It can be a nuisance at best or a red herring at worst. Accounting for autocorrelation can drastically improve the precision of estimates and elucidate underlying patterns in ecological data. Chapter 13 presents an easy-to-understand introduction to these complex topics.

Numerical Methods

Bayesian analyses and associated complementary approaches such as Markov Chain Monte Carlo techniques are widely used in a variety of circumstances ranging from analysis of mark-recapture data to genomic research in molecular genetics. In many situations, researchers are unable to make assumptions about the shape of the distribution of a data set, which can cause instability in precision estimates. This is especially the case when searching molecular genetics data for patterns among thousands—and in some cases millions—of base pairs in a sample of DNA. Chapter 14 rolls back the curtain to show how Bayesian analyses are used for the analysis of molecular genetic data and illuminates why these techniques have become widespread in the analysis of such data.

Machine-learning techniques are becoming more popular as ecologists and wildlife scientists tackle complex questions. Chapter 15 demonstrates how boosted regression trees and random forest methods, in particular, can provide strong inferences when predicting explanatory variables on a response is a priority. These techniques have no underlying assumptions and can fit many data types (e.g., categorical, continuous, binomial, and multinomial). They fit nonlinear relationships automatically and internally cross-validate their predictions. These techniques have been used for species distribution modeling, predictive inference of relationships, and classification modeling. While no machine-learning technique is effective in every application, such tools can be used in a wide variety of analyses with great success.

Chapter 16 addresses the role of expert knowledge in constructing models of wildlife-habitat and stressor relationships by comparing objectives and results of guided model creation with machine-learning and statistical model construction for various modeling objectives. It looks critically at definitions of expertise and how expert knowledge and experience can be codified and verified. Also acknowledging pitfalls and uncertainties in the use of expert knowledge, the chapter considers how to ensure the credibility and validity of expert knowledge-based models, as well as the kinds of constructs best used to represent knowledge and expert understanding, including mind mapping, influence diagrams, and Bayesian networks. Synthesizing machine-learning and expert-guided model structuring is also considered, as is the use of automated and statistical approaches to constructing models from empirical data, including rule-induction algorithms and structure equation modeling.

LITERATURE CITED

Buckland, S. T., E. A. Rexstad, T. A. Marques, and C. S. Oedekoven. 2015. Distance sampling: Methods and applications. Springer International Publishing, Switzerland.

Burnham, K. P., and D. R. Anderson. 1998. Model selection and inference. Springer, New York.

Burnham, K. P., and D. R. Anderson. 2002. Model selection and multimodel inference. Springer-Verlag, New York.

Burnham, K. P., D. R. Anderson, and J. L. Laake. 1980. Estimation of density from line transect sampling of biological populations. Wildlife Monographs 72:1–202.

Errington, P. L. 1945. Some contributions of a fifteen-year local study of the northern bobwhite to a knowledge of population phenomena. Ecological Monographs 15:1–34.

Guthery, F. S., L. A. Brennan, M. J. Peterson, and J. J. Lusk. 2005. Information theory and wildlife science: Critique and viewpoint. Journal of Wildlife Management 69:457–465.

Johnson, D.H. 1999. The insignificance of significance testing. Journal of Wildlife Management 63:763–772.

Mosteller, F., S.E. Fienberg, and R. E. K. Rourke. 1983. Beginning statistics with data analysis. Addison Wesley, Reading, MA.

Verner, J., M. L. Morrison, and C.J. Ralph. 2006. Wildlife 2000: Modeling habitat relationships of terrestrial vertebrates. University of Wisconsin Press, Madison.

Vogt, W. P., E. R. Vogt, D. C. Gardner, and L. M. Haeffele. 2014. Selecting the right analyses for your data: Quantitative, qualitative, and mixed methods. The Guilford Press, New York. 500 pp.

PART I GENERAL STATISTICAL METHODS

2

D A V I D B. W E S T E R

Regression

Linear and Nonlinear, Parametric
and Nonparametric

> The statistician knows . . . that in nature there never was a normal distribution,
> there never was a straight line, yet with normal and linear assumptions, known to be
> false, he can often derive results which match, to a useful approximation, those
> found in the real world. — George E.P. Box (1976:792)

Introduction

Regression is a versatile statistical tool with many applications in wildlife science. For example, an upland game-bird biologist may wish to study the relationship between bird numbers and growing season rainfall with a simple linear model; the data set might include bird numbers and rainfall recorded at 100 spatially separated study sites. This can be extended to a multiple linear model by adding a second explanatory variable, for example, a measure of grazing intensity at each site. If the same study sites are monitored over multiple years, the data set assumes a repeated measures quality that can involve nonindependent observations. A wildlife parasitologist may be interested in studying the relationship between abundance of eye worms in quail as affected by body weight and gender of a bird and the interaction between these two variables. In this example, body weight is a continuous variable and gender is categorical; and the response of parasite numbers (which likely are not normally distributed given body weight and gender) to body weight may depend on gender. A big-game biologist might be interested in home-range size of translocated deer as a function of time since translocation; for this analysis, a negative exponential decay model might be appropriate. Each of these three research questions can be addressed with regression.

Regression is an expression and analysis of the relationship between two or more variables. One of these variables, the *dependent* (or response) variable (Y_i), often is assumed to follow a specified distribution (e.g., normal, Bernoulli or binomial, Poisson, negative binomial) at each level of one or more *independent* variables (X_i) that are usually measured on an interval or ratio scale but may be categorical. The variables Y_i and X_i are related to each other through a specified function such as (1) a simple linear model $Y_i = \beta_0 + \beta_1 X_{i1} + \varepsilon_i$; (2) a multiple linear model $Y_i = \beta_0 + \beta_1 X_{i1} + \beta_2 X_{i2} + \varepsilon_i$; or (3) a simple nonlinear model $Y_i = \beta_0 + e^{\beta_1 X_{i1}} + \varepsilon_i$, where β_i are parameters to be estimated and ε_i is a random variable reflecting the distribution of Y_i at X_i. Models (1) and (2) are *linear* models because they are linear in their parameters; model (3) is *nonlinear* in its parameters. When a particular distribution cannot be assumed for a response variable, distribution-free methods are available. Common goals of regression include *describing the relationship* between Y_i and X_i (is it linear? curvilinear?), *predicting a value* of Y_i for specified X_i, and *identifying variables* that can be used to predict Y_i—these goals can involve both parameter estimation and hypothesis testing.

This definition of regression, coupled with the above examples, illustrates several concepts that are explored in this chapter: (1) dependent variables often follow distributions other than the normal distribution; (2) independent variables can be continuous or categorical, may interact with each other, and likely are not independent of each other; (3) in time-extensive data sets nonindependence can be expected; and (4) relationships between dependent and independent variables often are nonlinear. These issues are presented in the context of a known model; a final consideration addresses model specification.

Preliminary Considerations

In a typical application of simple linear regression, Hernández et al. (2013) studied quail population trends by regressing the number of quail (the dependent variable) on year (the independent variable). Their data set, which included a datum representing quail numbers for each year between 1960 and 2010, illustrates a key concept in our definition of regression and highlights a common feature of most observational regression studies: namely, whereas each level of the independent variable defines a population for the dependent variable, it is often the case that this population is sampled without replication (Mood et al. 1974:484–485; Scariano et al. 1984; Sokal and Rohlf 2012:476)—that is, there is a sample of $n = 1$ observation on Y_i for a given value of the independent variable. And yet, even in the absence of replication, we can make predictions, estimate variances, and offer α-level protected inferences. Our ability to do so hinges on making assumptions about the distribution of Y_i conditional on the independent variables that are impossible to test without simultaneously assuming that we have appropriately specified our model. With true replication, it is possible not only to directly test assumptions but also to perform lack-of-fit tests. True replication of Y_i at X_i is therefore recommended whenever possible.

Distributional Considerations

Understanding the distribution of the response variable ("dependent variable") is critical in regression analyses. Many textbooks illustrate the classical regression model using ordinary least squares (OLS) estimation with examples that assume a normal distribution for the dependent variable (conditional on X_i); another assumption is that the variance of the response variable (conditional on X_i) is constant across levels of the independent variable(s). Although variables such as home-range size, average number of coveys per acre, or body weight might be assumed to have a normal distribution, sometimes it may be less certain that variation in these variables is constant across the range of the independent variables. Fig. 2.1a shows the relationship between quail density (Y_i) in winter as a function of quail density in fall, X_1. Whether informally assessed with residual plots (Fig. 2.1b) or formally tested (e.g., with a Breusch-Pagan test; see Gujarati and Porter 2009), it is evident that as fall bird density increases so does variation in winter bird density—a violation of an assumption in classical regression. With heteroscedasticity, OLS estimates of regression coefficients are still linear and unbiased; further, they are consistent as well as asymptotically normally distributed. However, these estimators are not efficient—they are not minimum-variance estimators, and thus they are not "best linear unbiased estimators" (or "BLUE")—and therefore heteroscedasticity adversely impacts significance tests.

Heteroscedasticity often is expressed in patterns where the variance in Y_i is related to X_i. Fig. 2.2a illustrates homoscedasticity: the variance of Y_i is constant across levels of X_i. Heteroscedasticity is illustrated in Figs. 2.2b–d, where the variance of Y_i is proportional to X_i or powers of X_i.

Options to improve hypothesis tests when the assumption of homoscedasticity is untenable include weighted least squares (WLS) or estimation of *heteroscedasticity-consistent* ("robust") standard errors. Additionally, data transformation using the logarithm

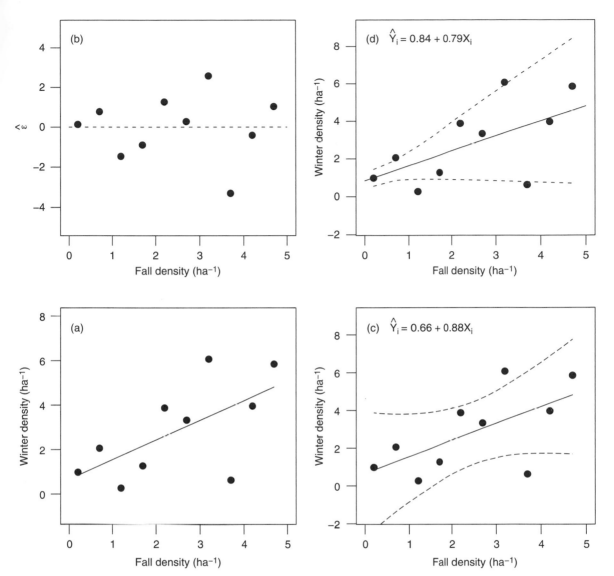

Fig. 2.1. (a) Winter bird density, Y_i, and fall bird density, X_{i1}, with an estimated regression equation using ordinary least squares; (b) A residual plot from the model $Y_i = \beta_0 + \beta_1 X_{i1} + \varepsilon_i$, where $\hat{Y}_i = \hat{\beta}_0 + \hat{\beta}_i X_{i1}$ and residuals $\hat{\varepsilon}_i = Y_i - \hat{Y}_i$ are plotted as a function of X_{i1}; (c) 95% confidence band around the estimated regression equation using ordinary least squares (OLS), which assumes homoscedasticity; note that the confidence band is narrowest when fall density, X_{i1}, is at its mean, $\bar{X}_1$, and widens symmetrically in both directions away from $\bar{X}_1$; (d) 95% confidence band from weighted least squares (WLS), which accounts for heteroscedasticity, reflected in a confidence band that widens as fall density counts increase.

of the independent variable (so that the scales of the regression are "linear-log"), the dependent variable ("log-linear" scales), or both ("log-log" scales) can be helpful. In WLS, observations with smaller variance are given more weight in the estimation process than observations with larger variance; when weights are known, the resulting estimators are BLUE. In most settings, however, weights are not known. In this case, White's method (Gujarati and Porter 2009) can be used to estimate robust standard errors; the method is asymptotically valid and may be helpful when there is little information about the nature of the heteroscedasticity. Some heteroscedasticity tests (e.g., Park's test; see Gujarati and Porter 2009) provide insight

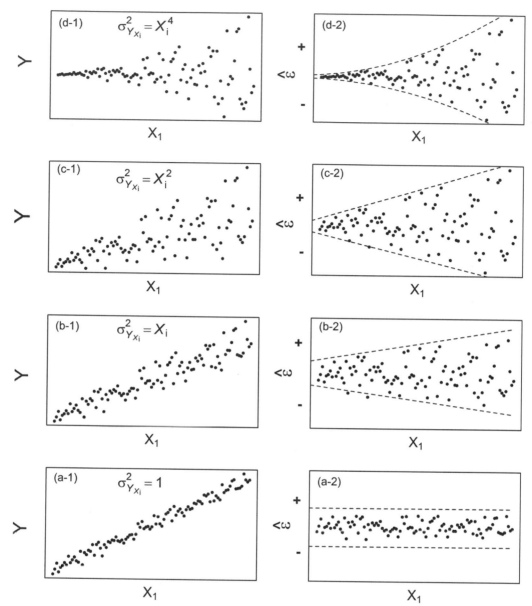

Fig. 2.2. Patterns of homo- or heteroscedasticity in simple linear regression. Plots (a-1), (b-1), (c-1), and (d-1) are of observations Y_i; plots (a-2), (b-2), (c-2), and (d-2) are corresponding residual plots from ordinary least squares (OLS) analysis where $\hat{Y}_i = \hat{\beta}_0 + \hat{\beta}_i X_{i1}$ and residuals, $\hat{\varepsilon}_i = Y_i - \hat{Y}_i$, are plotted as a function of X_{i1}. The statistical model is $Y_i = \beta_0 + \beta_1 X_{i1} + u_i$, where $u_i = X_{i1}^{\theta} e_i$, $e_i \sim NIID\ (0, 1)$, and $\theta = 0, 0.5, 1, 2$ for plots (a), (b), (c), and (d), respectively, corresponding to increasing heterogeneity of variance of Y_i as a function of X_{i1}.

into its nature and may reveal that variation in Y_i is a function of X_i; in these settings, transformations based on this knowledge may yield more efficient standard errors that are less dependent on large-sample considerations than White's method. WLS produces un-

biased estimators of regression coefficients that are more efficient than robust standard errors (Table 2.1, Fig. 2.3). Gujarati and Porter (2009) discuss data transformation and OLS when variation in Y_i is proportional to X_i (Fig. 2.2b) or to X_i^2 (Fig. 2.2c).

Table 2.1. Results from a Monte Carlo simulation ($N = 10{,}000$ simulations) of the simple linear model $Y_i = \beta_0 + \beta_1 X_{i1} + u_i$, where X_{i1} are integers between 1 and 10 inclusive, $u_i = X_{i1}^\theta e_i$, $e_i \sim NIID\,(0, 1)$, $\beta_0 = 5$, $\beta_1 = 1$, and $\theta = 0, 0.5, 1, 2$ so that variation in Y_i at X_{i1} is a function of X_{i1} when $\theta \neq 0$. "OLS" = ordinary least squares, "WLS" = weighted least squares, with weights $= 1/\sigma_{Y_{X_{i1}}}^2$; "OLS-HC$_3$" = "heteroscedasticity-consistent" (or "robust") standard errors (Gujarati and Porter 2009). True values for the standard error of the slope, $se(\beta_1)$ are given, together with Monte Carlo estimates of the mean, the standard error, minimum and maximum values of $se(\hat\beta_1)$; type I error rates are for $H_o : \beta_1 - 1$ vs $H_1 : \beta_1 \neq 1$. See Fig. 2.3.

θ	Estimation method	Standard error of the slope					Type I error rate
		True value	Mean	Standard error	Minimum	Maximum	
0	OLS	0.1101	0.1068	0.0268	0.0274	0.2097	0.0510
	OLS-HC$_3$	0.1101	0.1232	0.0422	0.0235	0.3236	0.0416
	WLS	0.1101	0.1068	0.0268	0.0274	0.2097	0.0510
0.50	OLS	0.2582	0.2487	0.0684	0.0407	0.5598	0.0499
	OLS-HC$_3$	0.2582	0.2838	0.1115	0.0450	0.9350	0.0480
	WLS	0.2190	0.2123	0.0534	0.0291	0.4145	0.0529
1	OLS	0.7377	0.6443	0.2011	0.1136	1.6343	0.0700
	OLS-HC$_3$	0.7377	0.7793	0.3572	0.1124	2.8780	0.0589
	WLS	0.4733	0.4591	0.1155	0.1011	0.9722	0.0511
2	OLS	6.8100	5.0096	1.8852	0.8912	15.3028	0.1170
	OLS-HC$_3$	6.8100	6.8684	3.5669	0.4843	27.6435	0.0724
	WLS	2.1109	2.0490	0.5163	0.6559	4.2467	0.0518

When OLS is used under heteroscedasticity, the "usual" estimator of the variance of $\hat\beta_1$, $\mathrm{var}(\hat\beta_1) = \sigma_\varepsilon^2 / \sum x_i^2$, may either over- or underestimate the actual variance, $\sum x_i^2 \sigma_\varepsilon^2 / (\sum x_i^2)^2$, the direction of bias depending on the pattern of heteroscedasticity as well as on values of X_i. When the variance of Y_i is proportional to X_i or to powers of X_i, the usual OLS estimator of the standard error of the slope, $se(\hat\beta_1)$, consistently underestimates the true standard error, and is generally smaller than the robust standard error; further, the robust standard error is generally closer to the true value (with increasing heteroscedasticity) than the OLS estimator (Table 2.1, Fig. 2.3). However, variation in the robust $se(\hat\beta_1)$ is higher than the usual OLS estimator at all levels of heteroscedasticity. Taken together, these considerations suggest that robust standard errors are little better than OLS estimators for heteroscedasticity patterns as shown in Table 2.1 (Dougherty 2011), and despite the fact that robust standard errors are asymptoti-

cally valid, a similar conclusion emerges with larger sample sizes.

This conclusion may not apply, however, when the goal is hypothesis testing and prediction. For example, we may be confident that there is a positive relationship between winter and fall bird densities. Hence, rather than testing what many consider an "obvious" null hypothesis, our interest may lie in testing a *particular* relationship: a slope greater than 1 may indicate reproduction and/or immigration during fall whereas a slope less than 1 may suggest mortality and/or emigration; either result has important management implications. Type I error rates (Table 2.1) show that WLS (with known weights) provides nominal α-level protection. In contrast, type I error rates with OLS are unacceptable with extreme heteroscedasticity, where robust standard errors are clearly preferable. Suppose our wildlife biologist wishes to estimate a confidence band around the estimated regression equation. When homogeneity

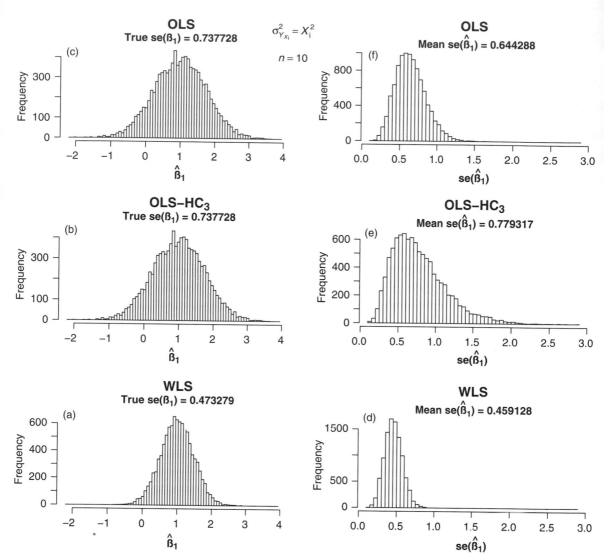

Fig. 2.3. Results from Monte Carlo simulation ($N = 10,000$ simulations) of the simple linear model $Y_i = \beta_0 + \beta_1 X_{i1} + u_i$, where X_i are integers between 1 and 10 inclusive, $u_i = X_{i1}^\theta e_i$, $e_i \sim NIID\,(0, 1)$, $\beta_0 = 5$, $\beta_1 = 1$, and $\theta = 1$, so that variation in Y_i at X_{i1} is proportional to X_{i1}^2. OLS = ordinary least squares, WLS = weighted least squares, with weights $1/\sigma^2_{Y_{X_{i1}}}$; HC$_3$ = heteroscedasticity-consistent (or "robust") standard errors (Gujarati and Porter 2009). (a–c) show the distribution of the slope estimator, $\hat{\beta}_1$, with true values for the standard error of the slope, $se(\beta_1)$; (d–f) show the distribution of the standard errors of slope, $se(\hat{\beta}_1)$, with Monte Carlo means for $se(\hat{\beta}_1)$. See Dougherty (2011) as well as Table 2.1 for additional results.

is satisfied, this band has the familiar appearance of being narrowest at $\bar{X}_1$ and widening *symmetrically* away from $\bar{X}_1$ (Fig. 2.1c). With heteroscedasticity, the variance associated with predictions is affected both by distance from $\bar{X}_1$ and by the variance of Y_i at X_i, and the estimated confidence band reflects both sources of variability, with obvious interpretational

consequences: uncertainty associated with predictions of winter quail density is larger with higher fall counts than with fewer fall counts (Fig. 2.1d).

Responses expressed as percentages (e.g., plant mortality or nest success) often are based on counts (e.g., three out of ten animals survive); these counts are not normally distributed but rather follow a

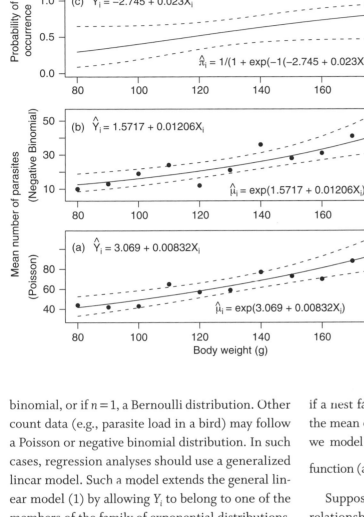

Fig. 2.4. Parasite numbers in quail as a function of body weight (g). (a) Parasite numbers follow a Poisson distribution at each body weight, with $\mu_Y = \sigma^2_{Y_{X_i}} = X_i/2$; (b) Parasite numbers follow a negative binomial distribution at each body weight, with $\mu_Y = k(1-p)/p$, $\sigma^2_{Y_{X_i}} = k(1-p)/p^2$, and $p = 0.8$ and $k = 1.5(X_i/2)$. (c) Logistic regression can be used to estimate probability of occurrence as a function of body weight. Estimated regression equations and a 95% confidence band around the estimated line are shown for each analysis.

binomial, or if $n = 1$, a Bernoulli distribution. Other count data (e.g., parasite load in a bird) may follow a Poisson or negative binomial distribution. In such cases, regression analyses should use a generalized linear model. Such a model extends the general linear model (1) by allowing Y_i to belong to one of the members of the family of exponential distributions, and (2) by using the right-hand side of the model ($\beta_0 + \beta_1 X_{i1}$ and now called the "linear predictor") to predict not the mean of Y_i but rather a function, g, of the mean. In other words, $g(\mu_i) = \beta_0 + \beta_1 X_{i1}$; $g(\mu_i)$ is called the "link" function because it links the mean of Y_i to the linear predictor. For example, if Y_i is Poisson-distributed, its mean is equal to its variance (and so heteroscedasticity is expected); in a generalized linear model with one independent variable, usually we assume that $\mu_i = e^{\beta_0 + \beta_1 X_{i1}}$, so that the link function is the log function: $\log_e(\mu_i) = \beta_0 + \beta_1 X_{i1}$. A log link is also used when the response variable follows a negative binomial distribution. If the response variable is Bernoulli-distributed (e.g., $Y_i = 0$

if a nest fails and $Y_i = 1$ if a nest is successful), then the mean of Y_i is π_i (the probability of success), and we model $\pi_i = 1/(1 + e^{-(\beta_0 + \beta_1 X_{i1})})$ so that the link function (a "logit") is $\log_e\left(\dfrac{\pi_i}{1-\pi_i}\right) = \beta_0 + \beta_1 X_{i1}$.

Suppose a wildlife parasitologist is studying the relationship between parasite numbers and body weight of quail. Results for Poisson and negative binomial regressions are shown in Fig. 2.4. Two concepts are noteworthy for generalized linear models. First, despite a *linear* predictor, there is a *curvilinear* relationship between estimated mean number of parasites and body weight. Second, care must be taken in interpreting regression coefficients. In a general linear model, we interpret the slope directly: $\hat{\beta}_1$ gives the unit change in the estimated mean of Y_i for a unit increase in X_i; thus, for Fig. 2.1d, where $\hat{\beta}_1 = 0.79$, we say that estimated mean winter density increases 0.79 birds ha^{-1} for each additional bird ha^{-1} in fall. For Poisson regression, however, we interpret $\hat{\beta}_1$ through the equation $(e^{\hat{\beta}_1} - 1)100\%$, which measures the percentage change in the estimated mean of Y_i for a unit increase in X_i. For example, in Fig. 2.4a, where $\hat{\beta}_1 = 0.00832$, $\hat{Y}_{X=80} = 3.735$ and $\hat{Y}_{X=81} = 3.743$, with corresponding estimated means $e^{\hat{Y}_i} = 41.87$ and 42.22; the percentage increase in estimated mean number of parasites from $X_i = 80$ to 81 is 0.8354%; in general, the percentage increase is $(e^{\hat{\beta}_1} - 1)\,100$ from X_i to X_{i+1}. A similar interpretation attends $\hat{\beta}_1$ in negative binomial regression. A generalized linear model with a logit link—logistic regression—is appropriate for count data generated by a binomial process. The estimated slope is interpreted in the context of "odds." Suppose the probability that a bird has parasites is $\pi = 0.75$; then the probability of not being parasitized is $1 - \pi = 0.25$, and the *odds* of being parasitized is $\pi / (1 - \pi) = 0.75/0.25 = 3$ (or "3 to 1"). For the incidence data in Fig. 2.4c, the estimated probability of parasite occurrence for an 80g bird is $\hat{\pi}_{X=80} = 1/(1 + e^{-(-2.745 + 0.023 \times 80)}) = 0.28802$, with odds $= 0.40454$; for a 81g bird, $\hat{\pi}_{X=81} = 0.29276$, with odds $= 0.41395$. Thus, the odds of parasitism increase 2.33% for a 1g increase in body weight; this value equals $(e^{\hat{\beta}_1} - 1)\,100\% = 2.33\%$ (in general, odds change $(e^{c\hat{\beta}_1} - 1)\,100\%$ for a c-unit increase in X_i). This is related to the concept of "odds ratio" ("OR") as follows: $\widehat{OR} = \widehat{odds}_{X=81} / \widehat{odds}_{X=80} = 1.0233$, and so $(\widehat{OR} - 1)\,100 = 2.33\%$. An odds ratio of 1 indicates no relationship between probability of parasitism and body weight.

Multiple Independent Variables

A multiple regression model has two or more independent variables. Suppose that parasite burdens in quail are affected not only by body weight but also by gender, a categorical variable. Or suppose that number of birds in an area is affected by breeding season precipitation and by cattle grazing intensity—both variables are measured on a ratio scale. *Interaction* is possible in both of these examples: perhaps parasite burdens are related to body weight for females but not for males; and it may be that the relationship between bird numbers and precipitation is different in heavily-grazed than in lightly-grazed settings. Multiple regression provides a mechanism to include several independent variables in a model. As before, we will assume that our models are correctly specified and focus on interpretational issues.

Just as a linear model with one independent variable fits a line in 2-dimensional space (Fig. 2.1), a linear model with two independent variables fits a regression plane in a 3-dimensional space (Fig. 2.5), and a model with three independent variables fits a hyper-plane in 4-dimensional space. The shape of the response surface—is it flat (Fig. 2.5) or curved (Fig. 2.6)?—is determined by the data and how independent variables are specified in the model.

Estimating plant biomass from morphological measurements is common in range and wildlife habitat studies. For example, nondestructive estimation of individual plant biomass is helpful in understanding effects of prescribed fire on forage production. Grass biomass is related both to plant basal area and to plant length (Sorensen et al., 2012): preliminary graphs and exploratory analyses of biomass with each variable indicated heteroscedasticity; when analyses were performed on a log-log scale, heteroscedasticity was not detected. Fig. 2.5 shows aboveground biomass as a function of both plant length and basal area. Although plant basal area and plant length are individually significantly related to biomass, and although *both* variables are significant when analyzed together, they do not interact with each other. Graphically, the response surface (Fig. 2.5) is not "level": estimated biomass increases both as plant length increases and as basal area increases; however, the absence of an interaction is revealed by a flat (but not level) response surface (on a log-log scale). For example, response of estimated biomass to increasing plant length *is the same* for plants with small basal area and plants with large basal area. When data are analyzed on a log-log scale, interpretation of estimated regression coefficients on the observed scale requires care (whether the model includes one or more independent variables). For example, the partial regression coefficient for log basal

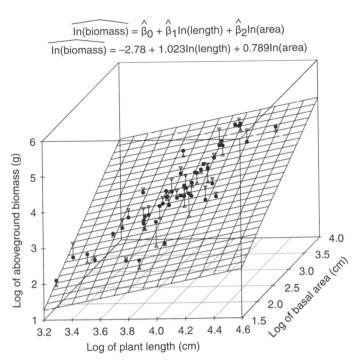

$$\widehat{\ln(\text{biomass})} = \hat{\beta}_0 + \hat{\beta}_1\ln(\text{length}) + \hat{\beta}_2\ln(\text{area})$$
$$\widehat{\ln(\text{biomass})} = -2.78 + 1.023\ln(\text{length}) + 0.789\ln(\text{area})$$

Fig. 2.5. Purple threeawn above-ground biomass (open triangles), estimated biomass (dots), and predicted response surface (hatched) as a function of plant length and basal area (all variables log-transformed). Each of $n = 50$ plants is represented by a dot (estimated biomass) and either an open triangle with point down (when observed biomass was greater than predicted biomass) or with point up (when observed biomass was less than estimated biomass). *Source: Reanalyzed from Sorensen et al. (2012).*

area ($\hat{\beta}_1 = 0.789$; Fig. 2.5) is interpreted as: "if plant length is held constant, then estimated biomass increases $100(e^{\hat{\beta}_1(\ln((100+p)/100))} - 1)\%$ for a $p\%$ increase in basal area." If log of plant length $= 4$, then, as basal area increases from 10 to 25 cm^2 (a 150% increase), estimated biomass increases from 22.85 to 47.07g, which is a $100(e^{0.789(\ln((100+150)/100))} - 1)\% = 106.06\%$ increase. The idea of holding one variable constant and examining changes in predictions as a second variable varies is the basis for the term "*partial* regression coefficient" and is most easily implemented in experimental settings.

Fig. 2.6 illustrates interaction between two continuous independent variables. Aboveground biomass of perennial snakeweed can be predicted by number and length of leafy stems (Tian and Wester, 1999). Although each variable is correlated with biomass, these variables interact: the response surface in Fig. 2.6 is neither level (both variables affect biomass) nor flat (response surface "edges" are not parallel); the response surface has straight edges but these edges are not parallel. For example, predicted biomass increases more rapidly with increasing number of leafy stems

when these stems are long compared to when these stems are shorter, and so the rate of change in predicted biomass to changes in one variable *depends* on the second variable. In models with interaction, a common interpretation of regression coefficients—the effect of one variable on estimated Y_i holding the second variable constant—is not appropriate.

Interactions can involve continuous and categorical variables. For example, night-time surveys to assess alligator populations often use estimated head size measurements to predict total body length. The relationship between head size and total body length, however, may differ between size classes of alligators. A linear model that (i) regresses total length (TL) on eye-to-nare length (ENL) and includes (ii) size class (as a dummy variable), as well as (iii) the interaction between ENL and size class, provides an opportunity to test these hypotheses: (1) Is there a relationship between TL and ENL? (2) Does this relationship differ between size classes? (3) Is a single model adequate or should size class–specific models be used? Fig. 2.7a shows a strong linear relationship between log TL and log ENL for each size class; this relation-

$$\widehat{biomass} = \hat{\beta}_0 + \hat{\beta}_1 number + \hat{\beta}_2 length + \hat{\beta}_3 (length)(number)$$
$$biomass = -2 + 0.04 number + 0.22 length + 0.028 (length)(number)$$

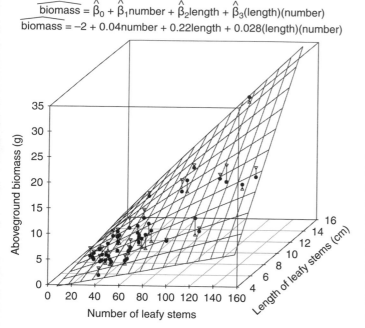

Fig. 2.6.   Perennial broomweed biomass (open triangles), estimated biomass (dots), and predicted response surface (hatched) as a function of number and length of leafy stems. Each of $n = 50$ plants is represented by a dot (estimated biomass) and either an open triangle with point down (when observed biomass was greater than predicted biomass) or with point up (when observed biomass was less than estimated biomass). *Source: Reanalyzed from Tian and Wester (1999).*

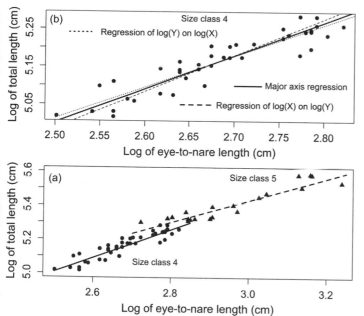

Fig. 2.7.   (a) Simple linear regression of log of total alligator length on log of eye-to-nare length for two sizes classes of animals. (b) Major axis regression (solid black line), OLS regression of log of total alligator length on log of eye-to-nare length (dotted line), and OLS regression of log of eye-to-nare length on log of total alligator length (dashed line). *Source: Adapted from Eversole et al. (2017).*

ship, however, differs between sizes—the slope for size class 4 is steeper than the slope for size class 5.

This example provides an opportunity to consider major axis (MA) regression and standard (or sometimes referred to as reduced) major axis (SMA) regression, commonly used in allometry (Smith 2009).

Recall that our definition of simple regression specified that each X_{i1} defines a population of Y_i. In this context, X_1 is not considered a random variable, whether or not its values are under the control of the investigator (Mood et al. 1974:484; Sokal and Rohlf 2012:477), and in either case, interest lies in predict-

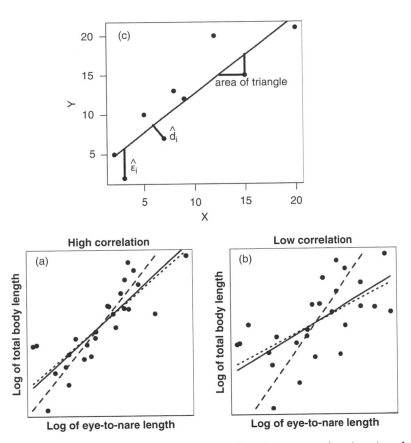

Fig. 2.8. Major axis regression (black solid line), OLS regression of log of eye-to-nare length on log of total length (dashed black line), and OLS regression of log of total length on log of eye-to-nare length (dotted black line) when log of total length and log of eye-to-nare length have (a) high ($r=0.83$) correlation or (b) low ($r=0.61$) correlation. (c) OLS regression minimizes the sum of the squares of $\hat{\varepsilon}_i$, the vertical distance between an observed value of Y_i and its corresponding predicted value on the regression line; major axis regression minimizes the sum of the squares of the orthogonal (perpendicular) distance, $\hat{d}_i$, between an observed value of Y_i and its corresponding predicted value on the regression line; and reduced major axis regression minimizes the sum of the areas of the triangles shown, which are defined by the horizontal and vertical distances between an observed value of Y_i and its corresponding points on the regression line.

ing the mean of Y_i at specified value(s) of X_1. It is obvious in the alligator example, however, that both TL and ENL are random variables that constitute a bivariate normal distribution. In many such settings it may be argued which is the dependent and which the independent variable: a regression of TL on ENL produces a different line than a regression of ENL on TL: in the former setting, interest focuses on the conditional distribution of TL at values of ENL, and in the latter interest focuses on the conditional distribution of ENL at values of TL. In a bivariate setting with random variables Y_1 and Y_2, conditional

inferences about Y_1 given Y_2 apply when (1) Y_{i1} are independent of each other and (2) Y_{i1}, when Y_2 is considered fixed or given, are normally distributed with conditional means $\beta_0 + \beta_1 Y_{i2}$ and homoscedasticity applies (Kutner et al. 2004:section 2.11). If, instead, interest lies simply in the relationship between TL and ENL, "there should be one line describing [their] pattern of covariation" (Smith 2009:478). In these cases, the analysis is called "model II" regression. Two model II regression techniques are common: (1) in MA regression, the sum of the squared perpendicular distances ($\hat{d}_i$ in Fig. 2.8c) is minimized; (2) in SMA regression, both vertical and horizontal differences between Y_i and the "best fit line" (triangular area in Fig. 2.8c) are minimized. Hypothetical data in Fig. 2.8 show that in a bivariate setting (1) OLS regression of Y_i on X_i [i.e., log(total length) on log(eye-to-nare length)] has the shallowest slope, (2) OLS regression of X_i on Y_i [i.e., log(eye-to-nare length) on log(total length)] has the steepest slope, and (3) MA regression has an intermediate slope. In fact, the divergence in slopes between the two OLS analyses increases as the correlation between the two random variables decreases (Fig. 2.8a and b). OLS and MA regressions for the alligator data are summarized in Fig. 2.7b, where the similarity between the two OLS regressions reflects high correlation between logs of total length and eye-to-nare length. In a bivariate setting, use of (1) OLS to estimate means of one random variable at selected levels of a second random variable, or (2) MA to investigate the relationship between the two random variables depends on the objectives of the analyst (Sokal and Rohlf 2012:536).

Higher-order interactions among independent variables are common in habitat studies. Wiemers et al. (2014) used multiple logistic regression to study resource selection by white-tailed deer as a function of time period (a categorical variable) and of operative temperature, vegetation height, and canopy cover (the latter three being continuous variables). One approach to data presentation involves holding two factors (e.g., time period and canopy cover) constant and showing responses to the remaining two factors. Fig. 2.9 shows how operative temperature and vegetation height affect estimated probability of use at 15% or 95% canopy cover during midday.

Multiple regression offers effective tools to analyze thresholds common in many ecological processes (Swift and Hannon 2010). Some thresholds can be approximated (i) by two straight lines with a transition *point* (Fig. 2.10a); other processes may involve (ii) two straight lines but a more gradual transition *region* (Fig. 2.10b); finally, (iii) a curvilinear model with a changing slope throughout may be appropriate (Fig. 2.10c). A common approach for (i) and (ii) is called "piecewise" regression, a method that includes both a continuous independent variable and a dummy variable, and a threshold point at $X_1 = c$. For (i), a common model (Gujarati and Porter 2009) is $Y_i = \beta_0 + \beta_1 X_{i1} + \beta_2 (X_{i1} - c) X_{i2} + \varepsilon_i$, where X_1 is a continuous explanatory variable and X_2 is a dummy variable ($X_2 = 1$ if $X_1 > c$ and $X_2 = 0$ if $X_1 \leq c$). When $X_1 \leq c$, the mean of Y_i is $E[Y_i] = \beta_0 + \beta_1 X_{i1}$; when $X_1 > c$, $E[Y_i] = (\beta_0 - c\beta_2) + (\beta_1 + \beta_2) X_{i1}$. Therefore, $\hat{\beta}_1$ and $(\hat{\beta}_1 + \hat{\beta}_2)$ estimate slopes when $X_1 < c$ and $X_1 > c$, respectively, and $\hat{\beta}_0$ and $(\hat{\beta}_0 - c\hat{\beta}_2)$ estimate intercepts when $X_1 < c$ and $X_1 > c$, respectively. This model specifies a transition *point* (Fig. 2.10a). For

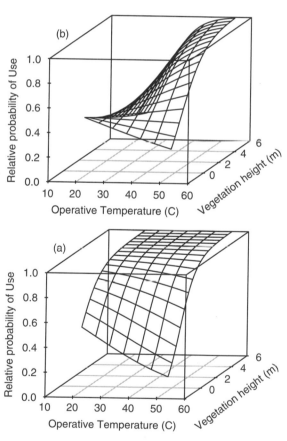

Fig. 2.9. Multiple logistic regression as a tool to assess resource selection by male white-tailed deer. Estimated probability of use as a function of operative temperature and vegetation height during mid-day with (a) 15% woody cover and (b) 95% woody cover. *Source:* From Wiemers et al. (2014).

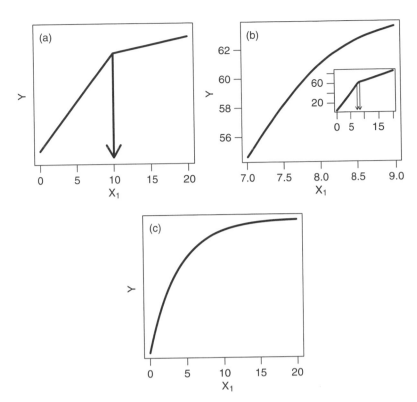

Fig. 2.10. Various approaches to modeling ecological thresholds. (a) An abrupt transition can be modeled with two simple linear regressions: $E[Y_i] = \beta_0 + \beta_i X_{i1} = 5 + 6 X_{i1}$ for $X_{i1} \leq 10$ and $E[Y_i] = \beta_0 + \beta_1 X_{i1} = 55 + 1 X_{i1}$ for $X_{i1} \geq 10$, with a threshold at $X_{i1} = 10$ where the slope changes. (b) Smoother transitions between two simple linear regressions can be modeled following Bacon and Watts (1971). In example shown, let $Y_i = \theta_0 + \theta_1 X_{i1} + \varepsilon_i$, $X_{i1} \leq X_0$ and $Y_i = \theta_0 + \theta_1 X_0 + \theta_2 (X_{i1} - X_0) + \varepsilon_i$, $X_{i1} \geq X_0$; then $Y_i = a_0 + a_1(X_{i1} - X_0) + a_2(X_{i1} - X_0) \tanh\{(X_{i1} - X_0)/\gamma\} + \varepsilon_i$ where $a_0 = \theta_0 + \theta_1 X_0$, $a_1 = \frac{1}{2}(\theta_1 + \theta_2)$ and $a_2 = \frac{1}{2}(\theta_2 - \theta_1)$; for this example, $X_0 = 8$; $\gamma = 0.999$; $\theta_0 = 5$; $\theta_1 = 7$; $\theta_2 = 2$; the parameter γ controls the abruptness of the transition, ranging from $\gamma = 0$, an abrupt "elbow" to $\gamma = 1$, a maximally rounded transition. (c) A nonlinear relationship across the range of X_{i1} defined by $E[Y_i] = \beta_0 + \beta_1 e^{(\beta_2 X_{i1})} = 10 - 10 e^{(-0.25 X_{i1})}$.

(ii), a smoother transition in the neighborhood of X_0 between two straight lines can be modeled with $Y_i = \alpha_0 + \alpha_1(X_{i1} - X_0) + \alpha_2(X_{i1} - X_0) \tanh\{(X_{i1} - X_0)/\gamma\} + \varepsilon_i$ (Bacon and Watts 1971); the parameter γ controls the abruptness of the transition, ranging from $\gamma = 0$, an abrupt "elbow," to $\gamma = 1$, a maximally rounded transition (Fig. 2.10b). When c is known, models can be fit with OLS; for example, in model (i), above, $Y_i = \beta_0 + \beta_1 X_{i1} + \beta_2(X_{i1} - c)X_{i2} + \varepsilon_i$, one simply includes $(X_{i1} - c) X_{i2}$ as a second independent variable. When c is not known, it can be included as a parameter to be estimated with nonlinear regression methods, or several candidate values for c can be used,

with final selection based on a measure of model fit (see Chapter 4). Finally, if the relationship is curvilinear throughout (see iii, above), a nonlinear model (e.g., $Y_i = \beta_0 + \beta_1 e^{(\beta_2 X_{i1})} + \varepsilon_i$; Fig. 2.10c) can be used. All three approaches are illustrated in Fig. 2.11.

Longitudinal Data

Some ecological processes are best studied with long-term data sets that include repeated measurements over time from multiple sampling locations. With longitudinal analysis (Fitzmaurice et al. 2004) one

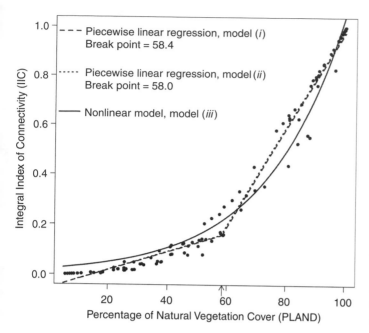

Fig. 2.11. Relationship between integral index of connectivity and percentage of natural vegetation cover as modeled with piecewise linear regression (dashed black line); a smooth transition between two linear regressions (dotted black line; Bacon and Watts 1971); an estimate of $\hat{\gamma} = 0.04$ indicates that approaches (*i*) and (*ii*) yield similar results: dashed and dotted lines are similar; and a nonlinear model, $Y_i = \beta_0 e^{\beta_1 X_{i1}} + \varepsilon_i$ (solid black line). *Source: Reanalyzed from Zemanova et al. (2017).*

can identify both spatial ("cross-sectional") and temporal ("longitudinal") effects as well as account for correlation among repeated measurements. For example, suppose the effect of woody cover on native plant species richness is studied by establishing permanent transects across a variety of sites and monitoring richness with yearly sampling. Transect-to-transect variability likely represents effects of initial woody cover and local site differences. Changes over time may reflect climatic as well as legacy effects. Suppose that s transects ($i = 1, 2, \ldots, s$) are monitored yearly for t years ($j = 1, 2, \ldots, t$). The statistical model is: $Y_{ij} = \beta_0 + \beta_1 X_{i1} + \beta_2 (X_{ij} - X_{i1}) + \varepsilon_{ij}$, where X_{ij} is woody cover on the *ith* transect in the *jth* year, β_1 estimates cross-sectional effects, and β_2 estimates longitudinal effects (Fitzmaurice et al. 2004:420). The following linear mixed model can be used (Fitzmaurice et al. 2004:213): $Y_{ij} = \beta_0 + \beta_1 X_{ij} + \beta_2 X_{i1} + b_{0i} + b_1 X_{ij} + \varepsilon_{ij}$; the correlation over time because of repeated measurements can be accounted for with an appropriate variance-covariance structure by including random intercepts and slopes for woody cover. In this linear mixed model, $\hat{\beta}_1$ estimates longitudinal effects (within-transect information that

arises because transects are measured repeatedly over time) and $(\hat{\beta}_1 + \hat{\beta}_2)$ estimates cross-sectional effects (between-transect information because transects vary in initial woody cover) (Fig. 2.12a). This model assumes that changes over time are independent of initial woody cover; it may be, however, that species richness changes more slowly on transects with less initial woody cover. A more realistic model includes $\beta_3 X_{i1} X_{ij}$ to account for interaction between temporal changes and "initial conditions"; the coefficient $\hat{\beta}_3$ estimates this effect (Fig. 2.12b).

Curvilinear Relationships: Linear and Nonlinear Models

The first step in any regression analysis should be to graph the response variable against independent variables. When this graph shows a curvilinear response, subject-matter insight may suggest potential relationships. For example, daily distance movements (Y_i) of translocated animals may be larger immediately following translocation and then decrease over time (X_1), reaching an approximate asymptote.

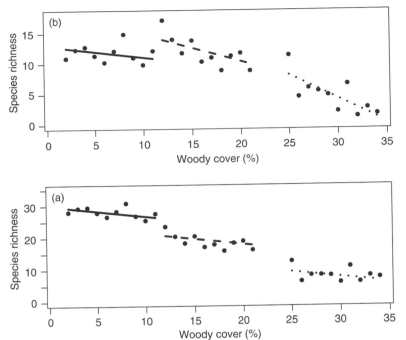

Fig. 2.12. Longitudinal regression analyses of herbaceous species richness (average of ten 0.25-m^2 quadrats on each transect) as affected by woody cover measured along three transects annually for 10 years. (a) The linear mixed model is $Y_{ij} = \beta_0 + \beta_1 X_{ij} + \beta_2 X_{i1} + b_{0i} + b_1 X_{ij} + \varepsilon_{ij}$, which does not account for interaction between initial woody cover and time: estimated regression equations for each transect are parallel. (b) $Y_{ij} = \beta_0 + \beta_1 X_{ij} + \beta_2 X_{i1} + b_{0i} + b_1 X_{ij} + \beta_3 X_{i1} X_{ij} + \varepsilon_{ij}$, a model that accommodates interaction between initial woody cover conditions and time: estimated slopes vary among transects. See associated text and Fitzmaurice et al. (2004).

a nonlinear negative exponential decay model, $Y_i = \beta_0 + \beta_1(1 - e^{\beta_2 X_{i1}}) + \varepsilon_i$, may be appropriate. Movements may also be more variable immediately following translocation and stabilize over time; WLS can accommodate heteroscedasticity (Fig. 2.13).

Although an exponential decay model may be appropriate for asymptotic responses, other settings call for different models that can involve data transformation (often with logarithms) or polynomials; in both settings, interpretation of partial regression coefficients requires care. For example, nutrient addition can enhance forage quality within a range of application, but at higher levels quality may decline; for this, a polynomial model, $Y_i = \beta_0 + \beta_1 X_{i1} + \beta_2 X_{i1}^2 + \varepsilon_i$ (which is, nevertheless, a linear model), may apply (Fig. 2.14). In this case, the effect of a unit change in X_1 on Y_i, $(\hat{\beta}_1 + 2\hat{\beta}_2 X_1)$, *depends* on X_1. Similarly, in a two-variable model with interaction, changes in $\hat{Y}_i$ to one variable depend on the level of the second variable (Fig. 2.6). Curvilinear relationships can sometimes be linearized by data transformation. For example, log transformation of Y_i commonly is used when "percentage changes in the dependent variable vary directly with changes in the independent variable" (Sokal and Rohlf 2012:526); in the model $\log_e (Y_i) = \beta_0 + \beta_1 X_{i1} + \varepsilon_i$, a c unit in X_1 multiplies the value of $\hat{Y}_i$ by $e^{c\hat{\beta}_1}$. It may be effective to use the model $Y_i = \beta_0 + \beta_1 \log_e (X_{i1}) + \varepsilon_i$ when proportional changes in X_1 produce linear changes in Y_i (Sokal and Rohlf 2012:527). In this case, the change in $\hat{Y}_i$ associated with a p% change in X_1 is $\hat{\beta}_1 \log_e [(100 + p)/100]$. Often, both dependent and independent variables are log-transformed in the model $\log_e (Y_i) = \beta_0 + \beta_1 \log_e (X_{i1}) + \varepsilon_i$ in the study of allometric relationships (Fig. 2.7). In this case, $\hat{Y}_i$ changes $100(e^{\hat{\beta}_1 (\log_e ((100+p)/100))} - 1)$% for a p% increase in X_1. It is important to appreciate, however, that although log transformation sometimes linearizes relationships, residuals associated with Y_i at X_1 may not be normally distributed.

Finally, when subject-matter insight does not inform a particular functional response, model selection for curvilinear settings can be based on visual assessment or a measure of goodness of fit (see Chapter 4). In Fig. 2.15, the number of *Baylisascaris procyonis* eggs in soil as a function of time varies with soil depth.

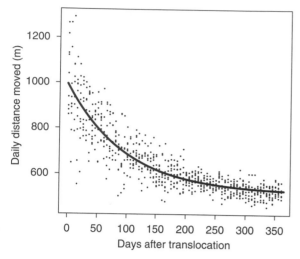

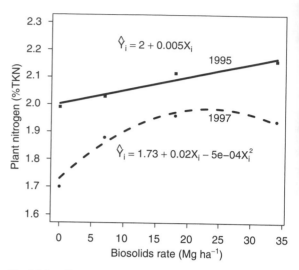

Fig. 2.13. Daily distance moved following animal translocation. The response that is modeled is $Y_i = \beta_0 + \beta_1(1 - e^{\beta_2 X_{i1}}) + \varepsilon_i$ with a weighted least squares analysis to accommodate heterogeneous variances in distance moved as a function of days after translocation.

Nonparametric Regression

Nonparametric regression involves two separate issues: (1) the distribution of Y_i at X_i, and (2) the form of the linear predictor. A nonparametric approach to the first issue might involve analyses on ranks of Y_i. For the second issue, a nonparametric approach replaces the linear parametric predictor (e.g., $\beta_0 + \beta_1 X_{i1}$ or some other function of X_i) with various smoothing functions with results conveyed graphically.

In Conover's (1999, 2012) presentation of nonparametric regression, both issues are addressed by using rank-based analyses that assume either a straight-linear relationship, or more generally a monotonic relationship between dependent and independent variables. This approach, however, is difficult to extend to multiple regression settings (Headrick and Rotou 2001) and is not widely used in wildlife applications.

Another nonparametric approach that is distribution-free but that specifies familiar statistical models involves permutation-based methods. Suppose we wish to study the relation between bird counts, Y_i, grazing intensity, X_1, and breeding season

Fig. 2.14. Plant nitrogen (% TKN Total Kjeldahl Nitrogen) as a function of biosolids application rate (Mg ha^{-1}) in 1995 ($\hat{Y}_i = 2 + 0.005X_{i1}$) and 1997 ($\hat{Y}_i = 1.73 + 0.02X_{i1} - 0.0005X_{i1}^2$) in study plots that were fertilized in 1994. This is also an example of an interaction between biosolids application rate, a continuous variable, and year of sampling, a categorical variable: the response of forage quality to nutrient addition depended on sampling year, with a linear response in 1995 and a quadratic response in 1997. *Source: Jurado and Wester (2001).*

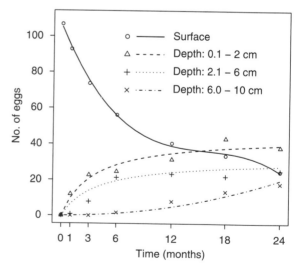

Fig. 2.15. Linear and nonlinear regression models used to predict number of parasite eggs at different soil depths in wet sand. Surface: $Y_i = \beta_0 + \beta_1 X_{i1} + \beta_1 X_{i1}^2 + \beta_1 X_{i1}^3 + \varepsilon_i$; depth 0.1–2 and 2.1–6 cm: $Y_i = \beta_1 X_{i1}/(\beta_2 + X_{i1}) + \varepsilon_i$; depth 6–10 cm $Y_i = \beta_1 X_{i1}^2 + \varepsilon_i$. *Source: Ogdee et al (2016)*

Assume Model: $Y_i = \beta_0 + \beta_1 X_{i1} + \beta_2 X_{i2} + \varepsilon_i$. For a test of $H_o : \beta_2 = 0$:

** Permutation of Y_i **

Step 1: Fit $Y_i = \beta_0 + \beta_1 X_{i1} + \beta_2 X_{i2} + \varepsilon_i$. Estimate β_2, $se(\beta_2)$ test $H_o : \beta_2 = 0$ with t_{calc}

Step 2: Randomly permute Y_i; denote these as Y_i^*. Do not permute X_1 or X_2.

Step 3: Fit $Y_i^* = \beta_0^* + \beta_1^* X_{i1} + \beta_2^* X_{i2} + \varepsilon_i$. Estimate β_2^*, $se(\beta_2^*)$ test $H_o : \beta_2^* = 0$ with t_{calc}^*

Step 4: Repeat Steps 2 and 3 many times, generating a distribution of t_{calc}^*

Step 5: Compare t_{calc} to t_{calc}^* to determine significance for test of $H_o : \beta_2 = 0$

** Permutation of residuals under 'reduced model' **

Step 1: Fit $Y_i = \beta_0 + \beta_1 X_{i1} + \beta_2 X_{i2} + \varepsilon_i$. Estimate β_2, $se(\beta_2)$ test $H_o : \beta_2 = 0$ with t_{calc}

Step 2: Fit $Y_i = \beta_0 + \beta_1 X_{i1} + \varepsilon_i$. Estimate β_0, β_1 and residuals $\hat{\varepsilon}_i$

Step 3: Residuals from Step 2 are randomly permuted and denoted $\hat{\varepsilon}_i^*$

Step 4: Using Steps 2 and 3, create $Y_i^* = \hat{\beta}_0 + \hat{\beta}_1 X_{i1} + \hat{\varepsilon}_i^*$

Step 5: Fit $Y_i^* = \beta_0^* + \beta_1^* X_{i1} + \beta_2^* X_{i2} + \varepsilon_i$. Estimate β_2^*, $se(\beta_2^*)$ test $H_o : \beta_2^* = 0$ with t_{calc}^*

Step 6: Repeat Steps 3-5 many times, generating a distribution of t_{calc}^*

Step 7: Compare t_{calc} to t_{calc}^* to determine significance for test of $H_o : \beta_2 = 0$

** Permutation of residuals under 'full model' **

Similar to 'reduced model' permutation except that residuals from Step 1 are permuted randomly, Steps 2 and 3 are removed, and Step 4 uses

$Y_i^* = \hat{\beta}_0 + \hat{\beta}_1 X_{i1} + \hat{\beta}_2 X_{i2} + \hat{\varepsilon}_i^*$.

Fig. 2.16. Steps involved in permutation-based multiple regression. *Source: Summarized from Anderson and Legendre (1999).*

rainfall, X_2: our model is $Y_i = \beta_0 + \beta_1 X_{i1} + \beta_2 X_{i2} + \varepsilon_i$. Suppose primary interest lies in testing hypotheses about the relationship between Y_i and X_2. Strategies include permutation of (1) observations, (2) residuals under the "reduced model," and (3) residuals under the "full model"; details are summarized in Fig. 2.16. Although all three approaches generally are asymptotically equivalent, control type I error rates, and have higher power than the usual OLS approach when errors are not normally distributed, reduced model permutation produces the most reliable and consistent results (Anderson and Legendre 1999).

In linear, logistic, Poisson, and negative binomial regression analyses, interest focuses on estimating the mean of the dependent variable as a function of the independent variable(s). Quantile regression (Koenker 2005) is a "semi-parametric" regression analysis that enables predictions of the response variable at locations of its distribution in addition to (or other than) the mean. For example, in a study of the thermal ecology of germinating seeds, one may wish to predict seedling emergence as a function of soil surface temperature (Fig. 2.17): whereas median seedling emergence (the 0.5 quantile) is not related to temperature, the 0.9 quantile is strongly related to temperature.

The foregoing nonparametric regression analyses require a model, and parameter estimation produces an equation useful for prediction. There is a host of additional nonparametric approaches that "do not have a formulaic way of describing the relationship between the predictors and the response" (Faraway 2006:211). These methods rely instead on graphical displays and so, while they "are less easily

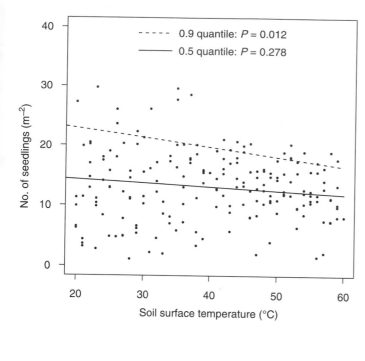

Fig. 2.17. An hypothetical example of simple quantile regression. The dependent variable is number of seedlings m⁻² and the independent variable is soil surface temperature. Quantile regression is a method that allows for inference about the estimated quantile of the response variable as a function of the independent variable. In this example, the 0.9 quantile of seedling emergence decreases ($P = 0.012$) with increasing temperature, whereas the 0.5 quantile (or median) seedling emergence is not related ($P = 0.278$) to soil surface temperature. See Koenker (2005).

communicated on paper" (Faraway 2006:212), they can be useful when little is known about an appropriate functional form of the model. Most of these methods fall into classes of techniques (Faraway 2006) that involve (1) kernel estimators (which, in their simplest form, are "just . . . moving average estimator[s]" [Faraway 2006:213]), (2) splines, and (3) local polynomials. Piecewise linear regression (Figs. 2.10, 2.11) is an application of splines where a linear function is applied between knots; more generally, however, splines divide the range of the independent variable(s) into segments and fit a polynomial of order d in each segment. Local polynomials combine "robustness ideas from linear regression and local fitting ideas from kernel methods" (Faraway 2006:221); a method called "loess" is common. To illustrate, many biological and ecological responses are periodic in nature, either in a spatial or a temporal sense (Batschelet 1981), and trigonometric models (Graybill 1976) are available. In the study of bird migration, for example, the percentage of a bird's body that is exposed to radar detection may be related to its orientation from the radar. Data in Fig. 2.18 were generated by a linear trigonometric

model; the solid line represents the known model and the dashed line represents a loess fit. Loess adequately portrays the relationship graphically but does "not provide an analytical expression for functional form of the regression relationship" (Kutner et al. 2004:140).

Model Specification, Specification Error, and Collinearity

Consider a wildlife biologist studying upland bird density. Based on an understanding of population biology and dynamics, informed by a thorough review of existing literature, and coupled with subject-matter insight and on-the-ground experience of the study area, she adopts a model that envisions bird density being affected by shrub density, breeding season rainfall, and predator abundance and specifies an appropriate model. If she assumes this model is correctly specified and assumptions are reasonably satisfied,[1] it is essential to appreciate an important philosophical and statistical implication: any variation in bird density not explained by her independent variables represents inherent variation in bird den-

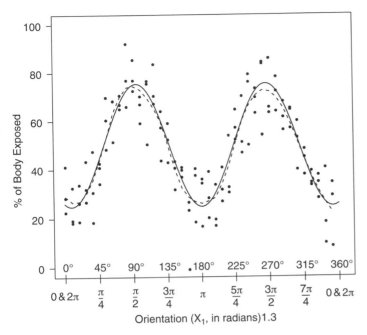

Fig. 2.18. The solid black line represents percent of a bird's body that is exposed to radar detection, Y_i, as a function of orientation from radar, with the linear trigonometric model

$$Y_i = 50 - 24\cos\left(\frac{2\pi X_{i1}}{3}\right) - 8\sin\left(\frac{2\pi X_{i1}}{3}\right) + \varepsilon_i$$

(Graybill 1976). The dashed black line represents a loess regression. *Source: Adapted from Yerrick (2016).*

sity. There is always *inherent variation* in the biological and ecological processes we study, and under the assumption that we have appropriately included relevant variables in our model, the "residual" term estimates this variation: including additional variables usually reduces error variance and may improve predictions (in a maximally overfit model, all the variation in Y_i can be explained) but overfitting seldom is encouraged (Gujarati and Porter 2009). Thus, *if the* model is correctly specified (if it includes all and only relevant variables), and accepting that biological/ecological processes can never be fully described with a finite model, it *may* be more appropriate to consider this error variation as "unexplain*able*" rather than "unexplain*ed*."

Of course, in most ecological settings it is naive to think that variation in what we study can be adequately explained by only one (or two or three) variable(s). Our biologist might suggest that, in addition to variables already included in her model, other variables (landscape metrics addressing habitat connectivity, assessment of disease, etc.) might be important. As soon as other independent variables are considered as candidates for our model, we real-

ize that we don't *know* what variables belong in our model—that the assumption that our model is the "true" model may be unrealistic. Box (1976:792) suggested that "since all models are wrong the scientist cannot obtain a 'correct' one by excessive elaboration." The use of the word "wrong" is strong; perhaps the real point is rather that our linear models probably are not exactly linear, that we may have failed to include relevant variable(s), that in fact our model is best considered an approximation of the true model (Tarpey 2009)—a reasonable view for many ecological applications. One way to explore this relies on the concept of "model specification error" (Rao 1971; Gujarati and Porter 2009; Dougherty 2011), which is best approached by assuming a known and true model and identifying consequences of misspecification—of using models that do not estimate the true model.

Suppose the model $Y_i = \beta_0 + \beta_1 X_{i1} + \beta_2 X_{i2} + \varepsilon_i$ is the true model; further, suppose that X_{i1} and X_{i2} are uncorrelated. Several important points follow: (1) in the estimated model $\hat{Y}_i = \hat{\beta}_0 + \hat{\beta}_1 X_{i1} + \hat{\beta}_2 X_{i2}$, parameter estimators are BLUE (when assumptions are met); (2) coefficients of determination, $r^2_{Y|X_i}$ (from regressing Y_i on each independent variable in separate simple regressions), will sum to $R^2_{Y|X_1, X_2}$, the coefficient of determination from the multiple regression equation; (3) the model $Y_i = \alpha_0 + \alpha_1 X_{i1} + \epsilon_i$ is referred to as a "misspecified model" (Rao 1971) because, by omitting X_{i2}, it does not estimate the true model; in this misspecified model, $\hat{Y}_i = \hat{\alpha}_0 + \hat{\alpha}_1 X_{i1}$, $\hat{\alpha}_1 = \hat{\beta}_1$: i.e., we obtain the same estimate of the regression coefficient for X_1 whether we fit the correct model (using both X_1 and X_2) or we fit the misspecified model (using only X_1). In *designed* experiments, it is often possible to have uncorrelated independent variables, but in observational studies this is rare.

In a more realistic setting for a two-variable model, independent variables are related to each other. In the estimated model, $\hat{Y}_i = \hat{\beta}_0 + \hat{\beta}_1 X_{i1} + \hat{\beta}_2 X_{i2}$, parameter estimators are BLUE just as they were when X_{i1} and X_{i2} were uncorrelated. Now, consider two kinds of model specification errors, represented by fitting (1) $\hat{Y}_i = \hat{\alpha}_0 + \hat{\alpha}_1 X_{i1}$, a misspecified model because it omits a relevant variable, and (2) $\hat{Y}_i = \hat{\beta}_0 + \hat{\alpha}_1 X_{i1} + \hat{\alpha}_2 X_{i2} + \hat{\alpha}_3 X_{i3}$, a misspecified model because it includes an irrelevant variable. These two specification errors have different consequences.

First, suppose a relevant variable, X_2, is omitted from the model. When this happens, the estimated regression coefficient for the included variable, $\hat{\alpha}_1$, is not BLUE for β_1 in the true model; in fact, $E[\hat{\alpha}_1] = \beta_1 + \beta_2 \gamma_{21}$, where γ_{21} is the slope when the omitted variable is regressed on the included variable: $\hat{X}_{i2} = \hat{\gamma}_0 + \hat{\gamma}_{21} X_{i1}$. Given that the true model includes both X_1 and X_2 but the estimated model omits X_2, if β_i are positive and X_{i1} and X_{i2} are positively correlated, $\hat{\alpha}_1$ will *overestimate* β_1; and if X_{i1} and X_{i2} are negatively correlated $\hat{\alpha}_1$ will *underestimate* β_1. Additionally, relevant variable omission biases the variance of estimators of included variables and the estimator of the residual variance, which vitiates significance tests. It is difficult to predict the impact of omitted variable bias mathematically when there are more than two explanatory variables (Dougherty 2011).[2]

This explains a common and puzzling occurrence in multiple regression. Much has been written about collinearity and its effects: Dormann et al. (2013:44) indicated that "collinearity problems cannot be 'solved,'" a conclusion echoed by Chen (2012), and that "interpretation of results must always be carried out with due caution." But even to view collinearity as "a problem" likely misses a more significant point. Leamer (1983:300) wrote: "There is no pair of words that is more misused both in econometrics texts and in the applied literature than the pair 'multicollinearity problem.'" Collinearity simply reflects the inherent and unavoidable complexity of the ecological systems we study. Much also has been written about what to do with estimated regression coefficients that "have the wrong sign." The above principle, however, provides a straightforward and effective explanation: when a relevant variable is omitted, estimated coefficients for the included variables (in addition to having smaller variances) actually reflect two things: their own influence on the dependent variable *and* the influence of omitted variables—obviously, it is a misinterpretation to conclude that a variable "has the wrong sign" only because of collinearity among variables included in the model (Chen 2012); it may be more likely the consequence of relevant variable omission. Therefore, in the simple regression of Y_i on X_1, X_1 is "acting partly as a variable in its own right and partly as a proxy for the missing" X_2 variable, and so the simple coefficient of determination reflects the "combined power of X_1 in both these roles" (Dougherty 2011:256).

Another consideration in the "wrong sign" issue likely involves a common tendency to interpret regression results in a cause-and-effect context. Many well-established texts are careful to explain that "no matter how strong is the statistical relation between X_i and Y_i, no cause-and-effect pattern is necessarily implied by the regression model" (Kutner et al. 2004:8) and that "the existence of a relationship between variables does not prove causality *or the direction of influence*" (Gujarati and Porter 2009:652–653; emphasis added). Freedman (2009) made a clear distinction between observational studies (where values of X_i are observed but not manipulated) and studies that manipulate values of X_i "by intervention," under which conditions causality inferences are stronger (see Sokal and Rohlf 2012:506ff).

The second kind of model misspecification—including an irrelevant variable—has less serious consequences. In fact, parameter estimates remain unbiased and consistent, and tests of hypothesis operate at nominal alpha-levels. This should not be interpreted to suggest that one should include all the variables one can imagine and measure: the problem will then be that these variables will likely not be independent of each other and this will increase the chance of collinearity, which in turn increases variances of parameter estimates and reduces the power of these tests (Gujarati and Porter 2009).

Interpretation of multiple regression coefficients can be complicated by two other considerations. First, interpreting a partial regression coefficient as the effect of this variable on the response variable "holding other variables constant" is inappropriate in polynomial models (Fig. 2.14) and in models that include interactions (Fig. 2.6). In a quadratic model it is not possible to vary X_1 without also varying X_1^2; and in a model with X_1, X_2, and X_1X_2, the partial regression coefficient for X_1 cannot be interpreted as the effect of X_1 on Y_i holding X_2 and X_1X_2 constant because both X_2 and X_1X_2 cannot be held constant if X_1 changes. But even in settings other than these, interpreting partial regression coefficients requires care. Kutner et al. (2004:284) provided a succinct and definitive explanation: "The important conclusion we must draw is: When predictor variables are correlated, the regression coefficient of any one variable depends on which other predictor variables are included in the model and which ones are left out. Thus, a regression coefficient does not reflect any inherent effect of the particular predictor variable on the response variable."

In fact, when the true regression model includes two or more correlated variables, the resulting collinearity (1) does not violate any assumptions of regression analysis (Achen, 1982:82), (2) does not represent model misspecification; (3) does not bias estimation of regression coefficients or their standard errors (under some conditions collinearity can reduce parameter variance estimates [Mela and Kopalle 2002] but more generally its effect is to "make it hard to get coefficient estimates with small standard error" [Achen 1982:82]); and (4) does not adversely affect prediction within the range of the data (although it can affect precision of predictions for some combinations of independent variables). Furthermore, (5) correlated variables are not always redundant (Hamilton 1987), and (6) estimated regression coefficients do not reflect any inherent effect of an independent variable on the dependent variable; rather, these coefficients (both their size and their sign) reflect the *combined* influence of the particular variable in question as well as omitted variables, which in turn means that, if a variable is thought to "have the wrong sign," the fault may be attributed to omitted variable bias, collinearity among included variables, or both, so that removing correlated variables—a common recommendation—may well "amount to throwing out the baby with the bathwater" (Hamilton 1987:132) and represents model misspecification because of variable omission. In short, once we admit that our models are incomplete—once we realize that another independent variable may matter but is not measured—then we cannot unambiguously describe the effect of an independent variable on the dependent variable. As Kutner et al. (2004) stated, "A regression coefficient does not reflect any inherent effect of the particular predictor variable on the response variable." Leamer (1983:301) wrote: "Better that we should rightly accept the fact that our non-experiments are sometimes not very informative about parameters of interest." An observation-based regression study (properly analyzed) produces a bottle that fits the data: it is best regarded as a curve-fitting exercise rather than discovery of cause and effect.

These principles can be demonstrated when one knows the correct model and explores consequences of specification error. But it is also true that one rarely knows the true model. A common interaction with colleagues goes something like this: "If I include distance-to-edge, rainfall, and predator abundance, my model doesn't make sense: it says that 'nest survival increases with predator abundance'; but if I remove rainfall, then the coefficients make more sense." At play here is a combination of effects of collinearity and variable omission in an everyday setting in wildlife science: our attempt to approximate reality with almost surely incomplete models that include related variables in observational studies.

Friedman (1953:5) suggested that "the only relevant test of the validity of a hypothesis is comparison of its predictions with experience," and in a regression context, the model represents our hypothesis. If the sample size is large enough, it is recommended to use a portion of the data set to estimate model parameters and the remainder of the data set to validate the model. In interpretating and applying regression results, therefore, it is helpful to (1) begin with an understanding of inferential scope and causality in observational and manipulative studies; (2) complement George E.P. Box's statement that "all models are wrong, but some are useful" with Tarpey's (2009) more hopeful perspective that "the truth is infinitely complex and a model is merely an approximation" of it; (3) appreciate what regression coefficients do (and do not) estimate; and then (4) evaluate regression results from observational studies in the light of our experience. When the goal is prediction, sizes and signs of regression coefficients are immaterial; and when the goal is understanding the influence of various independent variables on our response variable, it is simply a fact that, when our model includes correlated variables and suffers from omitted variables, variables that are included estimate their own influence as well as the influence of omitted variables.

Acknowledgments

The author would like to thank Christopher J. Nachtsheim, F. James Rohlf, and Jerrold H. Zar, as well as an anonymous reviewer, for providing thoughtful and constructive comments that improved an earlier version of this chapter.

NOTES

1. The idea here is based on Freedman (2009:210) who suggested that "there is no way to infer the 'right' model from the data unless there is strong prior theory to limit the universe of possible models"; whether there is "adequately strong theory" to support the inference is a consideration that belongs more to the ecological field than the statistical.

2. The foregoing assumed a true model and explored consequences of model misspecification—estimating a model different from the true model—as generally understood (Rao 1971; Freedman, 2009:149). As the example at the beginning of this section illustrates, however, it is unlikely that an ecologist would assume that the model being used includes all and only relevant variables. From this perspective it is important that, even though one's model is misspecified (relative to the true model), one still can have a correctly specified model for the conditional distribution of the response variable *given the particular independent variables being used* and that parameter estimates are BLUE *for the conditional model.* That is, an estimated regression model may be viewed as a model *for the conditional distribution of Y for a set of given Xs* (Tarpey 2009). Of course, when the ultimate goal of the regression analysis is a "description of a scientific law" wherein the functional relationship has a "clearly interpretable biological meaning"—when we have a "structural mathematical model" [*sensu* Sokal and Rohlf (2012:506-507); also see Freedman (2009, section 6.4)]—then it is important to use not just any set of Xs but Xs that can be interpreted accordingly.

LITERATURE CITED

Achen, C.H. 1982. Interpreting and using regression. Sage Publications, Beverly Hills.

Anderson, M. J., and P. Legendre. 1999. An empirical comparison of permutation methods for tests of partial regression coefficients in a linear model. Journal of Statistical Computation and Simulation 62:271–303.

Bacon, D. W., and D. G. Watts. 1971. Estimating the transition between two intersecting straight lines. Biometrika 58(3):525–534.

Batschelet, E. 1981. Circular statistics in biology (Mathematics in Biology). Academic Press, London.

Box, G. E. P. 1976. Science and statistics. Journal of the American Statistical Association 71(356):791–799.

Chen, G. J. 2012. A simple way to deal with multicollinearity. Journal of Applied Statistics 39(9):1893–1909.

Conover, W. J. 1999. Practical Nonparametric Statistics, 3rd ed. John Wiley and Sons, New York.

Conover, W. J. 2012. The rank transformation—an easy and intuitive way to connect many nonparametric methods to their parametric counterparts for seamless teaching introductory statistics courses. WIREs Computational Statistics 4:432–438. doi: 10.1002/wics.1216/.

Dormann, C. F., J. Elith, S. Bacher, C. Buchmann, G. Carl, G. Carré, J. R. García Marquéz, B. Gruber, B. Lafourcade, P.J. Leitão, T. Münkemüller, C. McClean, P.E. Osborne, B. Reineking, B. Schröder, A.K. Skidmore, D. Zurell, and S. Lautenbach. 2013. Collinearity: A review of methods to deal with it and a simulation study evaluating their performance. Ecogeography 36:27–46.

Dougherty, C. 2011. Introduction to econometrics. Oxford University Press, Oxford.

Eversole, C. B., S. E. Henke, D. B. Wester, B. M. Ballard, and R.L. Powell. 2017. Testing variation in the relationship between cranial morphology and total body length in the American alligator (*Alligator mississippiensis*). Herpetological Review 48(2):288–292.

Faraway, J. J. 2006. Extending the linear model with R: Generalized linear, mixed effects and nonparametric regression models. Chapman and Hall/CRC, Boca Raton, FL.

Fitzmaurice, G. M., N. M. Laird, and J. H. Ware. 2004. Applied longitudinal analysis. Wiley Interscience, Hoboken, NJ.

Freedman, D. A. 2009. Statistical models: Theory and practice, 2nd ed., Cambridge University Press, Cambridge, UK.

Friedman, M. 1953. The methodology of positive economics, pp. 3–43. In: Essays in positive economics. University of Chicago Press, Chicago.

Graybill, F. A. 1976. Theory and application of the linear model. Duxberry Press, North Scituate, MA.

Gujarati, D. N., and D. C. Porter. 2009. Basic econometrics, 5th ed., McGraw-Hill, New York.

Hamilton, D. 1987. Sometimes $R^2 > r_{yx_1}^2 + r_{yx_2}^2$: Correlated variables are not always redundant. American Statistician 41(2):129–132.

Headrick, T. C., and O. Rotou. 2001. An investigation of the rank transformation in multiple regression. Computational Statistics and Data Analysis 38(2):203–215.

Hernández, F., L. A. Brennan, S. J. DeMaso, J. P. Sands, and D. B. Wester. 2013. On reversing the northern bobwhite population decline: 20 years later. Journal of Wildlife Management 37(1):177–188.

Jurado, P., and D. B. Wester. 2001. Effects of biosolids on tobosagrass growth in the Chihuahuan desert. Journal of Range Management 54:89–95.

Koenker, R. 2005. Quantile regression. Cambridge University Press, Cambridge, UK.

Kutner, M. H., C. J. Nachtsheim, and J. Neter. 2004. Applied linear regression models, 4th ed. McGraw-Hill, Boston.

Leamer, E.E. 1983. Model choice and specification analysis. *In:* Handbook of Econometrics Vol. 1, edited by Z. Griliches and M.F. Intriligator (eds.), pp. 285-330. North Holland Publ. Co., Amsterdam.

Mela, C. F., and P. K. Kopalle. 2002. The impact of collinearity on regression analysis: The asymmetric effect of negative and positive correlations. Applied Economics 34(6):667–677.

Mood, A. M., F. A. Graybill, and D. C. Boes. 1974. Introduction to the theory of statistics, 3rd ed.. McGraw-Hill, Boston.

Ogdee, J. L., S. E. Henke, D. B. Wester, and A. M. Fedynich. 2016. Permeability and viability of *Baylisascaris procyonis* eggs in southern Texas soils. Journal of Parasitology 102(6):608–612.

Rao, P. 1971. Some notes on misspecification in multiple regressions. The American Statistician 25(5):37–39.

Scariano, S. M., J. W. Neill, and J. M. Davenport. 1984. Testing regression function adequacy with correlation and without replication. Communications in Statistics—Theory and Methods 13(10):1227–1237.

Smith, R. J. 2009. Use and misuse of the reduced major axis for line-fitting. American Journal of Physical Anthropology 140:476–486.

Sokal, R. R., and J. F. Rohlf. 2012. Biometry, 4th ed. W. H. Freeman and Co., New York.

Sorensen, G. E., D. B. Wester, and S. Rideout-Hanzak. 2012. A nondestructive method to estimate standing crop of purple threeawn and blue grama. Rangeland Ecology and Management 65:538–542.

Swift, T. L., and S. J. Hannon. 2010. Critical thresholds associated with habitat loss: A review of the concepts, evidence, and applications. Biological Reviews 85:35–53.

Tarpey, T. 2009. All models are right . . . most are useless. 2009 JSM Proceedings: Papers presented at the Joint Statistical Meeting. ISSN: 9780979174773. http://works.bepress.com/thaddeus_tarpey/44/.

Tian, S., and D. B. Wester. 1999. Aboveground biomass estimation of broom snakeweed (*Gutierrezia sarothrae*) plants. Texas Journal of Science 51(1):55–64.

Wiemers, D. W., T. E. Fulbright, D. B. Wester, J. A. Ortega-S., G. A. Rasmussen, D. G. Hewitt, and M. W. Hellickson. 2014. Role of thermal environment in habitat selection by male white-tailed deer during summer in Texas, USA. Wildlife Biology 20(1):47–56.

Yerrick, T. J. 2016. Evaluating an avian radar system in south Texas. Unpubl. MS Thesis, Texas A&M University-Kingsville.

Zemanova, M. A., H. L. Perotto-Baldivieso, E. L. Dickins, A. B. Gill, J. P. Leonard, and D. B. Wester. 2017. Impact of deforestation on habitat connectivity thresholds for large carnivores in tropical forests. Ecological Processes 6:21. https://doi.org/10.1186/s13717-017-0089-1.

3 Multivariate Models and Analyses

Erica F. Stuber,
Christopher J. Chizinski,
Jeffrey J. Lusk, and
Joseph J. Fontaine

The search for association among variables is a basic activity in all the sciences.
—Green (1978:38)

Working with Multivariate Data

The interrelated elements that shape biological systems create unique challenges for ecologists hoping to understand the causes and correlations of system processes. The diversity of species in a community, the multitude of environmental structures in a field, and the assortment of behaviors an individual expresses may be inexplicably related. By sampling broadly there is an opportunity to understand system complexity, but practical problems arise. To understand community diversity, for example, a system with 5 species requires consideration of 10 potential correlations, while 25 species may yield 300 potential correlations (Dillon and Goldstein 1984). Multivariate analysis techniques were developed to reduce the dimensionality of datasets and to explore or interpret how multiple response variables are related and influenced by explanatory variables.

Multivariate data are often presented as a matrix with columns representing the measured variables (e.g., species or environmental variables) and rows detailing the observations (e.g., sites or plots sampled). As with univariate analyses, variables can be binary, quantitative, qualitative, rank-ordered, or a mixture, with the form of the available data types guiding the choice of analysis. Because there are

many multivariate techniques, a decision tree can help biologists narrow down the possible analyses to implement on their data by considering study goals (exploratory versus testing), objectives (ordination versus classification), and data structures (Box 3.1).

Many multivariate datasets that contain noncategorical data can benefit by using data transformations for either statistical or ecological reasons. Choices made regarding outliers and transformation in data handling are critical in multivariate analyses and should be thoughtfully considered. Values of transformed variables will reflect the original data, but should be more amenable to particular analyses or ecological interpretation. Data transformations (e.g., log, square root, arcsine) help meet the statistical assumptions of the chosen sampling distribution (e.g., normality, linearity, homogeneity of variances) or make units comparable when the observations are measured on different scales. For example, an analysis of water quality might include temperature data that ranges from 0 to 70 °C, turbidity from 10 to 10000 NTU, and microcystin data from 0.35 to 32 µg/L, all of which need to be transformed to represent the same scale. Data transformations and standardizations can also be performed on the ecological distance measures commonly computed in clustering or ordination methods to reduce the influence

Box 3.1. Multivariate statistics decision tree.

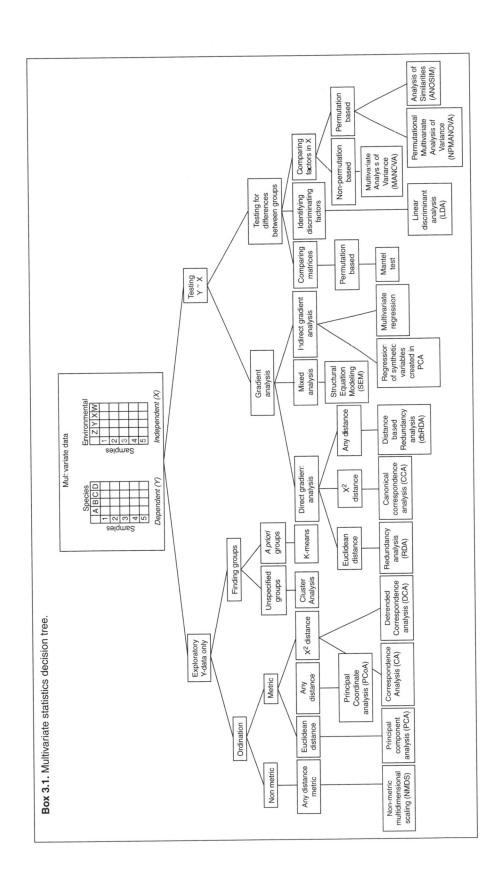

of absolute quantities or to equalize the relative importance of ubiquitous or rare species (Legendre and Legendre 1998; Legendre and Gallagher 2001). Some of the most common types of standardizations include relativized (dividing by row or column totals or maximums), z-score (subtracting the mean of the row or column and dividing by the standard deviation), Hellinger (for zero-inflated presence-absence data), and chord (weights data by rarity).

Ecological Distances

In most multivariate analyses, it is important to represent measurement (i.e., rows) and observations (i.e., columns) in your data set in terms of ecological resemblance (Fig. 3.1).

Similarity, which ranges from 1 (complete similarity) to 0 (no similarity), is the most common measure of resemblance and considers the number of measurements that observations have in common, divided by the total number of measurements taken. The complement to similarity is *dissimilarity* (i.e., dissimilarity = 1 − similarity). Alternatively, *distance* between measurements can also quantify ecological resemblance. The most common distance measure is Euclidean distance, which uses the Pythagorean theorem to measure distance between two points in multidimensional space (rather than geographic space) and is applicable to data of any scale. Some multivariate techniques are limited to a single distance measure (i.e., principal component analysis,

correspondence analysis) based on the type of data collected, but other techniques are more flexible (i.e., principal coordinate analysis, nonmetric multidimensional analysis). Each distance measure (Table 3.1) has its own strengths and weaknesses that must be considered before choosing an analysis technique (Legendre and Legendre 1998; McCune et al. 2002; Greenacre and Primicerio 2013).

Exploring Relationships in Multivariate Data
Principal Component Analysis

Principal component analysis (PCA) is one of the earliest methods developed to reduce the dimensionality of multivariate data (Hotelling 1933) and is widely used in ecology (James and McCulloch 1990). In many cases, PCA is used for exploratory data analysis (Ramette 2007) or for developing composite values for more complex methods (Janžekovič and Novak 2012). PCA reduces data into so-called components through linear combinations of variables that maximize the variance explained in the original data (James and McCulloch 1990; Legendre and Legendre 1998; Robertson et al. 2001; Ellison and Gotelli 2004). For example, a PCA of fish morphometric data might help determine which measurements to discontinue to reduce animal handling time (Fig. 3.2). Or, PCA could also be used to identify whether population parameters differ based on conservation or management practices such as no-take areas or species protected areas.

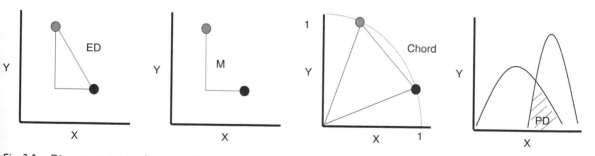

Fig. 3.1. Diagrammatic visualizations of common distance metrics. Distances between two points (e.g., species, or sites) are shown in black and gray. ED = Euclidean distance; M = Manhattan distance; Chord = Chord distance; PD = percent dissimilarity (hatched overlap), for example between abundance of two species along an environmental gradient.

Table 3.1. Comparison and formulation of distance measures commonly used in multivariate analyses. In each equation we are calculating the distance, in p dimensions, between sample units j and k from a matrix of observations of measured variables on sampling units (e.g., plots or individuals).

	Input data (x)	Range of distance value	Metric	Equations	Notes
Euclidean distance	any	nonnegative	y^1	$$ED_{jk} = \sqrt{\sum_{i=1}^{p}(x_{ij} - x_{ik})^2}$$	Pythagorean theorem applied to p dimensions; used in eigenvector ordination; literal distances, directly interpretable; sensitive to large outliers
Manhattan distance	any	nonnegative	y^1	$$M_{jk} = \sqrt{\sum_{i=1}^{p}|x_{ij} - x_{ik}|}$$	"City-block" distance; not compatible with Linear Discriminant Analysis, (LDA) Constrained Correspondence Analysis (CCA)
Chi-square distance	≥ 0	$d \geq 0$	y^1	$$X_{jk}^2 = \sqrt{\sum_{i=1}^{p}\frac{1}{x_{+i}}\times\left(\frac{x_{ij}}{x_{j+}} - \frac{x_{ik}}{x_{k+}}\right)^2}$$	Euclidean but standardized by the mean; used for species abundance data in CA and CCA
Chord distance	any	$0 \leq d \leq \sqrt{2}$	y^1	$$Chord_{jk} = \sqrt{\sum_{i=1}^{p}\left[\frac{x_{ij}}{\sqrt{\sum_{i=1}^{p}x_{ij}^2}} - \frac{x_{ik}}{\sqrt{\sum_{i=1}^{p}x_{ik}^2}}\right]^2}$$	Euclidean distance (ED) on a hypersphere of radius = 1; adjusted ED, by normalizing observations
Percent dissimilarity	≥ 0	$0 \leq d \leq 1^*$	n^2	$$PD_{jk} = 100 \times\left[\frac{\sum_{i=1}^{p}|x_{ij} - x_{ik}|}{\sum_{i=1}^{p}(x_{ij} + x_{ik})}\right]$$	Overlap between area under curves; not compatible with LDA, CCA
Bray-Curtis/ Sorensen distance	≥ 0	$0 \leq d \leq 1^*$	n^3	$$BC_{jk} = 100 \times\left[1 - \frac{2\sum_{i=1}^{p}\min(x_{ij}, x_{ik})}{\sum_{i=1}^{p}(x_{ij} + x_{ik})}\right]$$	Used for raw count data in Principal Components Analysis (PCA) and Nonmetric Dimensional Scaling (NMDS); sensitive to large outliers
Jaccard distance	≥ 0	$0 \leq d \leq 1^*$	Y^1	$$JD_{jk} = 100 \times\left[1 - \frac{\sum_{i=1}^{p}\min(x_{ij}, x_{ik})}{\sum_{i=1}^{p}(x_{ij} + x_{ik}) - \sum_{i=1}^{p}|x_{ij} - x_{ik}|}\right]$$	Proportion measure in Manhattan space

*: proportional measurements

1. Metric measurements satisfy four criteria: (1) minimum distance is 0 and occurs between identical observations; (2) distance will be positive if observations are not identical; (3) distances are symmetric; and (4) satisfies the triangle inequality

2. Non-metric measurements can take negative values

3. Semi-parametric measurements do not satisfy the triangle inequality
   j+ and k+ denote row sums
   +i denotes a sum of row sums

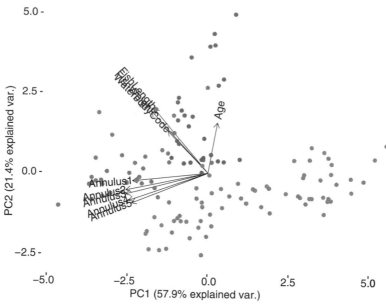

Fig. 3.2.   A principal component biplot depicting fish data collected from multiple waterbodies (see Table 3.2). Points correspond to measures of individual fish colored by species: black: largemouth bass; light gray: channel catfish; dark gray: bluegill. Points that are close together are individuals with similar scores for the measured variables (e.g., length and weight). Vectors represent loadings on the first two principal components, and each vector's length approximates the standard deviation in each original variable; all annulus measurements load strongly on PC1 (x-axis), and fish weight, length and age, as well as waterbody load on PC2 (y-axis). Taking the cosines of the angles between two vectors approximates the correlation between original variables. Generally speaking, if a point lies in the top right quadrant, it can be characterized by positive values of both components 1 and 2, while points that lie in bottom right quadrant would be characterized by positive component 1 values and negative component 2 values. The depiction demonstrates that fish species separate in multivariate space, as principal component scores do not overlap much between species. A fisheries biologist might use these results to classify unidentified fish species, or decide to stop collecting both fish length and weight when handling fish as they are highly redundant in this system.

Using Euclidean distances, each successive composite principal component is orthogonal to the last (reducing correlation) and explains less variation than the previous component (i.e., the first component explains the most variation). In many cases, most of the variation in the original data is explained by the first two dimensions and is displayed as a biplot (Fig. 3.2). The points on the biplot represent observations positioned in ordination space by their principal components such that component 1 is represented on the x-axis and component 2 on the y–axis (Jolicoeur and Mosimann 1960). The interpretation of ordination is dependent on whether a distance biplot (intersample relationships; scaling 1) or a correlation biplot (interspecies correlations;

scaling 2) is used, because observation scores are rescaled as a function of the scaling choice (Legendre and Legendre 1998; Ramette 2007).

A PCA can be performed on a correlation matrix or, if the data are collected on the same scale, on a variance-covariance matrix generated from the original data. An eigenanalysis of the variance-covariance matrix can calculate variances within each principal component, and the percent of variance explained by the components (Table 3.2; James and McCulloch 1990). The mathematics behind the eigenvectors and eigenvalues fall outside the scope of this chapter, but more detail is available in Rao (1952), Legendre and Legendre (1998), and most linear algebra textbooks.

Table 3.2. Variable loadings of measured variables on each principal component, which represent the coefficients of their linear combination as well as the standard deviation associated with each principal component. For example, the five measures of annulus all load strongly on component 1, but weakly on component 2, while fish length and weight do not load as strongly on component 1 as they do on component 2. We could assume that component 1 mostly reflects fish's annuli while component 2 described length and weight. The proportion of variance explained by each principal component represents how well any single principal component explains the variance in the multivariate dataset, while the cumulative proportion represents total variance explained by sequentially adding each principal component. In this case, fish biologists could safely reduce the variables they consider in analytical procedures from the nine initial variables to six principal components, without substantial loss of variance explained. Furthermore, because all five annulus measurements have very similar loadings, fish biologists might consider measuring only one annulus variable in the future.

Measurement variables	PC1	PC2	PC3	PC4	PC5	PC6	PC7	PC8	PC9
Waterbody	−0.20	0.35	0.62	−0.34	−0.56	0.04	−0.12	0.06	0.04
Fish length	−0.28	0.53	0.06	0.31	0.21	0.06	0.20	−0.67	0.10
Fish weight	−0.29	0.51	−0.09	0.36	0.18	−0.02	−0.18	0.66	−0.13
Age	0.04	0.42	−0.71	−0.49	−0.26	−0.02	0.02	−0.05	0.02
Annulus1	−0.38	−0.07	0.10	−0.57	0.58	0.41	0.10	0.09	−0.02
Annulus2	−0.41	−0.14	−0.03	−0.17	0.16	−0.72	−0.46	−0.18	0.00
Annulus3	−0.41	−0.18	−0.06	0.03	−0.19	−0.31	0.76	0.22	0.22
Annulus4	−0.40	−0.23	−0.16	0.14	−0.30	0.23	−0.01	−0.16	−0.76
Annulus5	−0.39	−0.25	−0.23	0.20	−0.25	0.41	−0.35	−0.02	0.59
Standard deviation	2.28	1.39	1.04	0.60	0.53	0.28	0.20	0.15	0.09
Proportion of variance	0.58	0.21	0.12	0.04	0.03	0.01	0.00	0.00	0.00
Cumulative proportion	0.58	0.79	0.91	0.95	0.98	0.99	1.00	1.00	1.00

When the data are not on the same scale, the correlation matrix must be used (Ramette 2007), but it is necessary to standardize the data, which removes the original variance and is therefore not reflected in the distance between variables. Thus, the interpretation of "variance explained," or the amount accounted for by each component, depends on whether a correlation matrix or variance-covariance matrix is used (James and McCulloch 1990). This is because correlation matrices are first standardized; thus, distances between principal component scores are independent from the scales of the original data. Using a correlation matrix or a variance-covariance matrix will lead to different principal components.

PCA has no formal assumptions (James and McCulloch 1990), but there are a few limitations:

- There are no rules determining the number of potential principal components. PCA generates orthogonal statistical relationships from the data which in some cases may have no biological interpretation.

- Principal component values have little meaning outside of the data being considered (e.g., a principal component value of 0.021 is not relevant for comparison with a different data set).

- PCA should be limited to data sampled on relatively short gradients representing linear relationships. Unless data are transformed, PCA cannot account for nonlinear relationships, so if data are clustered or nonlinear, other techniques such as correspondence analysis or nonmetric multidimensional scaling are more appropriate.

Principal Coordinate Analysis

Principal coordinate analysis (PCoA), or metric multidimensional scaling, is similar to a principal component analysis but can accept distance measures other than those that are Euclidean. Indeed, a PCoA using Euclidean distances produces the same results as a PCA, but there are instances where Euclidean distances are not appropriate (e.g., many double zeros in species presence/absence, abundance data [high beta diversity], genetic distances). For example, when studying community composition, Euclidean distance represents the distance between two sampling points in multidimensional space and does not account for species identity. As a result, two sampling points with the same species may be depicted being distinct when ordinated. Bray-Curtis distances, which account for species identity, may be more appropriate. Similar to principal components in PCA, principal coordinates in PCoA are combinations of original variables; however, principal coordinates are determined by the similarity or distance function selected (Legendre and Legendre 1998), which makes the interpretation of variable contributions distance-function-specific. Thus, the choice of dissimilarity function is critical and may require special attention when negative eigenvalues are produced (Legendre and Legendre 1998). Negative eigenvalues are indicative of convergence trouble, when PCoA cannot accurately represent a dissimilarity matrix, commonly due to the use of a nonmetric dissimilarity matrix.

Despite its ability to accommodate a broader array of distance functions, PCoA is used relatively infrequently in ecology. Many of the examples in the published literature focus on community composition or genetics across broad geographic ranges. For example, Montaño-Centellas and Garitano-Zavala (2015) used PCoA with a Gower distance (Gower 1966) to reduce the dimensionality of categorical environmental variables related to landscape type and human disturbance across an elevation gradient in the Andes. The synthesized variables from the first component and elevation were used in further analysis to predict avian species richness in the Andes.

As with PCA, there are few formal assumptions with PCoA. However, this methodology does have some limitations that should be noted:

- The choice of the dissimilarity index is important and can influence the interpretation of the data. It is important to research the strengths and weakness of the distance measure and how it will influence your results.
- As mentioned above, depending on the choice of dissimilarity matrix, there can be negative eigenvalues that will need to be remedied. In most applications, this does not influence the first few principal axes.
- Like PCA, PCoA assumes short linear gradients. Nonlinearity can result in an arc effect (see detrended correspondence analysis, discussed below, for flattening of curvature).

Correspondence Analysis

Correspondence analysis (CA) is a multivariate statistical technique originally developed to provide a graphical representation of contingency tables (e.g., standard two-way frequency tables), but are applicable for analyzing descriptors of any precision (Legendre and Legendre 1998). Consequently, ecologists frequently use CA to compare the similarity (correspondence) between sites (samples) and species abundances and represent these similarities in ordination space (Hill 1973, 1974). Correspondence analysis uses chi-square distances, which require that observations be positive and measured on the same scale (e.g., species abundances or percentage cover of vegetation types).

CA can employ numerous mathematical approaches to represent similarity, but the most common is reciprocal averaging, which follows five iterative steps: (1) a random number is assigned for all species (i.e., initial species scores); (2) for each site, a sample score is calculated as the weighted average of the initial species scores and the abundance of species; (3) for each species, a new species score is calculated as the weighted average of all site scores; (4) species and site scores are standardized again to get a mean of 0 and a standard deviation of 1; and (5) steps 2 through 4 are repeated until there are no further changes in the values.

Following a similar process as PCA and PCoA, CA decomposes the variance within the data into uncorrelated components that successively explain less variation. For each component, the overall correspondence between species scores and sample scores is summarized by an eigenvalue, which is equivalent to a correlation between species scores and sample scores (Gauch 1982). The results of CA are represented in a joint plot depicting both the site and species scores (Fig. 3.3), with the origin of the

plot indicating the centroid of all scores. As with the biplot for PCA, the choice of scaling influences the interpretation of the points in space. When the species composition changes along an environmental gradient, the sample positions may be displayed as an arc. In CA, this curve may be a mathematical consequence that both maximizes the difference between species and minimizes correlation between axes (ter Braak 1987). To remove this curvature, detrended correspondence analysis (DCA)

Fig. 3.3. Correspondence analysis of two-way contingency table of time series data of the number of fish caught representing 28 species (rows), from multiple waterbodies, over three field years (columns). Species frequencies (row values) and years (column values) are shown. The distance between species is a measure of similarity; species that are closer together have similar frequency profiles over the three years (e.g., white bass [*Morone chrysops*] and white sucker [*Catostomus commersonii*] have similar profiles, while striped bass [*Morone saxatilis*] and crappie [*Pomoxis annularis*] do not). Similarly, the distance between years is a measure of similarity. In this case, the sampling profiles of each field season are distinct. The distance between row and column points is not directly comparable; however, right angle projections of points onto vectors represent the value of that variable for each observation.

was developed. DCA flattens the distribution of the sites along the first CA axis without changing the ordination on the axis.

With CA, there are few statistical considerations to bear in mind:

- CA assumes that species have bell-shaped or Gaussian response curves to environmental variables such that a species is most abundant in the space that has the optimum environmental variables.
- CA uses the chi-square distance function, so the data must be scaled the same and be nonnegative.

Nonmetric Multidimensional Scaling

Nonmetric multidimensional scaling (NMDS) has increasingly found acceptance in ecology, particularly due to its capacity to deal with different types of data, including missing data. NMDS has the best performance characteristics of the unconstrained ordination methods in ecology (Minchin 1987).

While most ordination methods attempt to maintain the actual distance measures between points in multivariate space, there are situations where the exact distance is immaterial and the rank order of the relationships is sufficient—for example, when data are skewed, or multimodal, and may not neatly fit any statistical distribution. Nonmetric analyses, such as NMDS, allow for data without an identifiable distribution, and like the PCoA, NMDS is able to use a variety of distance measures. As opposed to other ordination methods, does not rely on eigenanalysis and does not maximize the variation explained by each axis. In fact, NMDS axes are arbitrary and can be flipped or rotated. We can compare the clustering suggestions of metric CA ordination with NMDS using the same data regarding fish species sampled from multiple waterbodies (Fig. 3.4).

Performing NMDS follows several iterative steps. The first step is to standardize the data. One of the most common methods is to standardize by the column maximum and then by the row totals (i.e., Wisconsin double standardization). Unlike the eigen-

analysis methods described in PCA, CA, and PCoA, the number of axes are described a priori in NMDS. The points are then oriented around the a priori specified number of axes at random locations (Oksanen et al. 2007). The distances from the random locations are compared to the distances in the original data using isotonic regression with a stress function (ranges between 0 and 1). The stress (i.e., measure of goodness of fit) indicates how different the ranks in the initial configuration are from the original data (Fig. 3.4). The points are iteratively moved in the direction of decreasing stress until the lowest stress value is found. The process can then be rerun with new random start locations to ensure the outcome represents the lowest stress value rather than a local minimum.

The results of NMDS are presented similar to those of PCA and CA, where species and sites are represented on a scatterplot and the proximity between observations represents the similarity between samples. However, the distances between observations do not correspond to the distances measured in the original data. Given this, the data can be rotated or flipped to help interpret the relationships between points.

Although NMDS makes few statistical assumptions, there are limitations to consider:

- Explaining the relationships between or among the points can require qualitative or subjective interpretations.
- NMDS uses an iterative process so the computation power to run the process can be substantial.

Cluster Analysis

Cluster analysis consists of a broad group of multivariate techniques that seeks to identify homogenous groups by maximizing between-group variation and minimizing within-group variation. The outcome is a reduction of observations into fewer groups (James and McCulloch 1990; Legendre and Legendre 1998). For example, a cluster analysis can spatially aggregate cases of disease on the landscape relative to noncases

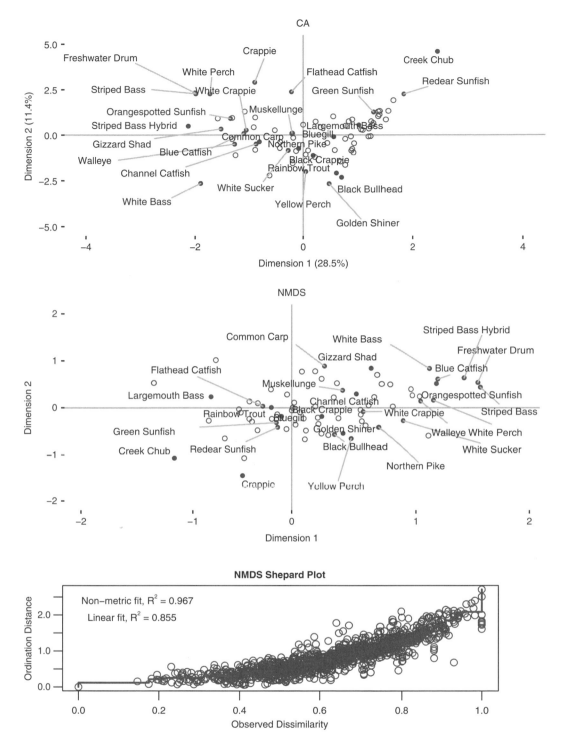

Fig. 3.4. Comparison visualizations of Correspondence Analysis (CA top) and Non-metric Multidimensional Scaling (NMDS middle) ordinations of fish species composition sampled from numerous waterbodies (open circles in ordination plots; unlabeled). In NMDS we specify two dimensions and use Bray-Curtis distances in our fish community × waterbody matrix. In CA, we do not specify the number of dimensions, but we show the first two components, which explain the greatest amount of variation in the data. Because the two analyses are based on different methods (e.g., metric vs. nonmetric, chi-square distance vs. Bray-Curtis), we might expect differences in the relationships. For example, the CA clusters species closer together around the origin than the NMDS, and we can see a weak arch effect suggesting that detrended correspondence analysis might be a better fit. The NMDS Shepard plot (bottom) depicts the degree of agreement between the final and original configurations. The low degree of scatter around the regression line, and low "stress score" (stress=0.18) suggest that reducing the dimensionality of the data to two adequately represents the original dissimilarities between communities.

in order to investigate whether landscape features can predict disease. Cluster analysis is often used in data-mining or other exploratory approaches, because it can identify previously undefined subgroups such as species assemblages and biogeographic patterns in data (Jackson et al. 2010). For example, it can classify a community based on similarities in species composition (Fig. 3.5). The technique works best when there are discontinuities in the data such as communities that occur only within discrete ecological boundaries (e.g., an agricultural field versus a forest), rather than a continuous gradient (e.g., a short-grass prairie that transitions into a tall-grass prairie) (Legendre and Legendre 1998). If there is continuous structure within the data, ordination techniques are generally preferred over classification as they assume groups respond independently to gradients, whereas cluster analysis may end up forcing the data into groups when

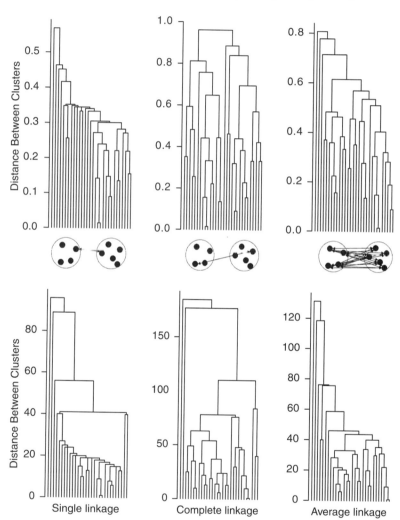

Fig. 3.5. Hierarchical agglomerative clustering of avian community sampled from 30 sites visualized as a dendrogram. Agglomerative clustering merges individual units and clusters that have the highest similarity using a linkage criterion, most commonly single, complete, or average. For example, field sites with the same species in the same abundances would have the highest similarity and be the first to cluster together. Clustering ends when all observations are merged into a single cluster (i.e., the top of the dendrogram). The composition within clusters is sensitive to the choice of clustering method and measures of distance (Orlóci 1978; Pielou 1984). For example, Bray-Curtis (top row), and Manhattan (bottom row) indices of distance produce disparate classifications and have different distance interpretations.

groups do not exist. For example, we might either use classification techniques to separate sampled individuals into one of three different subspecies or use ordination to represent the degree of similarity between the same individuals. Classification provides binary groupings, whereas ordination places subjects closer or farther from each other in multidimensional space.

Cluster analysis occurs in two general steps. First, a measure of similarity between observations is specified. For example, similarity in species abundance at different field sites (Table 3.3) is calculated in a distance matrix for pairwise combinations of field sites (rows) based on the species composition (columns) of each site. Second, using the distance measures, and after selecting a clustering rule, observations are clustered based on either a hierarchical or a partitioning technique (Kauffman and Rousseeuw 1990) (Box 3.2). When a new cluster is formed, distances between clusters and other observations (circular diagrams in middle row, Fig. 3.5) are calculated based on single linkage, or minimum distance (the distance between clusters is equal to the *shortest* distance from any member of one cluster to any member of the other); complete linkage, or maximum method (the distance between clusters is equal to the *greatest* distance from any member of one cluster to any member of the other); or average linkage (the distance between clusters is equal to the *average* distance from any member of one cluster to any member of the other).

Hierarchical techniques are useful as they can reveal relationships in a nested fashion, similar to how evolutionary relationships are depicted in a phylogenetic tree. Because hierarchical clustering requires a pairwise distance matrix between all observations, it is not very efficient for large sample sizes (e.g., >500 observations). Partitioning methods are typically applied when there are a priori reasons to group data,

Table 3.3. Example of grassland bird abundances sampled from 15 different field sites. An ecologist might be interested in differences between species (R-mode analysis) by calculating a matrix of distances between pairs of species (columns in community matrix), or differences between sites (Q-mode analysis) by calculating a distance matrix of distances between pairs of sites (rows in community matrix) (Legendre and Legendre 1998).

Site	Bobolink	Dickcissel	Eastern meadowlark	Field sparrow	Grasshopper sparrow	Horned lark	Lark bunting	Lark sparrow	Ring-necked pheasant	Upland sandpiper	Western meadowlark
1	0	6	0	2	0	0	0	0	0	0	0
2	4	6	0	0	0	0	0	0	0	0	0
3	2	0	0	0	0	2	0	1	2	2	0
4	0	2	3	0	2	2	0	0	0	0	10
5	0	2	0	0	0	0	0	0	0	14	2
6	0	2	0	0	0	0	0	0	0	0	0
7	0	0	0	0	2	4	0	0	0	0	6
8	0	4	0	0	0	0	0	0	2	0	2
9	0	14	0	0	0	0	0	0	2	0	21
10	0	0	0	0	6	2	43	0	6	0	6
11	0	0	2	0	0	2	4	0	4	0	2
12	0	0	0	0	4	6	0	0	0	0	8
13	0	6	0	0	0	2	2	0	4	0	6
14	0	4	0	0	2	4	17	0	2	0	6
15	0	2	0	0	0	0	0	0	6	0	0

Box 3.2. Overview and taxonomy of clustering process (based on Grabmeier and Rudolph 2002).

Glossary

Hierarchical- produces a hierarchy of clusters with decreasing similarity across groups as one moves down the hierarchy

Partitioning- split the data into a number of groups designated a priori by the researcher, or driven by the data

Agglomerative- individual observations are merged into similar pairs as we move up the hierarchy

Divisive- all observations begin as a single group that are subsequently divided into increasingly different groups

Monothetic- cluster membership is determined based on a single characteristic (e.g. sex)

Polythetic- cluster membership is determined by multiple criteria (e.g. sex and mass)

or when groups must be mutually exclusive (e.g., non-nested). For example, nonhierarchical cluster analyses can classify satellite imagery into land-use categories to investigate species-habitat relationships. Both defined (e.g., k-means) and data-driven (e.g., iterative self-organizing) partitioning methods follow four iterative steps: (1) randomly assign cluster centroids, (2) classify observations based on the closest centroid, (3) recalculate the group mean centroid after each observation is added, and (4) repeat steps 1–3 until the variation within cluster is minimized (Jensen 1986; Legendre and Fortin 1989). Unlike hierarchical clustering, partitioning methods do not require a data dissimilarity matrix and are generally computationally efficient, but partitioning methods are sensitive to outliers, and so outlier influence should be explored.

Interpretation of cluster analysis output is generally done either graphically (e.g., with dendrograms or geographic plots) to identify the number of groups contained in the data or with summary statistics (e.g., cluster means) that define the profile of each cluster. For example, when classifying satellite images into classes of land-use types, we can compare the means of each land-use type based on separation in multispectral light reflectance (Fig. 3.6). Separation in means of the clustering variables can highlight differences between groups and determine whether clusters are biologically distinguishable. In the case of classifying land-use types, it is possible to visualize the classifications and compare how the selection of the number of possible groups affects the classification of habitats on the landscape (Fig. 3.7). Once clusters are defined, the results can be visualized sequentially using biplots to explore how classified groups segregate over the variables measured.

Because cluster analysis is a hypothesis-generating tool, it is acceptable to use cluster analysis to explore data that do not fully meet assumptions for optimal statistical performance; however, there are limitations to consider:

• Partitioning methods are not suited to mixed data types (i.e., continuous and categorical data), and specific distance measures (e.g., Gower's distance) should be used on mixed data types in hierarchical analysis (Banfield and Raftery 1993).

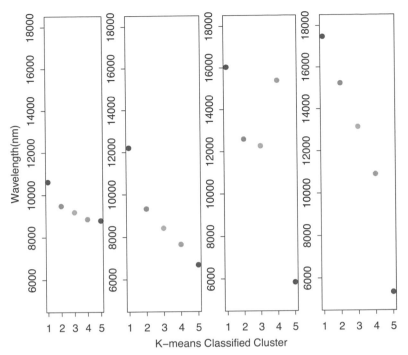

Fig. 3.6. Centroid values of five clusters segmented based on multiple clustering variables using k-means. Each panel represents a spectral band over specific ranges of light wavelengths (e.g., infrared, ultraviolet). A light reflectance of the earth measured over seven spectral bands was used to classify a satellite image into five distinct habitat types based on separation in the mean wavelengths across multiple bands. For example, observations in cluster 5 always have the lowest average wavelengths in each of the clustering variables.

- Cluster analysis assumes distance measures are based on independent variables that are normally distributed (continuous variables) or follow a multinomial distribution (categorical variables) such that clusters appear spherical, with similar variance.
- Clustering variables used are appropriate for group separation; otherwise, cluster solutions are not valid.
- Variables are mean-centered and scaled to standard deviation units (Gelman 2008). Because distance measures are strongly influenced by measurement units and magnitude, the variable with the greatest raw magnitude has an overwhelming impact on distance.
- Visual classifications are subjective, depending on the researcher's idea or expert opinion of how much distance constitutes a distinct group (Fig. 3.8).

Testing Relationships in Multivariate Data
Multivariate Regression Models

As an extension of univariate models introduced in Chapter 2 of this volume, multivariate regression models (Box 3.3) are particularly useful when the goal is to describe, explain, or decompose relationships between measures of flexible, or covarying attributes such as genes, behavior, physiology, or life-history traits (Searle 1961; Pigliucci 2003).

Two common multivariate regression models are the multivariate analysis of variance (MANOVA; an extension of the univariate ANOVA), when all predictor variables are categorical, and the multivariate analysis of covariance (MANCOVA; extension of univariate ANCOVA), when predictor variables are categorical and continuous. Contrary to other multivariate techniques, multivariate regression models are well suited to cases where the goal is not dimension reduction or for description of latent constructs or where it is not clear which variables should be considered a predictor versus a response (Dingemanse and Dochtermann 2013). For example, in community ecology the abundance of two species might correlate, but not be causally related, if an environmental condition such as precipitation is driving populations of both species independently.

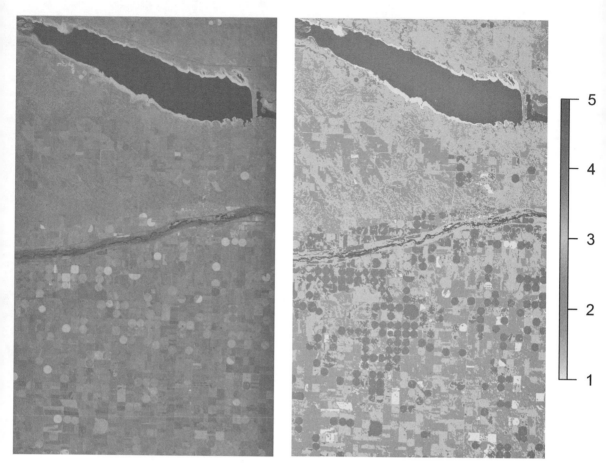

Fig. 3.7.   K-means classification of multispectral Landsat 8 satellite imagery at 30×30m spatial resolution. The left panel depicts an unclassified satellite image near Lake McConaughy, Nebraska. The right panel reflects the outcome of k-means clustering with an initial cluster value set to 5, representing hypothesized groups: water, corn fields, small grain fields, grassland, and "other." Typically, following such unsupervised classification, field data are used to "ground-truth" the classification. If data collected from the field indicate that the k-means classification did not perform well, it is necessary to investigate the characteristics of clusters that were misclassified by changing the number of groups, or including additional clustering variables to see if it is possible to increase accuracy. Once the classification reflects reality, the classification can be used for species distribution modeling or investigations of resource selection.

Multivariate regression is a powerful tool for examining the variance within and covariance among multiple interrelated measurements such as bidirectional relationships between multivariate datasets (Pigliucci 2003). Multivariate regression is particularly useful when relationships might be influenced by confounding variables (Fig. 3.9) or when we repeatedly collect multiple flexible response variables from subjects (between- and within-subject correlations may be estimated). For example, an individual may change its foraging strategy, antipredator response, or reproductive investment depending on its energy reserves. When considering highly flexible traits such as these, a simple correlation based on single measurements may not be informative because phenotypic associations are jointly shaped by correlations both between individuals and within individuals. Between-subject versus within-subject correlations can be decomposed by multivariate regression (Roff and Fairbairn 2007; Brommer 2013).

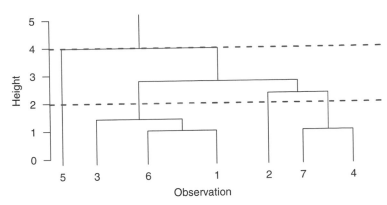

Fig. 3.8.   Schematic diagram of a hierarchical cluster analysis dendrogram. The dendrogram is read from bottom to top (in the typical agglomerative analysis), with solid horizontal lines depicting when clusters are formed and becoming increasingly nested toward the top. The position of the solid black horizontal lines shows the distance at which the observations were clustered (e.g., cluster 7-4 was joined at height = 2). Determining the number of groups represented by the dendrogram is subjective: if the similarity cutoff for defining a group (dashed lines) is 3, the dendrogram suggests four clusters in the data, whereas a less restricted cutoff of 5 would suggest two clusters. The cutoff criteria represent a trade-off between the accepted similarity of observations within defined clusters and the number of clusters considered. A lack of long stems in a dendrogram, as well as evidence of "chaining," when single observations are sequentially added, are evidence that clusters are not well-defined by the chosen clustering variables.

Box 3.3. Although the term multivariate regression is often used interchangeably with multivariable regression and multiple regression in the social science and ecological literature, they are distinct statistical concepts. Multivariable, and multiple linear regression refer to cases of a single dependent (response) variable, Y, but multiple independent (predictor) variables:

$$Y_i = \beta_0 + \beta_1 X_{i1} + \beta_2 X_{i2} + \varepsilon_i$$

but multivariate regression refers to cases with multiple dependent variables, Y, and Z:

$$Y_{ij} = \beta_{0y} + Subj_{0yj} + \varepsilon_{0yij},$$
$$Z_{ij} = \beta_{0z} + Subj_{0zj} + \varepsilon_{0zij}$$

The equations look like two univariate linear models; however, the residual error components (ε), and the random grouping effect in the case of repeated measures (*Subj*; e.g for repeated-measures on subjects) are estimated differently. Although both components (ε, *Subj*) are still assumed to have a mean of 0 as in univariate models, they are no longer considered independent across dependent variables. Therefore, we can use multivariate regression models to concurrently estimate the covariance between multiple dependent variables (Y, Z) measured simultaneously.

The results of a multivariate regression include estimated univariate models (one for each dependent variable) and the covariance or correlations between multivariate responses. The univariate results are interpreted like a simple linear regression model with coefficient estimates and associated uncertainty for each predictor on each dependent variable in the multivariate set. For example, as the temperature at a nest box increases by 1 °C, roosting birds delay sleep onset at night by 0.84 min (95% credible interval

[CI]: 0.19, 1.61min), awaken 0.75 min (95% CI 0.39, 1.05) earlier in the morning, and slightly increase the number of awakenings during the night (95% CI 0.01, 0.03); see Table 3.4). Running separate univariate regression models for each dependent variable would result in similar mean coefficients and errors (Table 3.4). In addition to the estimates the trivariate model provides pairwise correlations between responses after adjusting for fixed effects.

Similar to partitioning variance to within- and between-subject components, the covariance between multivariate responses can be partitioned within subjects or between subjects. For example, the variance in metabolic rate, daily energy expenditure, and exploratory tendencies of individual birds can be partitioned into between-individual (e.g., consistent differences) and within-individual (e.g., plasticity) components, thus making it possible to partition the covariance of traits into between-individual (e.g.,

syndrome), and within-individual (e.g., integration of plasticity) components.

As with univariate regressions, before drawing conclusions it is important to consider model fit, which is typically done through graphical residual analysis. Failure to fit the data can indicate where model assumptions were violated (Fig. 3.10).

Like univariate regression models, multivariate regression models assume:

- Observations are independent.
- The error term in our model is normally distributed with zero mean and constant variance.
- There is no pattern left unexplained by our predictors (Mertler and Reinhart 2016).

Violations of these assumptions can result in biased estimates or overly confident representations of uncertainty.

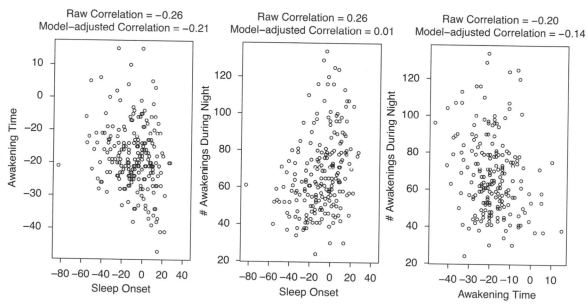

Fig. 3.9.   Relationships between three variables related to sleeping behavior: sleep onset time (minutes relative to sunset; normally-distributed), awakening time (minutes relative to sunrise; normally-distributed), and the number of awakening bouts during the course of the night (Poisson distribution). Sleep measures were taken on individual birds roosting in nest boxes in the wild. Although it is possible to calculate raw correlations (Pearson correlation) between pairwise combinations of sleep variables, the different sleep variables are affected by various environmental characteristics, including temperature, and intrinsic variables such as sex. A trivariate model will provide estimates of correlations after correcting for such confounding variables. *Source: Reanalyzed from Stuber et al. (2016).*

Table 3.4. Coefficient estimates and 95% credible intervals of the fixed effects of sex and temperature on three sleep variables measured from great tits (*Parus major*) roosting in nest boxes in the wild. Results were generated from a single tri-variate model simultaneously evaluating all three response variables (left) or three separate univariate models (right).

	Intercept (β_0)	Sex (β_1)	Evening Temperature (β_2)	Intercept (β_0)	Sex (β_1)	Evening Temperature (β_2)
Sleep onset	−13.21 (−16.77, −9.59)	4.59 (−0.11, 9.66)	0.84 (0.19, 1.61)	−13.35 (−17.50, −9.76)	4.58 (−0.52, 9.61)	0.81 (0.12, 1.56)
Awakening time	−15.21 (−16.85, −13.37)	−5.26 (−7.91, −2.93)	−0.76 (−1.05, −0.39)	−15.19 (−16.92, −13.37)	−5.19 (−7.67, −2.96)	−0.78 (−1.11, −0.46)
# Awakenings	4.11 (4.06, 4.17)	0.05 (−0.03, 0.13)	0.02 (0.01, 0.03)	4.11 (4.05, 4.17)	0.05 (−0.03, 0.13)	0.02 (0.01, 0.03)

Reanalyzed from Stuber et al. (2016).

Structural Equation Models

Understanding inherently complex ecological systems requires the ability to understand networks of relationships working simultaneously (Fig. 3.11) (Grace 2008). Structural equation modeling (SEM) provides a framework for encoding and testing hypotheses about how systems function as a whole, by analyzing the nature and magnitude of relationship webs between two or more predictor and response variables simultaneously. For example, a wildlife ecologist interested in the direct and indirect effects of avian predators on a community of prey may pose alternative hypotheses about species interactions by developing multiple possible community interaction webs based on previous work or prior knowledge of a system (Wootton 1994). The ability of SEM to evaluate support for interrelated webs of ecological processes is an analytical advance over sequential univariate analyses allowing ecologists to simultaneously evaluate support for biological hypotheses (Dochtermann and Jenkins 2011) and guiding explicit tests of proposed models through prediction and subsequent experimentation.

Path diagrams, a graphical representation of a hypothesis about sets of relationships, form the foundation of SEM (Fig. 3.11) and correspond to the underlying mathematical regression equations used for analysis (Box 3.4). Graphical path diagrams help clarify how multiple variables are related, considering measured (observed) and latent (unobserved) variables as well as direct and indirect relationships (Grace et al. 2010). Based on the path diagram, the inputs to SEM analysis include data (either raw data or correlation/covariance matrices), and the graphical paths that encode the *measurement* and *structural models* (Fig. 3.12).

The output of an SEM analysis is a set of estimated coefficients, or *loadings*, wherever a path is included in the graphical diagram. Variance associated with the measured variables, correlations between latent factors (Box 3.5), and a set of tests (e.g., Akaike information criteria) help to assess how well the proposed diagram fits the observed data (Mitchell 1992; Grace et al. 2010). Standardized path coefficients can be interpreted similarly to estimated regression coefficients (Fig. 3.13).

When latent variables (i.e., unmeasured constructs) are involved in the analysis, we will estimate factor loadings, which are similar to standardized regression coefficients, and represent the relationship between observed variables and their presumed underlying latent construct. Loadings also tell us how well any one measured variable could substitute for multiple variables when describing the latent factor.

Estimates of error associated with path estimates and factor loadings are also given, such that a p-value

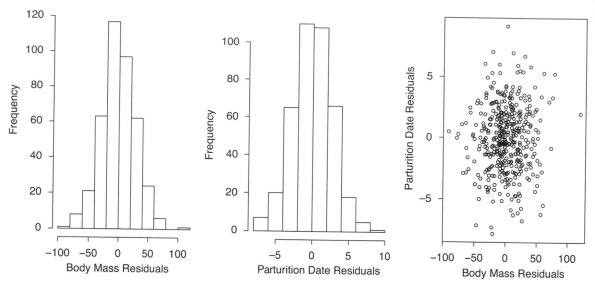

Fig. 3.10. Graphical residual analysis of multivariate regression. A biologist might be interested in the effects of supplemental feeding on size at reproduction and reproductive timing in ungulates. After collecting spring body mass and data of parturition from individual elk (*Cervus canadensis*) that were either provided supplemental feeding over-winter or not, we could model this data using a Multivariate Analysis of Variance (MANOVA) to look for mean differences in body mass and timing of reproduction between the treatment groups (supplemental feeding or control). Once residuals are extracted from our fitted model, we can plot histograms of residuals by each predictor variable, and bivariate scatterplots of residuals for each response variable. Histograms that are not symmetric and scatterplots that are not elliptical are indicative of data not sampled from a multivariate normal distribution, and we should consider transforming the data. Histograms of MANOVA residuals for elk body mass (left panel) and date of parturition (middle panel) are symmetric around zero, and appear multivariate-normal (mean=0, elliptical, uncorrelated; right panel). Because residual analysis does not show any indication of misspecification, we can draw conclusions from this model.

might be calculated for significance testing, and estimates of residual error variance are provided for observed variables. Double-headed arrows (e.g., Box 3.5, between latent variables) represent correlations between variables of interest (i.e., no directional causality is implied). For example, a behavioral ecologist might want to understand whether spatial food caching preferences are correlated with the amount of flexibility in caching behavior (Dochtermann and Jenkins 2007).

Overall SEM model fit (i.e., how well our hypothesized relationships fit the observed correlation structure in the network) is often assessed through chi-square tests, which evaluate the amount of difference between the expected and observed correlation matrices (Jöreskog and Bollen 1993; Kline 2015) or root mean squared error (RMSE), which represents

the amount of variance left unexplained by the proposed model (Steiger 1990; MacCallum et al. 1996).

If the hypothesized model does not adequately fit the observed data, it is possible to perform steps for model respecification, including identifying previously overlooked relationships or even formulating new alternative hypotheses. Once a SEM fits the data, the only way to test for causality is to perform experiments. For example, researchers have used SEM to understand and validate the relationships between winter storms and food web dynamics in kelp forests (Byrnes et al. 2011) by using estimates from the fitted SEM as predictions to test field experimentation.

In general, SEM is not as well suited for exploratory analysis because the causal relationships defined in the graphical diagram must be encoded a priori. Beyond this limitation, the structure of an

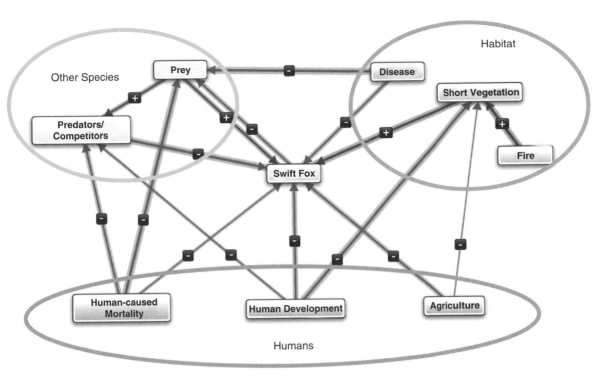

Fig. 3.11.   Hypothesized network of relationships showing the factors that affect swift fox (*Vulpes velox*) populations. Relationship webs, which are often used in community ecology, and wildlife science generally, are particularly amenable to analysis by structural equation modeling, which can simultaneously evaluate all hypothesized relationships, rather than sequentially through univariate modeling. The arrows represent direct relationships between ecological factors (e.g., other species, habitat, humans) and swift fox (the focal species) and relationships among factors that indirectly affect the focal species. The point of the arrow represents the direction of hypothesized causality of each relationship, line thickness represents expected strength of the relationship and positive/negative values qualifying relationships.

SEM also requires assumptions underlying the input data as well as the graphical models are met (Karlin et al. 1983), including:

Input data and estimation:

- Data are expected to conform to statistical assumptions of simple linear regression (see Chapter 2 on regression, this volume) as both regression and SEM are typically estimated through maximum-likelihood techniques.
- Observations are independent, or dependence is accounted for with a correlated error term.
- The joint distribution of constructs is multivariate normal; otherwise, a transformation or a non-maximum likelihood estimator is necessary.
- The model must be correctly specified.

Graphical diagrams:

- Indicators (observed variables) and constructs (latent variables) are correctly identified.
- Direction of the presumed effect or association is correctly specified.

Because regression-based structural equation models have many of the same assumptions as univariate and multivariate regressions, violations of these assumptions will result in similarly biased estimates of predictor effects and anti-conservative estimates of uncertainty.

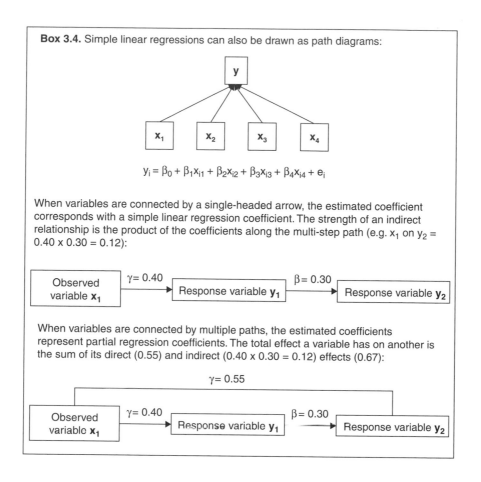

Box 3.4. Simple linear regressions can also be drawn as path diagrams:

$$y_i = \beta_0 + \beta_1 x_{i1} + \beta_2 x_{i2} + \beta_3 x_{i3} + \beta_4 x_{i4} + e_i$$

When variables are connected by a single-headed arrow, the estimated coefficient corresponds with a simple linear regression coefficient. The strength of an indirect relationship is the product of the coefficients along the multi-step path (e.g. x_1 on y_2 = 0.40 x 0.30 = 0.12):

When variables are connected by multiple paths, the estimated coefficients represent partial regression coefficients. The total effect a variable has on another is the sum of its direct (0.55) and indirect (0.40 x 0.30 = 0.12) effects (0.67):

Linear Discrimination Analysis

In ecology, groups are often predefined with questions focused on identifying what makes groups different. Linear discrimination analysis (LDA), like multinomial logistic regression, attempts to find linear combinations of continuous independent variables that best separate groups when predicting two or more dependent variables. For example, using truss-based morphometrics, Chizinski et al. (2010) used LDA to describe which morphometric measurements best differentiated the differences in body shape between stunted and nonstunted fish (Fig. 3.14). Features of the head were relatively larger in stunted fish, whereas the midbody was larger in nonstunted fish.

Like logistic regressions, LDA performs best for multivariate normal data when within group variance and covariance are similar (James and McCulloch 1990).

Permutational Multivariate Analysis of Variance

Permutational MANOVA (PERMANOVA; also sometimes referred to as "adonis," or AMOVA in multivariate molecular ecology) partitions sums of squares of a multivariate data set equivalent to MANOVA. Permutational MANOVA is an alternative to metric MANOVA and to ordination methods for describing how variation is attributed to different experimental treatments (e.g., LDA) or uncontrolled covariates (Anderson and Walsh 2013). Permutational MANOVA works with any distance function that is appropriate for the data being used. The per-

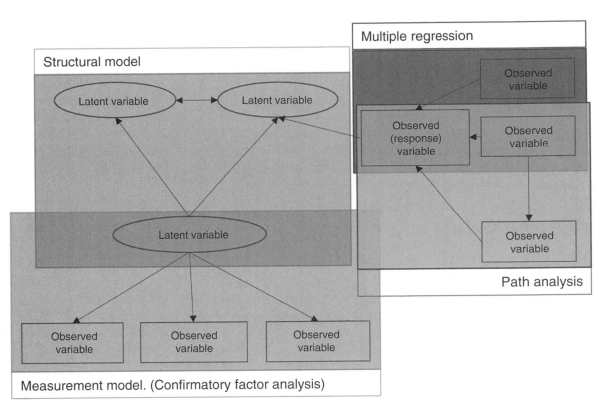

Fig. 3.12. Structural Equation Modeling (SEM) is an umbrella term encompassing many statistical techniques, including confirmatory factor analysis, path analysis, and multiple regression. *Structural models* represent the hypothesized causal relationships between dependent and independent measured (observed) and latent (unmeasured) variables, and *measurement models* depict relationships between observed variables and the expected latent variable construct.

mutations (generally 1,000 or more) of group membership are used to assess the statistical significance. Under permutation, a p-value is calculated as the proportion of the values of the test statistic that are equal to or more extreme than the observed value (Anderson and Braak 2003). In contrast to LDA, PERMANOVA is suited for predicting continuous dependent variables from categorical independent variables. For example, Edgar and Stuart-Smith (2009) used PERMANOVA to test whether two continuous variables, fish density and biomass, were significantly different inside versus outside of marine protected areas.

Because PERMANOVA is a permutation-based test, it does not have distribution-based assumptions (e.g., normality), and has only one formal assumption: observations (rows of the original data matrix) are exchangeable under a true null hypothesis

(Anderson 2001). PERMANOVA is sensitive to heterogeneity of variance within groups, particularly in unbalanced experimental design (see Table 3.5).

Mantel Test

The Mantel test compares two independent data sets for similar observations (Mantel 1967) using similarity matrices. Because the Mantel test compares similarity matrices, dependent and independent variables can be a mixture of continuous and categorical. The Mantel statistic is bounded between −1 and 1 and behaves like a correlation coefficient. The significance (p-value) of the Mantel statistic is tested by permutations of the groups as is done in PERMANOVA.

Another application of the Mantel test is for goodness of fit of ecological hypotheses against observed

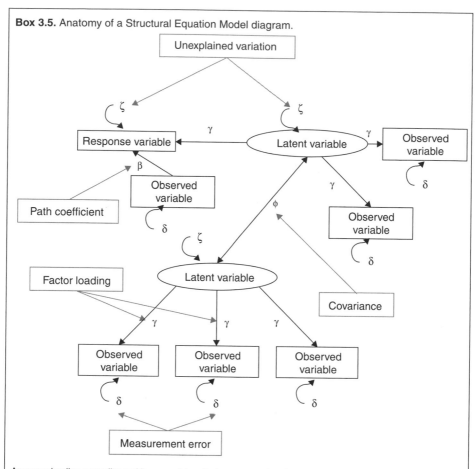

Box 3.5. Anatomy of a Structural Equation Model diagram.

An arrow implies causality and is an unelaborated summary of underlying causal mechanisms. Although little mechanistic information is contained, it represents the expected behavior of the system based on our previous knowledge. We learn about possible mechanisms by confronting our graphical models with data. Single-headed arrows represent a regression while double-headed arrows represent covariation. Fitting a structural equation model will provide us with estimates of path coefficients (β; like regression coefficients) for relationships between observed (dependent) variables and response variables, factor loading (γ; like standardized regression coefficients) for the relationship between observed variables and an underlying (unmeasured) latent variable, and correlations (ϕ) between response variables and/or latent variables.

data (Legendre and Legendre 1998). For example, Chizinski et al. (2006) proposed several hypotheses (e.g., equal occurrence, fish invasiveness, proximity, human preference) that would account for patterns of fish presence-absence in urban lakes. They used Mantel tests to compare the observed presence-absence patterns to matrices representing the nine hypotheses explaining the patterns. Results indicated that the model representing fish invasiveness best fit the observed incidence patterns.

The Mantel test has a few key assumptions:

- The relationships between the two matrices are linear.
- Observations are spatially independent (a widespread issue in ecological and evolutionary data). Spatial autocorrelation in data should be minimized during study design (Guillot and Rousset 2013) or addressed during analysis because it can produce spurious relationships (see Table 3.6).

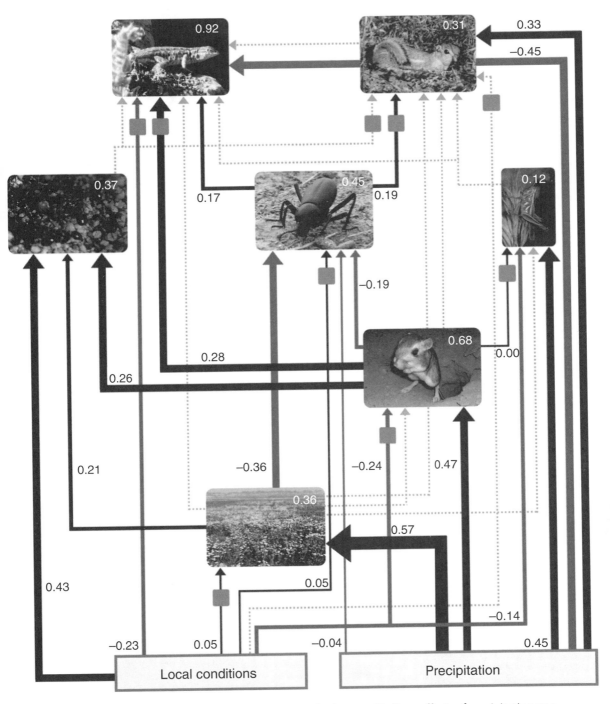

Fig. 3.13. Estimated structural equation model investigating the direct and indirect effects of precipitation on a grassland community. Precipitation can affect the animals of the community directly through modification of physiology or behavior or indirectly through effects on the plant community. Coefficients of estimated effects are interpreted similarly to regressions. For example, holding all other values constant, a 1 standard deviation increase in precipitation results in a 0.57 standard deviation increase in plant biomass. SEM model fit is assessed similarly to regression models. SEM output can return R^2 values for each variable in the model (e.g., numbers inside picture boxes). For example, local conditions and precipitation explain 68% of the variation in giant kangaroo rat (*Dipodomys ingens*) density. *Source: Deguines et al. (2017).*

(a)

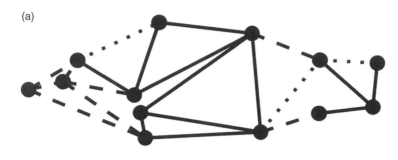

(b)

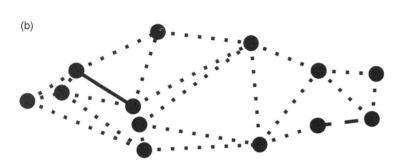

Fig. 3.14. Significant loadings and landmark coordinates between (a) stunted and nonstunted, and (b) adult and juvenile *Morone americana*. Solid lines depict morphological distances associated with greater distances in the groups, and dashed lines depict morphological distances associated with shorter distances in the groups. Dotted lines represent measures with no association between groups. *Source: Adapted from Chizinski et al. (2010).*

Table 3.5. PERMANOVA results comparing average thoracic ventilation of anuran species between control and atrazine treated plots. The anuran community (10 species) was sampled from 30 control and two sets of 30 treatment sites. Data from the first set of 30 treatment sites had the same variance as data collected from control sites (e.g., model assumption homogeneity of variance between groups was satisfied), whereas data from the second set of 30 treatment sites had much higher variance associated with it (e.g., model assumption homogeneity of variance between groups was *not* satisfied). Both sets of treatment sites had the same mean thoracic ventilation and differed only in the variance around the mean. While we were able to detect a difference in means between the control and the first set of treatment sites (left side), we were unable to detect the same difference in mean between the control and second set of treatment sites with unequal variance (right side). Based on simulated data.

	df	MS	Pseudo-F	P(perm)	df	MS	Pseudo-F	P(perm)
Treatment	1	3986	2.52	0.02	1	5389	0.64	0.75
Residuals	58	1581			58	8358		
Total	59				59			

df = degrees of freedom

MS = mean sum of squares

Pseudo-F = pseudo F statistic

P(perm) = p-value based on 999 permutations

Analysis of Similarities

Analysis of similarities (ANOSIM) is a nonmetric test for differences between two or more groups using any distance measure (Clarke 1993). ANOSIM compares the ranks of distances between groups with the rank distances within groups, thereby producing an R-statistic. The R-statistic varies from approximately 0 to 1, with 1 indicating large separation between the groups and 0 indicating little separation. The significance (p-value) of the R-statistic is tested by permutations

Table 3.6. Mantel test results testing the similarity between two matrices: simulated insect community species abundance from five species, and three simulated environmental covariates measured from 50 sampling sites. We generated environmental variables under conditions of spatial independence and spatial autocorrelation by adding an exponential variogram (sill = 1 pixel, range = 5 pixels) to the model of spatial independence. Community abundance data were only weakly dependent on the environmental characteristics generated. Using the Mantel test to determine whether two matrices are independent (community abundance matrix and environmental variables matrix, measured at 50 sampling sites), we are unable to reject the null hypothesis that the matrices are independent when data come from spatially uncorrelated sites. However, when environmental data are collected under conditions of spatial autocorrelation, the Mantel test suffers from inflated type I error (false positives) when the test's assumptions of independent samples is violated (Guillot and Rousset 2013). Indeed, in data generated with the same underlying characteristics but with spatial autocorrelation in the environmental predictors (a common occurrence in wildlife science data), the Mantel test produced a much smaller p-value.

	r	P(perm)
Spatially independent	−0.033	0.665
Spatially autocorrelated	0.089	0.086

of the groups as is done in permutational MANOVA.

ANOSIM has one key assumption: The ranges of dissimilarities within groups should be similar. Running ANOSIM on groups with very different within-group dispersions can lead to spurious results.

Indirect Gradient Analysis

While ordination techniques such as PCA and CA do not account for direct relationships between suites of species, it can be inferred that there are underlying environmental conditions that influence community composition. In PCA or CA, the synthesized observation scores and traditional statistical techniques (e.g., ANOVA, linear regression) can be used to assess the influence of an environmental variable on the synthesized scores. For example, before performing a regression analysis, Bai et al. (2016) first derived PCA scores describing the wing shape of multiple insect species. Subsequently, Bai et al. (2016) investigated how environmental characteristics including temperature, precipitation, and elevation predicted the morphometric PCA scores using linear regression. Similarly, we can use the ranks of the observations on each axis in PCoA

and NMDS and the ranks of environmental variables using Spearman's rank correlation (Legendre and Legendre 1998). For example, the plot-level NDMS scores from an NDMS analysis of a vegetation community from plots located across a broad spatial extent could be correlated with temperature, soil characteristics, and precipitation metrics across the study sites. Because this type of analysis uses the synthesized variables created from a multivariate technique and then compares them to environmental variables, it is termed "indirect gradient analysis."

Direct Gradient Analysis

As opposed to an indirect gradient analysis, which analyzes the relationship between ordination scores and environmental variables a posteriori, a direct gradient analysis uses the environmental variables directly in the ordination, forcing the scores to be maximally related to the environmental gradients. As such, only the variation in species composition explained by the environmental variables is analyzed, not all the variation in the species composition (Legendre and Legendre 1998). Thus, direct gradient analysis requires that the environmental gradient is known and represented by measured variables. For example, it is possible to perform an ordination procedure based on all of the data from a vegetation community rather than only on plant species composition data. Gradient analysis can be performed with any ordination technique, but because ordination computes optimal linear combinations of variables, there is always the assumption of linearity between ordination scores and gradients, and non-linear relationships will likely be undetected.

Redundancy Analysis

Redundancy analysis (RDA) is an extension of PCA, but is generally used for statistical testing, rather than exploration, where each canonical axis (similar to principal components) corresponds to a direction in the multivariate dependent variables (e.g., species), measured in Euclidean distance—which is maximally related to linear combinations of the multivariate independent variables (e.g., environmental conditions) (Rao 1952). Multiple linear regression is used to explain the variation between the independent and dependent variables, and through iteration, the best ordination of the observations is found. Redundancy analysis preserves the Euclidean distances between the *fitted* dependent variables (Legendre and Fortin 1989). To conduct an RDA requires a data frame corresponding to the abundance of species across sites and a data frame corresponding to environmental conditions at the sites (Fig. 3.15). RDA provides the species variation that can be explained by environmental variation, the cumulative percentage of variance of the species-environment relationship, and the overall statistical significance of the relationships between the species and environmental data frames (Legendre et al. 2011).

An extension of redundancy analysis is distance-based redundancy analysis (dbRDA), which uses non-Euclidean distance measures (Legendre and Anderson 1999). The general methodology of dbRDA is similar to RDA described above. However, dbRDA uses PCoA rather than PCA, which allows for the use

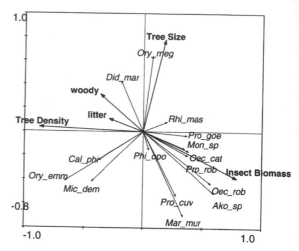

Fig. 3.15.   Biologists used redundancy analysis to describe habitat use (tree density, woody understory, litter depth, tree size) by a small mammal community and the relationships with resources (insect biomass). The angles between response (species) and explanatory vectors (tree size, tree density, woody, litter, and insect biomass) represent their correlations, and distances between species represent Euclidean distances in ordination space. Two axes of redundancy analysis explained 47% of variance in species abundance and largely agreed with multiple univariate regression analyses. *Source: Lambert et al. (2006).*

of other distance measures like Bray-Curtis. For instance, Geffen et al. (2004) used dbRDA to assess the variation of genetic distances in the gray wolf (*Canis lupus*), which is explained by temperature, rainfall, habitat type, water barriers, and climate.

Like PCA, RDA can be represented in a biplot of the species score and site scores, but RDA can also have environmental variables overlaying the scores (i.e., triplots). The environmental variables are represented as arrows for continuous variables or as additional shapes for qualitative or nominal variables (ter Braak 1994). Like the biplot projections in PCA, the choice of scaling (focus on the sites or the species) also influences how the points in space are interpreted (ter Braak 1994).

Canonical Correspondence Analysis

Canonical correspondence analysis (CCA) is similar to RDA, except that it is based on the unimodal species-environment relationship in correspondence analysis. The calculations involved in CCA, however, are more complex than RDA and can be approached using three different techniques, details of which are found in ter Braak (1986, 1987) and Legendre and Fortin (1989). Like CA, CCA preserves the chi-square distances among observations rather than the Euclidean distances in RDA (Legendre and Fortin 1989). CCA is well adapted to relate environmental data to species abundance and occurrence even when a species is absent at some sites. However, CCA is particularly sensitive to rare species that occur in species-poor samples (Legendre and Legendre 1998). To conduct a CCA, it is necessary to compare a matrix corresponding to the abundance of species across sites with another matrix of environmental data corresponding to the sites in the species data frame. The analysis provides the species variation explained by environmental variation, the cumulative percentage of variance of the species–environment relationship, and the overall statistical significance of the relationships between the species and environmental data frames.

The results of a CCA can also be represented in the biplot or triplot format, of which the interpretation of the points in multidimensional space are dependent on the type of scaling chosen. The same interpretation of the relationships between sample and species points is found in CA and CCA. Species scores are often interpreted as species' environmental optima, but this assumes that abundance or probability of occurrence follows a symmetric, unimodal distribution of position over the environmental gradients considered.

Conclusion

Many interesting questions in wildlife science regard complex patterns in ecological communities and how these patterns are influenced by interactions with multifaceted environments. Multivariate methods provide a formal framework to evaluate datasets with large numbers of often correlated variables, which are common to wildlife biology and ecology. The statistical techniques outlined in this chapter provide methods for summarizing redundancy (e.g., for variable reduction in subsequent studies) or combining variables (e.g., those that reflect unmeasured latent variables or describe composite variables) to allow us to detect and quantify multivariate patterns. This chapter has provided an overview of some of the more popular multivariate statistical techniques in wildlife ecology. If you are interested in more detail regarding one of the processes discussed or other techniques available to you, you are encouraged to read books that treat the material in much greater depth (McGarigal et al. 2000; McCune et al. 2002; Borcard and Legendre 2011; Legendre and Legendre 2012). Further, because most modern statistical programs (e.g., R, MATLAB, SAS) can conduct most types of multivariate statistics, you are free to explore these techniques and processes in much more detail in the statistical environment you are most comfortable in.

Multivariate analyses are often descriptive in nature, acting as important hypothesis-generating mechanisms, but are also used to make inferences when appropriate study design is followed (e.g., probabilistic samples are drawn from a population) and analytical assumptions are met. Although some multivariate techniques often require the investigator to make decisions regarding the level of similarity defining a group or the interpretation of latent or composite variables, they can also be viewed as highly flexible mechanisms for exploring alternative hypotheses about ecological systems. Regardless of method, multivariate or univariate, inference based on experimental manipulation is the gold standard for science. Once understood, multivariate methods are powerful tools that the wildlife scientist can use to describe the multivariate nature of wildlife systems and test complex interrelationships.

Acknowledgments

The authors are grateful to Leonard Brennan for inviting us to contribute this chapter, and all the editors for their patience and encouragement during the writing process. They would like to thank Mark

Burbach and two anonymous reviewers for their insightful comments on an earlier draft, and Lutz Gruber for help in compiling mathematical equations for Table 3.1. Christopher J. Chizinski was supported by Hatch funds through the Agricultural Research Division at the University of Nebraska-Lincoln and from Federal Aid in Wildlife Restoration project W-120-T, administered by the Nebraska Game and Parks Commission. Any use of trade, firm, or product names is for descriptive purposes only and does not imply endorsement by the U.S. government. The Nebraska Cooperative Fish and Wildlife Research Unit is supported by a cooperative agreement among the U.S. Geological Survey, the Nebraska Game and Parks Commission, the University of Nebraska, the U.S. Fish and Wildlife Service, and the Wildlife Management Institute.

LITERATURE CITED

Anderson, M. J. 2001. A new method for non-parametric multivariate analysis of variance. Austral Ecology 26(1):32–46.

Anderson, M. J., and C. T. Braak. 2003. Permutation tests for multi-factorial analysis of variance. Journal of Statistical Computation and Simulation 73(2):85–113.

Anderson, M. J., and D. C. Walsh. 2013. PERMANOVA, ANOSIM, and the Mantel test in the face of heterogeneous dispersions: What null hypothesis are you testing? Ecological Monographs 83(4):557–574.

Bai, Y., J. J. Dong, D. L. Guan, J. Y. Xie, and S. Q. Xu. 2016. Geographic variation in wing size and shape of the grasshopper Trilophidia annulata (Orthoptera: Oedipodidae): Morphological trait variations follow an ecogeographical rule. Scientific Reports 6 (32680).

Banfield, J. D., and A. E. Raftery. 1993. Model-based Gaussian and non-Gaussian clustering. Biometrics 49(3):803–821.

Borcard, D., F. Gillet, and P. Legendre. 2011. Introduction. In Numerical Ecology with R. pp. 1–7 Springer.

Brommer, J. E. 2013. On between-individual and residual (co) variances in the study of animal personality: Are you willing to take the "individual gambit"? Behavioral Ecology and Sociobiology 67(6):1027–1032.

Byrnes, J. E., D. C. Reed, B. J. Cardinale, K. C. Cavanaugh, S. J. Holbrook, and R. J. Schmitt 2011. Climate-driven increases in storm frequency simplify kelp forest food webs. Global Change Biology 17(8):2513 2524.

Chizinski, C., C. Higgins, C. Shavlik, and K. L. Pope. 2006. Multiple hypotheses testing of fish incidence patterns in an urbanized ecosystem. Aquatic Ecology 40(1):97–109.

Chizinski, C., K. L. Pope, G. Wilde, and R. Strauss. 2010. Implications of stunting on morphology of freshwater fishes. Journal of Fish Biology 76(3):564–579.

Clarke, K. R. 1993. Non-parametric multivariate analyses of changes in community structure. Australian Journal of Ecology 18(1):117–143.

Deguines, N., J. S. Brashares, and L. R. Prugh. 2017. Precipitation alters interactions in a grassland ecological community. Journal of Animal Ecology 86(2):262–272.

Dillon, W. R., and M. Goldstein. 1984. Multivariate analysis: Methods and applications. Wiley Publishing, New York, NY.

Dingemanse, N. J., and N. A. Dochtermann. 2013. Quantifying individual variation in behaviour: Mixed-effect modelling approaches. Journal of Animal Ecology 82(1):39–54.

Dochtermann, N. A., and S. H. Jenkins. 2007. Behavioural syndromes in Merriam's kangaroo rats (Dipodomys merriami): A test of competing hypotheses. Proceedings of the Royal Society of London B: Biological Sciences 274(1623):2343–2349.

Dochtermann, N. A., and S. H. Jenkins. 2011. Developing multiple hypotheses in behavioral ecology. Behavioral Ecology and Sociobiology 65(1):37–45.

Edgar, G. J., and R. D. Stuart-Smith. 2009. Ecological effects of marine protected areas on rocky reef communities— A continental-scale analysis. Marine Ecology Progress Series 388:51–62.

Ellison, G. N., and N. Gotelli. 2004. A primer of ecological statistics. Sinauer, Sunderland, MA.

Gauch, H. G. 1982. Multivariate analysis in community ecology. Cambridge University Press. New York, NY

Geffen, E., M. J. Anderson, and R. K. Wayne. 2004. Climate and habitat barriers to dispersal in the highly mobile grey wolf. Molecular Ecology 13(8):2481–2490.

Gelman, A. 2008. Scaling regression inputs by dividing by two standard deviations. Statistics in Medicine 27(15):2865–2873.

Gower, J. C. 1966. Some distance properties of latent root and vector methods used in multivariate analysis. Biometrika 53:325–338.

Grace, J. B. 2008. Structural equation modeling for observational studies. Journal of Wildlife Management 72(1):14–22.

Grace, J. B., T. M. Anderson, H. Olff, and S. M. Scheiner. 2010. On the specification of structural equation models for ecological systems. Ecological Monographs 80(1):67–87.

Greenacre, M., and R. Primicerio. 2013. Multivariate data analysis for ecologists. Foundation BBVA, Bilbao, Spain.

Guillot, G., and F. Rousset. 2013. Dismantling the Mantel tests. Methods in Ecology and Evolution 4(4):336–344.

Hill, M. O. 1973. Reciprocal averaging: An eigenvector method of ordination. Journal of Ecology 61(1):237–249.

Hill, M. O. 1974. Correspondence analysis: A neglected multivariate method. Applied Statistics 23(3):340–354.

Hotelling, H. 1933. Analysis of a complex of statistical variables into principal components. Journal of Educational Psychology 24(6):417.

Jackson, D. A., S. C. Walker, and M. S. Poos. 2010. Cluster analysis of fish community data: "new" tools for determining meaningful groups of sites and species assemblages. Pages 503–527 in K. B. Gido and D. A. Jackson, editors. Community ecology of stream fishes: concepts, approaches, and techniques. American Fisheries Society, Symposium 73, Bethesda, Maryland.

James, F. C., and C. E. McCulloch. 1990. Multivariate-analysis in ecology and systematics—Panacea or Pandora's box?. Annual Review of Ecology and Systematics 21:129–166.

Janžekovič, F., and T. Novak. 2012. PCA—A powerful method for analyzing ecological niches. In Principal component analysis—Multidisciplinary applications, edited by P. Sanguansat, pp. 127–142. Elsevier Science, Amsterdam, the Netherlands.

Jensen, J. R. 1986. Introductory digital image processing: A remote sensing perspective. Prentice-Hall, Englewood Cliffs, NJ.

Jolicoeur, P., and J. E. Mosimann. 1960. Size and shape variation in the painted turtle. A principal component analysis. Growth 24:339–354.

Jöreskog, K. G. 1993. Testing structural equation models. In Testing Structural Equation Models, edited by K.A. Bollen, and J. S. Long pp.257–294Sage Publications. Newbury Park, CA.

Karlin, S., E. Cameron, and R. Chakraborty. 1983. Path analysis in genetic epidemiology: a critique. American Journal Of Human Genetics 35(4):695.

Kauffman, L., and P. Rousseeuw. 1990. Finding groups in data. In An introduction to cluster analysis. John Wiley & Sons, Inc., Hoboken, NJ.

Kline, R. B. 2015. Principles and practice of structural equation modeling. Guilford Press, New York, NY.

Lambert, T. D., J. R. Malcolm, and B. L. Zimmerman. 2006. Amazonian small mammal abundances in relation to habitat structure and resource abundance. Journal of Mammalogy 87(4):766–776.

Legendre, P., and M. J. Anderson, 1999. Distance-based redundancy analysis: Testing multispecies responses in multifactorial ecological experiments. Ecological Monographs 69(1):1–24.

Legendre, P., and M. J. Fortin. 1989. Spatial pattern and ecological analysis. Vegetatio 80(2):107–138.

Legendre, P., and E. D. Gallagher. 2001. Ecologically meaningful transformations for ordination of species data. Oecologia 129(2):271–280.

Legendre, P., and L. Legendre. 1998. Numerical ecology. Elsevier, Amsterdam.

Legendre, P., and L. Legendre. 2012. Numerical ecology. Elsevier, Amsterdam.

Legendre, P., J. Oksanen, and C. J. ter Braak. 2011. Testing the significance of canonical axes in redundancy analysis. Methods in Ecology and Evolution 2(3): 269–277.

MacCallum, R. C., M. W. Browne, and H. M. Sugawara. 1996. Power analysis and determination of sample size for covariance structure modeling. Psychological Methods 1(2):130.

Mantel, N. 1967. The detection of disease clustering and a generalized regression approach. Cancer Research 27(2 Part 1):209–220.

McCune, B., J. B. Grace, and D. L. Urban. 2002. Analysis of ecological communities. MjM Software Design, Gleneden Beach, OR.

McGarigal, K., S. Cushman, and S. G. Stafford. 2000. Multivariate statistics for wildlife and ecology research. Springer, New York, NY.

Mertler, C. A., and R. V. Reinhart. 2016. Advanced and multivariate statistical methods: Practical application and interpretation. Routledge, New York, NY.

Minchin, P. R. 1987. An evaluation of the relative robustness of techniques for ecological ordination. Vegetatio 69:89–107.

Mitchell, R. 1992. Testing evolutionary and ecological hypotheses using path analysis and structural equation modelling. Functional Ecology 6(2):123–129.

Montaño-Centellas, F. A., and Á. Garitano-Zavala. 2015. Andean bird responses to human disturbances along an elevational gradient. Acta Oecologica 65:51–60.

Oksanen, J., F. G. Blanchet, M. Friendly, R. Kindt, P. Legendre, D. McGlinn, P. R. Minchin, R. B. O'Hara, G. L. Simpson, P. Solymos, M. H. H. Stevens, E. Szoecs, and H. Wagner 2018. vegan: Community Ecology Package R package version 2.5–2.

Orlóci, L. 1978. Ordination by resemblance matrices. In R.H. Whittaker (ed.) Ordination of plant communities, pp. 239–275. Springer, Dordrecht, the Netherlands.

Pielou, E. C. 1984. The interpretation of ecological data: A primer on classification and ordination. John Wiley & Sons, New York, NY.

Pigliucci, M. 2003. Phenotypic integration: Studying the ecology and evolution of complex phenotypes. Ecology Letters 6(3):265–272.

Ramette, A. 2007. Multivariate analyses in microbial ecology. Federation or European Microbiological Societies Microbiology Ecology 62(2):142–160.

Rao, C. R. 1952. Advanced statistical methods in biometric research. Wiley, Oxford, UK.

Robertson, M., N. Caithness, and M. Villet. 2001. A PCA-based modelling technique for predicting environmental suitability for organisms from presence records. Diversity and Distributions 7(1–2):15–27.

Roff, D., and D. Fairbairn. 2007. The evolution of trade-offs: Where are we? Journal of Evolutionary Biology 20(2):433–447.

Searle, S. 1961. Phenotypic, genetic and environmental correlations. Biometrics 17(3):474–480.

Steiger, J. H. 1990. Structural model evaluation and modification: An interval estimation approach. Multivariate Behavioral Research 25(2):173–180.

Stuber, E. F., C. Baumgartner, N. J. Dingemanse, B. Kempenaers, and J. C. Mueller. 2016. Genetic correlates of individual differences in sleep behavior of free-living great tits (*Parus major*). G3: Genes, Genomes, Genetics 6(3): 599–607.

ter Braak, C. J. 1986. Canonical correspondence analysis: A new eigenvector technique for multivariate direct gradient analysis. Ecology 67(5):1167–1179.

Ter Braak C.J.F. (1987) The analysis of vegetation-environment relationships by canonical correspondence analysis. In: Prentice I.C., van der Maarel E. (eds) Theory and models in vegetation science. Advances in vegetation science, vol 8. Springer, Dordrecht, the Netherlands.

Ter Braak, C.J. 1994. Canonical community ordination. Part I: Basic theory and linear methods. Ecoscience 1(2):127–140.

Wootton, J. T. 1994. Predicting direct and indirect effects: An integrated approach using experiments and path analysis. Ecology 75(1):151–165.

4 — Comparing Ecological Models

Mevin B. Hooten and
Evan G. Cooch

Model selection based on information theory is a relatively new paradigm in the biological and statistical sciences and is quite different from the usual methods based on null hypothesis testing.
—Burnham and Anderson (2002:3)

Why Compare Models?

Statistical models provide a reliable mechanism for learning about unknown aspects of the natural world. By their very nature, however, statistical models are placeholders for true data-generating processes. Because true data-generating processes are unknown, statistical models represent our mathematical understanding of them while accounting for inherent randomness in how the underlying ecological process operates and how the data we may observe arise. In designing statistical models, we may have multiple perspectives about the mechanisms that give rise to the data. Therefore, a critical component of the scientific process involves the assessment of model performance, often in terms of predictive ability. We value statistical models that provide accurate and precise predictions because they excel at mimicking the data-generating mechanisms. We refer to an assessment of the predictive ability of a statistical model as *validation*.

We are also generally interested in model interpretation, which is focused on the question of which variables are more or less important in predicting the response. In a high-dimensional problem, it is likely that some subset of predictor variables is not strongly associated with the response variable or, in the extreme, not associated at all (such that the estimates for these variables are zero). This interest in variable selection is often intrinsically linked to questions about relative scoring among a set of candidate models fit to a set of data.

In this chapter, we introduce the concept of model scoring for parametric statistical models, how we calculate model scores, and what we do with the scores. We place the concept of scoring in a broader discussion about parsimony and its utility for prediction. We begin with likelihood-based methods and then shift to Bayesian methods, providing examples throughout based on a generalized linear model (GLM) for avian species richness. We also discuss the relationship between model scoring and variable selection.

Scoring Models
Deviance

Given that prediction is an important indicator by which to compare models, we often rely on a quantitative metric (i.e., a score) for assessing the predictive ability of models. Prediction is a form of learning about unobserved random quantities in nature. In statistics, prediction typically refers to learning about unobserved data (e.g., at future times or new locations). For example, suppose there are two sets of data, one you collect (**y**, an $n \times 1$ vector) and one you do not collect ($\mathbf{y}_u$) (i.e., data that are unobserved). Statistical prediction involves learning about $\mathbf{y}_u$ given **y**. A point prediction results in our single best understanding of the unobserved data $\hat{\mathbf{y}}_u$ given the observed data (**y**) and statistical model M. Then, to assess our prediction, we might consider a score that measures the distance between our prediction and truth (Gneiting 2011).

Numerous issues arise when scoring statistical models based on predictive ability. First, we typically do not know the unobserved data ($\mathbf{y}_u$), so a score cannot be calculated. Second, if we did have access to the unobserved data, the score would depend on the way we measured distance. We can address the first issue by collecting two data sets—one for fitting the model and one for validating the model. If the validation data set is large enough, it will provide an accurate representation of the predictive ability. The second issue is impossible to resolve without setting some ground rules. Thus, statisticians have traditionally recommended scoring functions that are based on distances inherent to the type of statistical model being used for inference. Such scores are referred to as *proper scores* (Gneiting and Raftery 2007).

The *deviance* is a proper score (also referred to as the "logarithmic score"; Gneiting and Raftery 2007). It is proper because it involves the likelihood associated with the chosen statistical model (i.e., hypothesized mechanism that gives rise to the data). The deviance is usually expressed as $D(\mathbf{y}) = -2\log f(\mathbf{y} \mid \beta)$, which involves a function f representing the likelihood evaluated for the data based on the model parameters $\beta \equiv (\beta_1, \ldots, \beta_p)'$. We use the -2 multiplier in the deviance to be consistent with historical literature, and it implies that smaller scores indicate better predictive ability. There are other types of proper scoring functions, but because the deviance is one of the most commonly used scores, we focus on t throughout.

Validation

It is tempting to use the within-sample data **y** to score a model based on the deviance $D(\mathbf{y})$. However, the score will be "optimistic" about the predictive ability of the model because we learned about the parameters in the model using the same set of data (Hastie et al. 2009). "Optimism" is a term commonly used by statisticians to refer to an artificially inflated estimate of true predictive ability. This idea is easily understood by considering a data set with n data points, to which we wish to fit models containing as many as p predictor variables. Using a familiar least-squares approach, we seek to minimize residual sums of squares (RSS), a commonly used objective function in linear regression. However, the RSS will decrease monotonically as the number of parameters increases (often characterized by an increase in the calculated R^2 for the model). While this may seem like a positive outcome, there are two important problems. First, a low RSS (or high R^2) indicates the model has low error with respect to the within-sample data (sometimes referred to as the "training" data), when our interest is in choosing a model that has a low error when predicting out-of-sample data (validation, or "test" data). Coefficient estimates will be unbiased, and, if n is much larger than p, coefficient estimates will have low variance. However, as $p \to n$, there will be a substantial increase in variance. In the extreme, where $p \geq n$, the variance of the coefficient is infinite, even though $R^2 \approx 1$. Second, if the criterion for model selection is based solely on minimizing RSS in the training data (in the least squares context; equivalently, minimizing deviance in a likelihood framework for linear models), then the model containing all p parameters would always be selected. In fact, we want to select a model with good ability to predict the response outside the sample (i.e., we seek low test error).

Formally, the preceding relates to what is known as the *bias-variance tradeoff*. As a model becomes complex (more parameters), bias decreases, but variance of the estimates of parameter coefficients increases. Following Hastie et al. (2009), we illustrate the basis for this relationship by proposing that a response variable y can be modeled as $y = g(\mathbf{x}) + \varepsilon$. The corresponding expected prediction error is written as $E((y - \hat{g}(\mathbf{x}))^2)$. If $\hat{g}(\mathbf{x})$ is the prediction based on the within-sample data, then

$$E((y - \hat{g}(\mathbf{x}))^2) = \sigma^2 + (E(\hat{g}(\mathbf{x})) - g(\mathbf{x}))^2 + \text{var}(\hat{g}(\mathbf{x}))$$
$$= \text{irreducible error} + \text{bias}^2 + \text{variance}$$

The first term, irreducible error, represents the uncertainty associated with the true relationship that cannot be reduced by any model. In effect, the irreducible error is a constant in the expression. Thus, for a given prediction error, there is an explicit trade-off between minimizing the variance and minimizing the bias (i.e., if one goes down, the other goes up).

Out-of-sample validation helps control for optimism by using separate procedures for fitting and scoring models. Out-of-sample validation corresponds to calculating the score for the out-of-sample data. In the case where two data sets (i.e., training and validation data) are available, the deviance can be calculated for each model using plug-in values for the parameters based on the point estimates $\hat{\boldsymbol{\beta}}$ from a model fit to $\mathbf{y}$. Thus, for a set of models $M_1, \ldots, M_l, \ldots, M_L$, we calculate $D_l(\mathbf{y}_u)$ and compare to assess predictive performance (lower is better).

A completely independent second data set is not often available to use for validation. In that case, cross-validation can be useful to recycle within-sample data for scoring. In cross-validation, the data set is split into two parts—a temporary validation data set $\mathbf{y}_k$ and a temporary training data set $\mathbf{y}_{-k}$ (where the $-k$ subscript refers to the remaining set of data from $\mathbf{y}$ after the kth subset is held out). For each "fold" of training data $\mathbf{y}_{-k}$, we fit model l and calculate the score $D_l(\mathbf{y}_k)$ for the validation data set. After we have iterated through all K folds, we compute the joint score as $\sum_{k=1}^{K} D_l(\mathbf{y}_k)$ and compare among the L models to assess predictive ability.

Cross-validation is an appealing method because it automatically accounts for optimism and can be used in almost any setting without requiring a separate set of validation data. However, it is based on a finite set of data and includes a circular procedure because, presumably, the best predicting model identified in the comparison would then be fit to the entire data set for final inference and/or additional prediction. Furthermore, cross-validation can be computationally intensive if the associated algorithms require substantial computing resources. In modern computing environments, the computational burden is less of an issue than it was decades ago, but the cross-validation procedure (i.e., fitting the model for a set of folds and models) can require slightly more of an overhead programming investment.

SPECIES RICHNESS: CROSS-VALIDATION

As a case study, we consider continental U.S. bird species richness (Fig. 4.1) as a function of state-level covariates throughout this chapter. Suppose that we wish to model the bird counts (y_i) by U.S. state based on a set of state-level covariates x_i, for $i = 1, \ldots, n$ where $n = 49$ continental states in the United States (including Washington, DC) and the covariates are: state area (sq. km/1,000), average temperature (average degrees F), and average precipitation (average inches per year). Because the data y_i are nonnegative integers, a reasonable starting place for a data model for y_i is the Poisson distribution such that

$$y_i \sim \text{Pois}(\lambda_i). \tag{4.1}$$

We link the mean richness (λ_i; also known as "intensity") to the covariates (x_i) and regression coefficients ($\beta_0, \ldots, \beta_p$) using a log link function

$$\log(\lambda_i) = \beta_0 + \beta_1 x_{1,i} + \cdots + \beta_p x_{p,i} \tag{4.2}$$

for a set of covariates ($x_{j,i}, j = 1, \ldots, p$). It is common to see the regression part of the model written as $\log(\lambda_i) = \beta_0 + x_i' \boldsymbol{\beta}$ or $\log(\lambda_i) = x_i' \boldsymbol{\beta}$, depending on whether the intercept is included in $\boldsymbol{\beta}$ (in the latter case, the first element of vector x_i is 1).

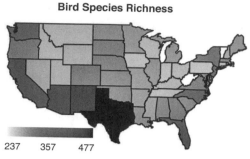

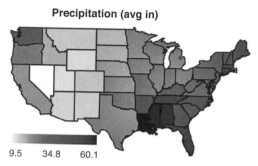

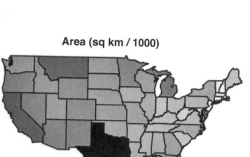

Fig. 4.1. Bird species richness in the continental United States and covariates: state area, average annual temperature, average annual precipitation.

The set of models we seek to compare are:

1. null model with only an intercept (no covariates)
2. intercept and area as covariate
3. intercept and temperature as covariate
4. intercept and precipitation as covariate
5. intercept and area and temperature as covariates

We excluded models with both state area and precipitation because they are strongly negatively correlated ($r = -0.63$). We also excluded models containing both temperature and precipitation because of moderate collinearity ($r = 0.48$). Substantial multicollinearity among covariates can cause regression models to be unstable (i.e., "irregular," more on this in what follows) and result in misleading inference (Christensen 2002; Kutner et al. 2004). Furthermore, we scaled the covariates to have mean zero and variance one before conducting all analyses. Standardizing covariates like this can reduce collinearity in some models and also puts the regression

coefficients on the same scale so that they can be more easily compared.

In this case, because we had 49 observations, we used 7-fold cross-validation, breaking the data up into seven subsets. For each fold, we fit the models to 6/7ths of the data and computed the validation score (i.e., deviance) for the remaining 1/7th of the data (summing over all folds). The resulting deviance score, calculated using cross-validation $\sum_{k=1}^{7} D_l(y_k)$, for each of our five models $l = 1, \ldots, 5$) was 762.8, 597.1, 687.8, 755.4, and 551.5, respectively. The cross-validation scores indicated that models 5 and 2, both containing the "area" covariate, perform best for prediction because they have the lowest cross-validation scores.

Information Criteria

The inherent challenges associated with scoring models based on out-of-sample validation and cross validation inspired several developments that control for optimism based only on within-sample data. In the maximum likelihood paradigm for specifying and fitting statistical models to data, two scoring approaches have been popularized: Akaike's information criterion (AIC; e.g., Akaike 1983; Burnham and Anderson 2002) and Bayes information criterion (BIC; e.g., Schwartz 1978; Link and Barker 2006). They are both similar in that they depend on the deviance as a primary component of the score, and they account for optimism in the predictive ability by penalizing the deviance based on attributes of the model or data collection process. Because smaller deviance indicates a better score, AIC and BIC penalize the score by adding a positive term to the deviance. For AIC, we calculate the score as $D(\mathbf{y}) + 2p$, where p is the number of unknown parameters β in the model. The score for BIC is calculated similarly as $D(\mathbf{y}) + \log(n)p$, with n corresponding to the dimension of the data set $\mathbf{y}$ (i.e., the sample size). Note that as $\log(n)$ becomes larger than 2, the penalty will have more influence on the score in BIC than AIC.

Within-sample scores are referred to as *information criteria* because the derivation of their penalties corresponds to certain aspects of information theory. *Information theory* arose from early work in signal processing and seeks to account for the information content in data. Given that statistics allows us to model data using probability distributions, information theory is concerned with how close the probability distribution we used to model the data is to the truth (Burnham and Anderson 2002). While it is impossible to calculate the distance between our model and the truth when the truth is unknown, the penalty used in AIC allows us to compare among a set of models to assess which is closest to the truth (with the lowest score indicating the closest). Conveniently, a separate derivation of the AIC penalty showed that it can also identify the best predicting model in certain circumstances (Stone 1977). Critically, the AIC penalty ($2p$) is a function of the number of unknown model parameters. Thus, complex models are penalized more than simpler ones. The concept of Occam's razor indicates that there is a

sweet spot in model complexity that provides the best out-of-sample predictive ability (Madigan and Raftery 1994), with highly parameterized models being poorer predictors in limited data situations. Thus, it is often said that information criteria seek to balance model fit (to within-sample data) with parsimony (reducing model complexity to control for optimism in predictive ability).

The BIC score was derived with a different goal in mind than that of AIC. Under certain conditions, BIC identifies the data-generating model out of a set of models that includes the truth. It is also naturally a good score to use for calculating weights for model averaging, again, under certain conditions (Link and Barker 2006). Both AIC and BIC tend to rank models similarly when there are large gaps in the model performance, but AIC will select more complex models in general when the differences among models are small.

SPECIES RICHNESS: INFORMATION CRITERIA

Recall that AIC is defined as AIC $= D(\mathbf{y}) + 2p$ (based on the plug-in point estimates $\hat{\boldsymbol{\beta}}$), where p is the number of model parameters (p is equal to 1, 2, 2, 2, 3 for our five models, respectively). Similarly, BIC is defined as BIC $= D(\mathbf{y}) + \log(n)p$, where $n = 49$ for our case study. For our models, the deviance is calculated as

$$D(\mathbf{y}) = -2\sum_{i=1}^{n}\log(\text{Pois}(y_i \,|\, \exp(\hat{\beta}_0 + \hat{\beta}_1 x_{1,i} + \cdots + \hat{\beta}_p x_{p,i}))), \tag{4.3}$$

where "Pois" stands for the Poisson probability mass function. We calculated AIC and BIC for each of our five models.

The results in Table 4.1 indicate that AIC and BIC are similar and agree on the ranking of the models. The information criteria also agree with the results of the cross-validation, in that models 5 and 2 are the top two models for our data. Across all models, the intercept was fairly consistent and the estimated coefficients for the "area" and "temp" predictor variables were positive while that for "precip

Table 4.1. Likelihood-based information criteria and coefficient point estimates.

Model	AIC[1]	BIC[2]	Intercept	Area	Temperature	Precipitation
1	741.1	743.0	5.761	–	–	–
2	571.2	575.0	5.755	0.100	–	–
3	669.2	673.0	5.758	–	0.069	–
4	706.1	709.8	5.759	–	–	0.049
5	526.7	532.4	5.754	0.092	0.055	–

[1]Akaike information criterion

[2]Bayes information criterion

was negative. These results imply that increases in state area and average temperature predict higher bird species richness, while an increase in average precipitation predicts lower bird species richness.

Regularization

The concept of penalizing complex models to account for optimism and improve predictive ability is much more general than the way in which it is used in AIC and BIC. *Regularization* is a type of penalization that allows users to choose a penalty that suits their goals and meshes well with their perspective about how the world works. A general scoring expression is $D(\mathbf{y}) + a\sum_{j=1}^{k}|\beta_j|^b$, where a and b are *regularization parameters* that affect the type and strength of penalty. Because the user can set b, there are infinitely many regularization forms, but the two most commonly used are the "ridge" penalty ($b=2$, equation 4a; Hoerl and Kennard 1976) and the "lasso" penalty ($b=1$, equation 4b; Tibshirani 1996):

$$D(\mathbf{y}) + a\sum_{j=1}^{k}|\beta_j|^2 \quad \text{ridge penalty,} \quad (4.4a)$$

$$D(\mathbf{y}) + a\sum_{j=1}^{k}|\beta_j| \quad \text{lasso penalty.} \quad (4.4b)$$

It is intuitive to view these penalties geometrically. For example, in the case where a model has two parameters, we can view the penalty as a shape in two-dimensional parameter space. Consider all possible values that the parameters β_1 and β_2 can as-

sume in the space depicted in Fig. 4.2. When the model is fit to a particular data set without any penalization, the point estimate $\hat{\beta}$ will fall somewhere in this space (the lower right quadrant in the example shown in Fig. 4.2). When penalized, the estimates will be "shrunk" toward zero by some amount controlled by the penalty.

The ridge penalty ($b=2$) is represented as a circle with its radius being a function of the regularization parameter a. The lasso penalty ($b=1$) is represented as a diamond where the area is a function of the regularization parameter a. For more than two parameters, the constraint shapes become higher dimensional (e.g., spheres and boxes for $p=3$). Note that the lasso constraint has corners whereas the ridge constraint is smooth. These features play an important role in the penalization. The penalized parameter estimates are snapped back to a point on the shape (i.e., where the error ellipses intersect with the shape of the particular constraint function).

The ridge and lasso coefficient estimates represent a compromise between the model fit to the available data and a penalty to account for optimistic predictive ability. The strength of the penalty is induced by the regularization parameter a. As a increases, the shapes shrink in size (i.e., distance between the edge and the origin decreases), taking the penalized parameter estimates with them. Notice that as the regularization parameter a increases past a certain point, the lasso-penalized estimate will shrink to exactly zero for one of the parameters, whereas the ridge-penalized estimate

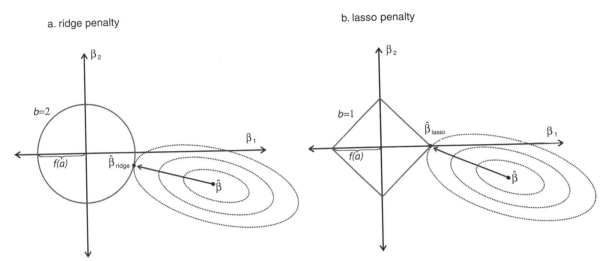

Fig. 4.2. Ridge penalty (gray circle, $b=2$) and lasso penalty (gray diamond, $b=1$) with three point estimates in parameter space for a model with two parameters (β_1, β_2): the unpenalized estimate ($\hat{\beta}$), the ridge estimate ($\hat{\beta}_{ridge}$), and the lasso estimate ($\hat{\beta}_{lasso}$). The arrows indicate the shrinkage induced by the penalty in each case. The ellipses represent contours of the distribution for the unpenalized estimate (β).

will shrink toward zero but at an asymptotic rate never reaching zero exactly until $a \rightarrow \infty$. These trajectories are a result of simple geometry. The sharp corners of the lasso penalty imply that the lasso estimates will more likely fall on a point of the diamond shape, which means that one of the two parameters will be estimated as zero (effectively removing that effect from the model, as in Fig. 4.2, where $\beta_{lasso,2} = 0$). By contrast, the ridge penalty will shrink all parameters in the model, but not to zero unless $a \rightarrow \infty$.

The application of these differences in shrinkage trajectory represents the user perspective about the world. In the case of lasso, when certain parameters are set to zero by the penalty, they are effectively removed from the model, making the model discretely less complex. By contrast, the ridge estimates leave all parameters in the model, but reduce their influence appropriately. Some have argued that the ridge penalty better mimics the real world because everything is affected by everything else, even if only by an infinitesimal amount, although this may complicate the interpretability of the model. However, lasso regularization has other beneficial properties,

such as retaining sparsity in the parameter space (by forcing some parameters to be zero) and it has become popular (Tibshirani 1996).

Regularization can also help alleviate the effects of multicollinearity on inference. When covariates are highly correlated (i.e., collinear), the associated coefficient estimates will often oppose each other (i.e., one gets large and the other gets small) because they are effectively fighting over the same type of variability in the data. In extreme collinearity cases, the parameter estimates can oppose each other strongly. Regularization shrinks the parameter estimates toward zero, thereby reducing the effects of collinearity. The resulting regularized estimates are technically biased, but have much lower variability. Ridge regression was developed for precisely this purpose. The term "regularization" is so named because it induces regularity in models (Hoerl and Kennard 1976). Because regular models have fewer parameters than data and do not have highly collinear predictor variables, regularization helps with both cases.

The catch with regularization is that the user has to choose the parameters a and b. The shape parameter b is often chosen based on the goals of the study and the desired type of shrinkage, but the strength parameter a is not as easy to set. In principle, it would be most satisfying to formally estimate a along with the other model parameters β. However, the within-sample data do not carry enough information to estimate a by themselves. Thus, a is typically set based on how well it improves the predictive ability of the model for out-of-sample data using the same validation or cross-validation techniques described in the previous section. We illustrate the cross-validation approach to regularization and finding an optimal value for a in what follows. The shape parameter b can also be chosen using cross-validation alone, or it can be selected a priori depending on whether the user wants some parameter estimates to be set to zero if necessary.

SPECIES RICHNESS: REGULARIZATION

Regularization allows for the comparison of an infinite set of models because we can include all covariates and let the penalty shrink them toward zero based on cross-validation. We fitted both ridge and lasso Poisson regression to the bird richness data set using values of a ranging from 0 to 100. We included a simulated covariate ("sim") that is strongly correlated with the "area" covariate to demonstrate the differences among coefficient estimate trajectories for the real versus simulated covariates. The resulting trajectories and cross-validation scores are shown in Fig. 4.3. Both types of regularization, ridge and lasso, shrink the coefficient estimate for the simulated covariate ("sim") to zero faster than others. The simulated covariate is shrunk exactly to zero by lasso, immediately resulting in the optimal model for prediction. With the ridge penalty, the coefficient estimate for the simulated covariate actually changes sign (from negative to positive) but ends up near zero in the optimal model for prediction. The regularization results in larger effects for the real covariates for both ridge and lasso. Between the two types of penalties, the resulting optimal score for the ridge penalty was better (542.7) than the score for lasso (544.2).

Scoring Bayesian Models
Posterior Predictions

Bayesian statistics are similar to likelihood-based statistics in that a parametric probability distribution is chosen as a model for the data. The difference is that parameters are treated as unobserved random variables in Bayesian models, as opposed to fixed variables in non-Bayesian models (Hobbs and Hooten 2015). We use conditional probability statements to find the probability distribution of unknown variables (i.e., parameters and predictions) given known variables (i.e., data). Thus, if we treat our model parameters β as random variables with distribution $f(\beta)$ before the data are observed, our Bayesian goal is to find the posterior distribution $f(\beta|\mathbf{y})$ after the data are observed. Using conditional probability, we find that the posterior distribution is $f(\beta|\mathbf{y}) = f(\mathbf{y}|\beta)f(\beta)/f(\mathbf{y})$, which is a function of likelihood $f(\mathbf{y}|\beta)$, the prior $f(\beta)$, and the marginal distribution of the data $f(\mathbf{y})$ in the denominator (Hobbs and Hooten 2015). The marginal distribution of the data is often the crux in solving for the posterior distribution because it usually involves a complicated integral or sum. Therefore, we use numerical approaches such as Markov chain Monte Carlo (MCMC) algorithms for approximating the posterior distribution and associated quantities (Gelfand and Smith 1990).

The Bayesian mechanism for prediction is the posterior predictive distribution $f(\mathbf{y}_u|\mathbf{y})$. Bayesian point predictions can be calculated by $\hat{\mathbf{y}}_u = \sum_{t=1}^{T} \mathbf{y}_u^{(t)}/T$ (i.e., the mean of the posterior predictive distribution) using posterior predictive samples $\mathbf{y}_u^{(t)}$ arising from the MCMC model-fitting algorithm. However, the posterior predictive distribution provides much more information about the unobserved data as well, such as the uncertainty in our predictions. Thus, while it is tempting to use only the point predictions, we can obtain a much deeper understanding of what we know, and do not know, about the things we seek to predict using additional characteristics from the posterior predictive distribution.

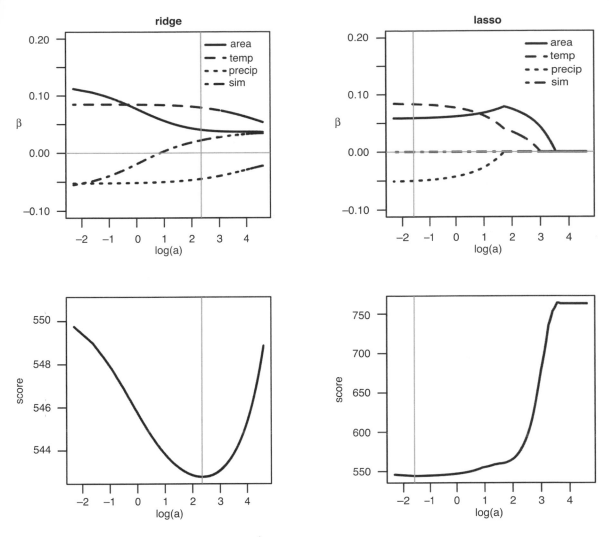

Fig. 4.3. Poisson regression coefficient estimate ($\hat{\beta}$) trajectories (top panels) based on the ridge penalty (left panels; $b=2$) and lasso penalty (right panels; $b=1$); regularization score trajectories (bottom panels) based on cross-validation. Vertical gray lines represent the values for the strength parameter a based on the optimal scores.

Scoring Bayesian Models

In principle, the same concept of scoring described in previous sections can be applied to Bayesian point predictions (Vehtari and Ojanen 2012). In this case, the deviance can be calculated using the same likelihood as in the non-Bayesian models but with Bayesian point estimates as plug-in values for the parameters β. Alternatively, we can leverage one of the key benefits that comes for free with MCMC: namely, the ability to obtain inference for any function of

model components (e.g., data, predictions, and parameters). Therefore, a natural Bayesian score would be the mean posterior predictive deviance $\bar{D}(\mathbf{y}_u)$, which can be calculated using MCMC samples as $\sum_{t=1}^{T} D(\mathbf{y}_u^{(t)})/T$, where t corresponds to a MCMC iteration and T is the total number of MCMC iterations. The posterior predictive distribution provides the Bayesian mechanism for obtaining predictions from a model. Thus, because the deviance can be calculated using predictions, we treat it as a derived quantity (i.e., a statistic that does not affect model fit, but that is a function of predictions or parameters in the model). MCMC makes it easy to obtain an estimate of this statistic, which becomes the score for our model. Calculating this new score using Bayesian methods, we can compare models as before based on out-of-sample data or cross-validation to account for optimism.

Bayesian models may require more time to fit than non-Bayesian models. Although Bayesian models can provide richer forms of inference, cross-validation may become infeasible for very large data sets and/or complex Bayesian models. Thus, within-sample scoring methods for Bayesian models have been developed. The most commonly used within-sample score for Bayesian models is the *deviance information criterion* (DIC; Spiegelhalter et al. 2002; Celeux et al. 2006). DIC takes the same form as AIC, with the deviance based on a plug-in Bayesian point estimate for β, plus a penalty. The DIC penalty is $2p_D$, with p_D representing the effective number of parameters. It is not possible to count parameters discretely in Bayesian models because the prior provides some information about model parameters. Instead, we can think of a measure for model complexity (i.e., the optimism) as the difference in score between the deviance that accounts for the uncertainty in model parameters ($\bar{D}$) and the deviance based on only the plug-in parameter estimates ($\hat{D}$), resulting in $p_D = \bar{D} - \hat{D}$. As it turns out, for simple models, p_D is close to the number of parameters p when priors are less informative.

An alternative Bayesian score based on within-sample data uses the posterior predictive distribution directly (Richardson 2002; Watanabe 2010). The so-called Watanabe-Akaike information criterion (WAIC; Watanabe 2013) substitutes in the logarithm of the posterior predictive density for the deviance and uses a different calculation for the penalty. Thus, WAIC is specified as $-2 \log f(\mathbf{y}|\mathbf{y}) + 2p_D$. Using the MCMC sample, WAIC can be calculated as $-2\sum_{i=1}^{n} \log \sum_{t=1}^{T} f(y_i|\beta^{(t)})/T + 2p_D$, where $p_D = \sum_{i=1}^{n} \text{var}(\log f(y_i|\beta))$ and the "var" corresponds to the variance over the posterior distribution for β. Heuristically, WAIC balances fit with parsimony to

improve predictive ability because the uncertainty will increase as the model complexity increases. Thus, as the posterior variance of the deviance increases, the model is penalized more.

SPECIES RICHNESS: BAYESIAN INFORMATION CRITERIA

Recall that DIC is defined as $\text{DIC} = \hat{D} + 2p_D$, for $p_D = \bar{D} - \hat{D}$. These different forms of deviance can be computed using MCMC output from our model using

$$\hat{D} = -2\sum_{i=1}^{n} \log(\text{Pois}(y_i|\hat{\lambda}_i)), \tag{4.5}$$

and

$$\bar{D} = -2 \frac{\sum_{t=1}^{T} \sum_{i=1}^{n} \log\left(\text{Pois}(y_i|\exp(\beta_0^{(t)} + \beta_1^{(t)}x_{1,i} + \cdots + \beta_p^{(t)}x_{p,i}))\right)}{T}, \tag{4.6}$$

where $\hat{\lambda}_i$ is the posterior mean of λ and $\beta_j^{(t)}$ is the jth coefficient on the tth MCMC iteration (for $j = 1, \ldots, p$ and a MCMC sample of size T).

Similarly, the Watanabe-Akaike information criterion is

$$\text{WAIC} = -2\sum_{i=1}^{n} \text{lppd} + 2p_D, \tag{4.7}$$

where "lppd" stands for log posterior predictive density for y_i and can be calculated using MCMC as

$$\text{lppd}_i = \log\left(\frac{\sum_{t=1}^{T} \text{Pois}(y_i|\exp(\beta_0^{(t)} + \beta_1^{(t)}x_{1,i} + \cdots + \beta_p^{(t)}x_{p,i}))}{T}\right), \tag{4.8}$$

and where Gelman and Vehtari (2014) recommend calculating p_D as

$$p_D = \sum_{i=1}^{n} \left(\frac{\sum_{t=1}^{T} (\log(\text{Pois})_i^{(t)} - \sum_{t=1}^{T} \log(\text{Pois})_i^{(t)}/T)^2}{T}\right), \tag{4.9}$$

where

$$\log(\text{Pois})_i^{(t)} = \log(\text{Pois}(y_i \mid \exp(\beta_0^{(t)} + \beta_1^{(t)} x_{1,i}$$
$$+ \cdots + \beta_p^{(t)} x_{p,i}))).$$

To specify a Bayesian model for the bird species richness data, we used the same Poisson likelihood as previously and then specified priors for the parameters. A reasonable prior for unconstrained regression coefficients is Gaussian (because the support for β_j includes all real numbers). Thus we specify

$$\beta_j \sim N(\mu_j, \sigma_j^2) \quad \text{for} \quad j = 1, \ldots, p, \quad (4.10)$$

as priors with means $\mu_j = 0$ and variances $\sigma_j^2 = 100$. Using these priors, we fit each of our models to the bird richness data and calculated DIC and WAIC.

While the values in Table 4.2 for DIC and WAIC are different, the ordering of models remains the same and is consistent with the non-Baycsian information criteria (although that may not always be true). Again, we see that models 5 and 2 provide the best predictive ability among our set of five models.

Bayesian Regularization

Comparing Bayesian models is not limited to use with out-of-sample scoring and information criteria. The concept of regularization also naturally transfers to the Bayesian setting (Hooten and Hobbs 2015). In fact, the regularization penalty already exists in

Table 4.2. Bayesian information criteria.

Model	DIC[1]	WAIC[2]
1	741.2	748.4
2	571.3	577.4
3	669.2	680.1
4	706.1	720.3
5	526.7	533.8

[1] Deviance information criterion

[2] Watanabe-Akaike information criterion

Bayesian models as the prior. To see this connection, notice that the logarithm of the numerator in conditional probability is $\log(f(\mathbf{y} \mid \beta)) + \log(f(\beta))$. Thus, multiplying by -2, we have the same regularization expression as in previous sections, but with the penalty equal to $-2 \log(f(\beta))$. Therefore, the penalty is a function of the prior.

In regression models, the most common prior for the coefficients is a normal distribution. If we let the prior for the intercept be $\beta_0 \sim N(0, \sigma_0^2)$ and the prior for the slope coefficients be $\beta_j \sim N(0, \sigma_\beta^2)$, then the regularization shape parameter is $b = 2$ (as in ridge regression) and the strength of the penalty a is proportional to the reciprocal of prior variance $(1/\sigma_\beta^2$; Hooten and Hobbs 2015). Thus, to induce a stronger penalty, we simply make the prior variance for the slope coefficients small, hence shrinking the posterior for the slope coefficients toward zero. To choose the optimal value for σ_β^2, we can perform cross-validation (or out-of-sample validation), as before, to improve predictive ability of the model (Watanabe 2010).

The Bayesian regularization procedure can be used with a suite of different model specifications that involve regression components (e.g., logistic regression, Poisson regression, occupancy models, capture-recapture models; Hooten and Hobbs 2015). Different regularization penalties can be imposed by using different priors. For example, using a double exponential prior for β_j instead of a Gaussian prior results in a Bayesian lasso penalty (Park and Casella 2008; Kyung et al. 2010). The only potential disadvantage is that a Bayesian regularization procedure may involve more computation than fitting the model a single time when cross-validation is applied and the regularization is tuned. Even so, when fitting many types of models on modern computers, Bayesian regularization is feasible. Alternatively, strong priors that are set in the traditional way, using pre-existing scientific knowledge about the process, may be enough to facilitate a natural Bayesian regularization without requiring an iterative model-fitting procedure (Seaman et al. 2012).

Discussion

We presented several different approaches for comparing parametric statistical models in this chapter. Our focus was mainly on comparing models with respect to predictive ability, but not all of the methods are designed to be optimal for prediction in the same sense. For example, BIC is more closely related to model averaging, and model averaged predictions outperform the predictions from any one model alone. Thus, multimodel inference can refer to a comparison of models based on predictive ability or an explicit combination of models to improve desired inference.

Any probability distributions (e.g., predictive distributions) can be averaged to form a new distribution, but the weights with which to average them are not unique, and any one set of weights is optimal only under certain circumstances. Bayesian methods provide the most coherent justification for model averaging because the optimal weights have been shown to equal the posterior model probabilities, $P(M_l|\mathbf{y})$, for $l=1, \ldots, L$ (Hoeting et al. 1999; Link and Barker 2006; Hooten and Hobbs 2015). The posterior model probability blends information from the model and data with a prior understanding of model suitability. Posterior model probabilities can be calculated easily for some Bayesian models, but they are intractable for others. Thus, BIC was developed for computing the optimal model averaging weights under certain conditions (equal prior model probabilities and flat prior distributions; Schwarz 1978).

The model averaging weights associated with BIC are proportional to $e^{-BIC_l/2}$, where BIC_l is the Bayesian information criterion calculated for model M_l. Burnham and Anderson (2002) suggested replacing BIC_l with AIC_l and using it in a non-Bayesian context to model average parameter estimates and predictions, a practice that has become popular in wildlife biology and ecology. However, model averaging should be performed only for quantities that do not change in their interpretation across models

(Cade 2015; Banner and Higgs 2017). Regression parameters have different interpretations among models unless the predictor variables are uncorrelated; they are interpreted conditional on the other parameters in the model. However, predictions of data always have the same interpretation in models, so they can be safely averaged for final inference (Burnham and Anderson 2002; Burnham and Anderson 2004).

To illustrate model comparison based on predictive ability, we employed a Poisson GLM for count data and compared a set of five models including a variety of covariates. Across all methods we demonstrated, state area always appeared in the best predicting model. However, all of the covariates improve predictive ability beyond the null model (i.e., intercept only). In the models we considered, the information criteria provided similar insights as cross-validation, and the non-Bayesian information criteria agreed with the Bayesian approaches. This occurred because we used relatively vague priors for the regression coefficients, and thus, the information content from the data was approximately the same across models.

The regularization approaches also provided similar results for our data. However, while lasso immediately shrunk the simulated covariate to zero, ridge regression shrunk this covariate, and the others, more slowly. In this particular case, the smoothness of the trajectory allowed the regularization procedure to find a better predicting model (out of infinite possibilities) under the ridge penalty.

The algorithms required to fit the specific models we presented in this chapter, and its relatively small example data set, yielded nearly immediate results in all cases. However, for larger data sets, the cross-validation and Bayesian approaches may require more time to implement. Still, we used readily available software to fit all models and nested model-fitting commands within "for loops" to perform cross-validation when necessary. With modern computing resources and easy parallel computing, cross-validation can be sped up substantially with minimal extra effort, so the added computational burden is not nearly as limiting now as it was in the past. However, while cross-validation automatically accounts for optimism in scoring predictive ability, it still depends on the initial data set and is limited in representing true out-of-sample predictive ability.

Finally, Ver Hoef and Boveng (2015) argue that there are valid situations that call for the use of a single, well-designed model that best represents the scientist's understanding of the ecological mechanisms and data collection process. In such cases, the emphasis is not on predictive ability, but rather on gaining a better understanding of the model components. A model component may be a simple population mean that is unknown, such as the average biomass in a survey plot, or it could be the true animal abundance in a closed study area. In these cases, we may have no need for model comparison because the desired inference is clear and the study design can be customized to answer these questions.

Acknowledgments

The authors thank David Anderson, Ken Burnham, Steve Ellner, Tom Hobbs, Jennifer Hoeting, Bill Link, Jay Ver Hoef, and Gary White. Hooten also acknowledges support from NSF EF 1241856, which helped fund this work. Any use of trade, firm, or product names is for descriptive purposes only and does not imply endorsement by the U.S. Government.

LITERATURE CITED

Akaike, H. 1983. Information measures and model selection. International Statistical Institute 44:277–291.

Banner, K. M., and M. D. Higgs 2017. Considerations for assessing model averaging of regression coefficients. Ecological Applications 27:78–93.

Burnham, K. P., and D. R. Anderson. 2002. Model selection and multimodel inference, 2nd ed. Springer-Verlag. New York, NY.

Burnham, K. P., and D. R. Anderson. 2004. Multimodel inference: Understanding AIC and BIC in model selection. Sociological Methods and Research 33:261–304.

Cade, B. S. 2015. Model averaging and muddled multimodel inferences. Ecology 96:2370–2382.

Celeux, G., F. Forbes, C. P. Robert, and D. M. Titterington. 2006. Deviance information criteria for missing data models. Bayesian Analysis 1:651–674.

Christensen, R. 2002. Plane answers to complex questions. Springer, New York.

Gelfand, A. E., and A. F. M. Smith. 1990. Sampling-based approaches to calculating marginal densities. Journal of the American Statistical Association 85:398–409.

Gelman, A., J. Huang, and A. Vehtari. 2014. Understanding predictive information criteria for Bayesian models. Statistics and Computing 24:997–1016.

Gneiting, T. 2011. Making and evaluating point forecasts. Journal of the American Statistical Association 106:746–762.

Gneiting, T., and A. E. Raftery. 2007. Strictly proper scoring rules, prediction, and estimation. Journal of the American Statistical Association 102:359–378.

Hastie, T., R. Tibshirani, and J. Friedman. 2009. Elements of statistical learning: Data mining, inference, and prediction, 2nd ed. Springer. New York.

Hoerl, A. E., and R. W. Kennard. 1976. Ridge regression: Biased estimation for nonorthogonal problems. Technometrics 12:55–67.

Hoeting, J. A., D. Madigan, A. E. Raftery, and C. T. Volinsky. 1999. Bayesian model averaging: A tutorial. Statistical Science 14:382–417.

Hobbs, N. T., and M. B. Hooten. 2015. Bayesian models: A statistical primer for ecologists. Princeton University Press, Princeton, NJ.

Hooten, M. B., and N. T. Hobbs 2015. A guide to Bayesian model selection for ecologists. Ecological Monographs 85:3–28.

Kutner, M. H., C. J. Nachtsheim, and J. Neter. 2004. Applied linear regression models. McGraw-Hill/Irwin, New York.

Kyung, M., J. Gill, M. Ghosh, and G. Casella. 2010. Penalized regression, standard errors, and Bayesian lassos. Bayesian Analysis 5:369–412.

Link, W. A., and R. J. Barker. 2006. Model weights and the foundations of multimodel inference. Ecology 87:2626–2635.

Madigan, D., and A. E. Raftery. 1994. Model selection and accounting for model uncertainty in graphical models using Occam's window. Journal of the American Statistical Association 89:1535–1546.

Park, T., and G. Casella. 2008. The Bayesian lasso. Journal of the American Statistical Association 103:681–686.

Richardson, S. 2002. Discussion of the paper by Spiegelhalter et al. Journal of the Royal Statistical Society, Series B 64:626–227.

Schwarz, G. E. 1978. Estimating the dimension of a model. Annals of Statistics 6:461–464.

Seaman, J. W. III, J. W. Seaman Jr., and J. D. Stamey. 2012. Hidden dangers of specifying noninformative priors. American Statistician 66:77–84.

Spiegelhalter, D. J., N. G. Best, B. P. Carlin, and A. van der Line. 2002. Bayesian measures of model complexity and fit. Journal of the Royal Statistical Society, Series B 64:583–639.

Stone, M. 1977. An asymptotic equivalence of choice of model cross-validation and Akaike's criterion. Journal of the Royal Statistical Society, Series B 36:44–47.

Tibshirani, R. 1996. Regression shrinkage and selection via the lasso. Journal of the Royal Statistical Society, Series B 58:267–288.

Vehtari, A., and J. Ojanen. 2012. A survey of Bayesian predictive methods for model assessment, selection and comparison. Statistics Surveys 6:142–228.

Ver Hoef, J. M., and P. L. Boveng. 2015. Iterating on a single model is a viable alternative to multimodel inference. Journal of Wildlife Management 79:719–729.

Watanabe, S. 2010. Asymptotic equivalence of Bayes cross-validation and widely applicable information criterion in singular learning theory. Journal of Machine Learning Research 11:3571–3594.

Watanabe, S. 2013. A widely applicable Bayesian information criterion. Journal of Machine Learning Research 14:867–897.

PART II ESTIMATION OF ABUNDANCE AND DEMOGRAPHIC PARAMETERS

5

GARY C. WHITE

Estimation of Population Parameters Using Marked Animals

Population-level questions are of obvious interest to wildlife managers and scientists.
—Dinsmore and Johnson (2012:349)

Preliminaries

Certain words have specific meaning when used in this field. A *parameter* is a quantity of a population, such as its size (N) or its survival rate (S). Generally, the value of a parameter—i.e., exactly how many animals are in the population—is unknown. The goal is to assess this value using an *estimator*. For example, the Lincoln-Petersen estimator of population size is $\hat{N} = \dfrac{n_1 \times n_2}{m_2}$, where n_1 is the number of animals marked in the first sample from a population, n_2 is the number of animals sampled in a second sample, with m_2 of them marked during the first sample (i.e., m_2 animals were captured in both samples). This formula is an estimator, specifically an analytical estimator, because it can be written down analytically. If 100 animals are captured in the first sample, and 150 in the second sample, with 50 of them marked from the first sample, then the estimate of population size is 300; i.e., $\hat{N} = \dfrac{100 \times 150}{50} = 300$. Most of the estimators in this chapter will be numerical estimators, produced by a computer using numerical methods. No specific formula can be constructed, which is to say that analytical methods cannot generate a formula for the estimator.

Various methods are used to obtain estimators and hence estimates of population parameters.

Maximum likelihood estimators are derived from a likelihood function, formally a joint probability density function of the sample data. Under likelihood theory, all the information available from the data is contained in the likelihood function. Some likelihood functions can be solved to provide analytical estimators and hence simple estimates, but generally numerical methods are needed to solve the likelihood function to obtain estimates as well as the variance of an estimate. In addition, *confidence intervals* are obtained for the parameter estimate to provide a measure of how precise the parameter was estimated.

Multiple methods are available to obtain confidence intervals. Beginning statistical classes teach constructing confidence intervals based on a normal distribution (±2SE). When the distribution of the parameter is thought to be not normally distributed, the parameter is transformed to a different numerical scale to generate an approximate normal distribution. The ±2SE interval is computed on the transformed scale and then converted back to the original numerical scale. The preferred method, albeit numerically intensive, profile likelihood confidence intervals, are computed directly from the log-likelihood function. Chapter 2 of White et al. (1982) provides a more complete discussion of these topics.

Bayesian estimators also use the same likelihood function, but place prior distribuitons on the parameters and combine the prior distributions with the likelihood function to obtain the posterior distribution of a parameter. By sampling the posterior distribution, we can obtain an estimate of the parameter, as well as interval estimates, termed "credibility intervals." Link and Barker (2010) discuss the differences in interpretation of Bayesian estimation versus frequentist methods (i.e., methods based only on the likelihood function).

This chapter does not get into the details of and differences between frequentist and Bayesian estimation. Both approaches use the same likelihood function. For typical problems, the mode of the posterior distribution is exactly the maximum likelihood estimate if "uninformative" priors are used. Thus, the two approaches are not unrelated; both have the same model for the data. The advantage of maximum likelihood estimation is that it takes considerably less computer time to produce estimates, on the order of 10 to 100 times less. The advantage of Bayesian estimation is that prior information can be incorporated through the prior distribution (although this feature can also be viewed as a handicap), and hyperdistributions can be placed on sets of parameters (hierarchical modeling). As an example of a hyperdistribution, suppose we have 50 years of survival estimates from a population such as North American mallards (*Anas platyrhynchos*). Whereas reporting 50 separate estimates is not particularly enlightening, the mean and variance of the hyperdistribuiton summarizing these estimates is a more useful result. See Royle and Dorazio (2008) for a thorough treatment of hierarchical modeling.

Sampling

Population biologists are often untrained in *sampling theory*. To properly obtain a representative sample from a population, each individual must be enumerated and then a representative (typically a random) sample is drawn. Thus, a sampling frame (i.e., the biological population) is created, and individuals are randomly selected from this sampling frame. From this sample, inference is made to all the individuals in the sampling frame.

This sampling theory is not appropriate for estimating the size (N) of a biological population because if we knew exactly how many animals are in the population (i.e., we know N), then there is no need to make an estimate. As a result, methods to estimate N are assumed to not be random samples, and the various models attempt to estimate N based on a nonrandom sample.

However, sampling is important when estimating parameters of a population such as fecundity, fidelity, survival, and so on. Suppose we want to estimate the survival rate, S, of the population by attaching radios on a sample of animals and monitoring their survival for some interval. We obtain an estimate, $\hat{S}$, of the survival rate of this sample. But, we desire an estimate appropriate for the entire population, not just the sampled animals. To make this inference, we must assume that our sample is representative of the entire population. For this assumption to be true, our methods of capturing animals to attach radios must not select for larger animals from the population or select in any way that produces a survival rate of the sampled animals that is different from that of the population. This assumption can never really be evaluated; we can only speculate about possible problems. The point is to be aware that how you "sample" a population to mark animals for survival estimation is critical to the validity of the inferences you make back to the population.

Population Closure

The simple assumption we make when estimating the size of a population, N, at a specific point in time, is that the N parameter is fixed, and the population is not changing during the interval while the mark-recapture surveys are taking place. Two forms of population closure are considered: demographic and geographic. *Demographic closure* requires that no

Fig. 5.1. Illustration of the lack of geographic closure with a 7 × 12 trapping grid.

tions, an unbiased estimator of D is obtained. This may be all that is needed for an analysis. However, consider a case where a harvested species is being studied, and in order to set a harvest figure for a population management unit, we really need to know N. If D is constant across the entire area of the population being considered—i.e., the grid we sampled has the same density of animals as the rest of the population management unit that will be harvested from—then computing $\hat{N} = \hat{D} \times A$ may be appropriate, where A is now the area of the population management unit. However, such a strong assumption is seldom true, so beware of the deceptive simplicity of the N and D relationship.

Major Reviews

Several reviews of the topics covered in this chapter are worth noting. Seber (1982) provides a treatise on these methods, albeit now dated. Williams et al. (2001) discuss all the methods covered here and provide a more detailed explanation of them than can be presented in this chapter. Borchers et al. (2002) cover estimation of closed populations. Amstrup et al. (2005) also provide explanations of most of the methods presented in this chapter.

More recent reviews are provided by Royle and Dorazio (2008), Link and Barker (2010), King et al. (2010), Royle et al. (2014), and Kéry and Royle (2016). These authors' volumes feature Bayesian estimation methods, although often maximum likelihood methods are discussed as well.

population parameters are changing—i.e., there are no births or deaths. *Geographic closure* requires that entering (immigration) or leaving (emigration) the area being sampled does not occur. Consider the scenario depicted in Fig. 5.1, where a 7 × 12 trapping grid is used to capture animals depicted by the elliptical home ranges. Some animals are entirely on the grid, whereas other are mostly off the grid. The trapping grid is not geographically closed.

Population Size versus Density

The relationship between population size (N) and population density (D, animals per area) seems straightforward at first glance: $D = \dfrac{N}{A}$, where A is the area being sampled. Unfortunately, the relationship is not that simple when populations are not geographically closed. When the situation looks like Fig. 5.1, the population estimate is going to be the number of animals using the grid. So, if A is taken as the area of the grid, density is biased high, because not all the $\hat{N}$ animals are always on the grid.

Spatially explicit capture-recapture methods (Royle et al. (2014) attempt to account for this lack of geographic closure. Under appropriate assump-

Population Estimation

All population estimates are based on a simple relationship: a count (C) of animals divided by an estimate of the probability of detecting these animals ($\hat{p}$), $\hat{N} = \dfrac{C}{\hat{p}}$. This simple relationship applies to distance sampling (discussed in Chapter 6 of this volume), change-in-ratio estimators, and mark-recapture or mark-resight estimators. Consider how the Lincoln-Petersen analytical estimator can be re-arranged to this form (n_1 = number captured on first occasion, n_2 = number captured on second occasion, and m_2 = number captured on both occasions):

$$\hat{N} = \frac{n_1 \times n_2}{m_2} = \frac{n_2}{\frac{m_2}{n_1}} = \frac{n_2}{\hat{p}}.$$

We have estimated the detection probability for the count of animals in the second survey (n_2) as the proportion of marked animals detected in the second survey $\left(\hat{p} = \frac{m_2}{n_1} \right)$. Thus, population estimation is the process of estimating a detection probability.

Biologists have been extremely innovative in obtaining counts of animals, considering that the field basically started with simple trapping grids where the number of animals captured was the count of animals. DNA greatly enhanced our ability to individually identify animals, i.e., "mark" them. DNA samples are collected via hair snares, rub trees, scat samples, or physical capture. Combinations of these various methods can greatly improve population estimates in the face of individual heterogeneity of animal detection probabilities because different methods detect different kinds of animals and therefore increase the overall count of animals observed. Camera trapping (O'Connell et al. 2011) has provided means of detecting animals that are identifiable by their individual characteristics, such as tigers, as well as animals visually marked by the investigator, e.g., ear tags. Passive integrated transponder (PIT) tags provide a less invasive mark that can be detected passively once the tag is inserted.

Closed Populations

Otis et al. (1978) proposed a set of eight models with detection probabilities varying by occasion (time, t), behavioral effects from capture (b), and individual heterogeneity (h) (Fig 5.2). Model M_0 uses a zero subscript for zero heterogeneity. Reviews of these models are provided by Borchers et al. (2002) and

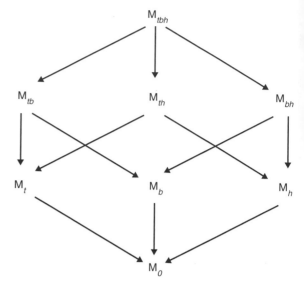

Fig. 5.2.   Set of eight models presented to model the effects of time (t), behavioral effect of capture (b), and individual heterogeneity (h) on detection probabilities.

Amstrup et al. (2005), and treatment under the Bayesian paradigm is provided by Link and Barker (2010), Royle and Dorazio (2008), and King et al. (2010). I will discuss the models without h first, because they are the easiest to understand.

Consider a scenario with three occasions, in which animals are individually identified when captured on each of the three occasions. There are eight possible encounter histories, as shown in Table 5.1. The probability of initial capture of an animal for each occasion is p_i, and after an animal has been captured once, its recapture probability becomes c_i. Thus, the effect of behavioral response to capture, either positive (trap happy) or negative (trap shy), is reflected in the recapture probabilities. There is no c_1 because an animal can never be recaptured on the first occasion. The number of animals observed one or more times for the observable encounter histories for t occasions is $M_{t+1} = \sum_{i=1}^{2^t-1} n_i$, leaving $N - M_{t+1}$ animals that are not observed (i.e., never captured or detected). The notation M_{t+1} comes from the number of marked animals at time, $t + 1$, i.e., the number of animals seen during the t occasions. Information

Table 5.1. Possible encounter histories and their associated probabilities for model M_{tb} with three occasions. The terms $Pr(x_i)$ are used to build the likelihood described in the text, where the definitions of the notation are also provided.

Index i	Encounter history x_i	$Pr(x_i)$	Number of animals
1	111	$p_1 c_2 c_3$	n_1
2	110	$p_1 c_2 (1-c_3)$	n_2
3	101	$p_1 (1-c_2)c_3$	n_3
4	011	$(1-p_1) p_2 c_3$	n_4
5	100	$p_1 (1-c_2)(1-c_3)$	n_5
6	010	$(1-p_1) p_2 (1-c_3)$	n_6
7	001	$(1-p_1) (1-p_2) p_3$	n_7
8	000	$(1-p_1) (1-p_2) (1-p_3)$	$N-M_{t+1}$
Sum		1	N

in Table 5.1 provides the necessities, specifically the $Pr(x_i)$ definitions, to build the likelihood function for the full likelihood where N is included:

$$L(N; p_1,\ldots,p_t; c_2,\ldots,c_t \mid x_i, n_i)$$

$$= \binom{N}{M_{t+1}} Pr(x_{2^t})^{N-M_{t+1}} \prod_{i=1}^{2t-1} Pr(x_i)^{n_i}$$

The combinatorial $\binom{N}{M_{t+1}} = \dfrac{N!}{M_{t+1}!(N-M_{t+1})!}$ is

necessary to account for the number of ways that M_{t+1} animals can be sampled from a population of N animals. Typically, the likelihood function is converted to the logarithmic scale, and then numerical optimization is used to estimate the log L for the parameters $N, p_1, \ldots, p_t$ and $c_2, \ldots, c_t$, because the log function is a strictly increasing function of the likelihood function, and the maximums for each of the parameters coincide for the two functions.

Because the term $Pr(x_{2^t})^{N-M_{t+1}}$ is included in the likelihood, individual covariates measured at the time an animal is captured cannot be used in this parameterization. Huggins (1989, 1991) produced a simpler likelihood function by removing the term for the animals never encountered. Then, the probabilities for each of the observable encounter histories are divided by 1 minus the probability of an

animal never being seen (i.e., $1 - \prod_{i=1}^{t}(1-p_i)$), generally denoted as $p*$ in the literature), so the probabilities of observable encounter histories sum to 1. The estimate of N is derived as the count of animals observed divided by the probability that an animal is seen one or more times just as described above for all population estimation methods:

$$\hat{N} = \frac{M_{t+1}}{1 - \prod_{i=1}^{t}(1-p_i)} = \frac{M_{t+1}}{p*}.$$

When individual covariates are measured on the captured animals, the initial detection and recapture probabilities are now specific to each individual j:

$$\hat{N} = \sum_{j=1}^{M_{t+1}} \frac{1}{1 - \prod_{i=1}^{t}(1-p_{ij})}.$$

Just for clarification, the parameter set $N, p_1, \ldots, p_t$ and $c_2, \ldots, c_t$ for model M_{tb} is not identifiable without a constraint. That is, it is theoretically not possible to estimate all of these parameters, even with an infinite number of observations. Typically, the constraint $\text{logit}(c_i) = \text{logit}(p_i) + \alpha$ is used, where α provides a constant offset on the logit scale, with the logit function $\log\left(\frac{\theta}{1-\theta}\right)$. The reason for placing this constraint on the logit scale is to make the constraint additive. Only a single parameter (α) is needed to model all the c_i values from the p_i values. Model M_t is obtained by setting $p_i = c_i$, M_b by $p_. = p_i$ and $c_. = c_i$, and M_0 by $p_. = p_i = c_i$. Seber (1982:164) noted that there is nothing gained by using model M_0 instead of M_t, but it is included here for the sake of completeness.

Performance of the full likelihood parameterization with N in the likelihood and the Huggins conditional likelihood are generally nearly identical, albeit the full likelihood parameterization might provide a bit more precise estimate of N (but not always). Generally, almost nothing is lost by using the Huggins parameterization, and much is gained because of the ability to use individual covariates in the models and the logit-normal random effects models that are described below.

When individual heterogeneity is considered in these models, all kinds of complexities arise. First, note that each animal cannot have its own unique encounter history in a model, because the model would not be identifiable. Hence, we must use a reduced model that provides a distribution of encounter histories. I will consider only likelihood-based models here to maintain consistency with the likelihood-based models already presented.

Individual heterogeneity is the case where each animal has its own detection probability. Ordinary estimators of population size are biased low when individual heterogeneity is ignored. Animals with a high detection probability are seen more often and bias $\hat{p}$ high, which lowers $\hat{N}$. In effect, we do not obtain a representative sample of the population. To address this, we use model-based inference to "correct" for the missing observations. Three strategies—finite mixtures, logit-normal mixture, and beta-binomial mixture—have been used; see Royle and Dorazio (2008:196–198) for a concise discussion of the three approaches. I will discuss only the most common approaches (i.e., finite mixtures and

logit-normal mixture); however, various models can fit a set of data identically, or nearly so, and provide widely different estimates of N (Link 2003). These differences arise because different models extrapolate to the unobserved animals differently (Fienberg 1972). Thus, it is critical that individual heterogeneity models are based on biological explanations of what might be happening, rather than blindly following automated model selection procedures.

Pledger (2000) formalized the use of finite mixtures for closed population estimation. With the finite mixtures approach, a mixture of two (or more) kinds of animals is hypothesized to exist in the population. Each type has its own p and c parameters, so that the cell probabilities of Table 5.1 are generalized. Thus, consider the history 1001 to illustrate the process with two mixtures, a and b, for model M_{tbh}:

$$\Pr(1001) = \begin{pmatrix} \pi_a p_{a1}(1-c_{a2})(1-c_{a3})(c_{a4}) \\ + \\ (1-\pi_a)p_{b1}(1-c_{b2})(1-c_{b3})(c_{b4}) \end{pmatrix}.$$

The parameter π_a is the probability of an animal being in the a mixture, with $1-\pi_a$ the probability of it being in the b mixture. To make this model identifiable, some constraints are required. Typically, an offset on the logit scale is used, $\text{logit}(p_{bi}) = \text{logit}(p_{ai}) + \alpha_p$, with one additional offset for differences between p and c, $\text{logit}(c_{ai}) = \text{logit}(p_{ai}) + \alpha_c$, $\text{logit}(p_{bi}) = \text{logit}(p_{ai}) + \alpha_p$, and $\text{logit}(c_{bi}) = \text{logit}(p_{ai}) + \alpha_p + \alpha_c$. However more complex models can be constructed and still be identifiable. Models M_h, M_{th}, and M_{bh} are possible with finite mixtures by imposing appropriate constraints. Further, by fixing π_a to 1 and constraining the mixtures equally, the remaining models are possible with appropriate constraints imposed.

Finite mixtures work with either the full likelihood parameterization with N in the likelihood, or the Huggins conditional likelihood. Dorazio and Royle (2003) and Pledger (2005) provide simulations that compare the performance of the finite mixture model to competitors. Kendall et al. (2009) employed the finite mixture estimator and used three sampling methods, hair trap, bear rub, and physical capture, in an extensive study to estimate grizzly bear (*Ursus arctos*) population size in northwestern Montana.

The logit-normal mixture, first proposed by Coull and Agresti (1999) for capture-recapture models, has also been widely used. The Huggins random effects estimator is based on the capture-recapture estimator, which is conditioned on the animals captured one or more times (Huggins 1989, 1991). The estimator is extended by including an individual random effect for the capture probability (p_{ik}) of each animal i, constant across occasions $k = 1, \ldots, t$ on the logit scale following McClintock et al. (2009b) and Gimenez and Choquet (2010); i.e.,

$$\text{logit}(p_{ik}) = \beta_k + \varepsilon_i,$$

with β_k a fixed effect modeling time and ε_i a normally distributed random effect with mean zero and unknown variance σ_p^2. Hence

$$p_{ik} = \frac{1}{1 + \exp\left(-(\beta_k + \sigma_p Z_i)\right)},$$

where $Z_i \sim N(0, 1)$. The estimate of population size, N, is obtained following Huggins (1989) as the summation across animals captured one or more times, $\hat{N} = \sum_{i=1}^{M_{t+1}} \frac{1}{p_i^*}$. Royle and Dorazio (2008:201) provide R code to perform this process for the full likelihood parameterization. White and Cooch (2017) provide details on the structure of the estimator and simulations to document the performance of the Huggins parameterization.

All the models described are available in Program MARK (White and Burnham 1999). In addition, Royle and Dorazio (2008) provide R code for many of these models under a frequentist paradigm at http://www.mbr-pwrc.usgs.gov/pubanalysis/roylebook/chapters.htm.

Bayesian Analysis

All the models described so far in this chapter can also be treated in a Bayesian paradigm. Link and Barker (2010:227) provide code for Model M_t using WinBUGS (a Windows package that computes Bayesian statistical estimators). Royle and Dorazio (2008) take a different tactic and use data augmentation in the Bayesian paradigm to estimate population size. Data augmentation relates closed population estimation to occupancy estimation (see Chapter 7, this volume). In occupancy estimation, we estimate the proportion of sites that are occupied, Ψ. So, the idea is to augment the observed encounter histories, one per animal, with a bunch of all-zero encounter histories, so that there are more encounter histories being considered than there are N animals in the population. So, assume that we have M encounter histories after augmentation, each representing just one animal (where the augmented histories might be hypothetical animals). Then, the proportion of the augmented encounter histories that are "occupied" with probability Ψ provides an estimate of N as $\hat{N} = \hat{\psi} \times M$.

WinBUGS code is provided by Royle and Dorazio (2008:185) for an implementation of M_0. Data augmentation with individual covariates is shown on pages 214–216. They also discuss issues with using WinBUGS on page 206. All the code from Royle and Dorazio (2008) is available at http://www.mbr-pwrc.usgs.gov/pubanalysis/roylebook/chapters.htm.

Program MARK (White and Burnham 1999) also provides some Bayesian estimation of the closed models, but does not allow the flexibility of coding individual effects as a hierarchical model. However, at this time, there is not an inclusive package for the analysis of closed models in the Bayesian paradigm. For advanced Bayesian analyses, users are required to use R and code their own samplers, or use WinBUGS, OpenBUGS, or JAGS, possibly in combination with R. But I have no doubt that Bayesian models will be more common as biologists become more experienced and gain technical knowledge of computing (probably mainly in R), application of Markov Chain Monte Carlo (MCMC) sampling, and probability calculus.

Mark-Resight Models

Mark-resight models assume that a sample of animals representative of the population is marked, and that the detection probability of these individually marked animals can be used to correct counts of unmarked animals. Thus, marking may occur only once, with multiple resighting surveys. These models were originally developed for use with radio-marked animals, e.g., deer or mountain sheep, where aerial surveys provide resighting data. For example, suppose 100 sheep are radio-marked, and then multiple resighting surveys are conducted where encounter histories for the 100 radio-marked sheep are constructed. For each resighting survey, the count of unmarked sheep is made. These counts are then corrected with the detection probabilities estimated from the marked sheep. However, radio-marked animals are not required. These models have been applied to camera surveys where only some of the animals are individually identifiable (e.g., only some of the elephants have tears and holes in their ears to make them individually identifiable).

McClintock et al. (2009a, 2009b) and McClintock and White (2009) have developed a set of three models: logit-normal abundance model where animals can only be observed once on a resighting survey, logit-normal model extended for immigration and emigration from the study area, and Poisson-log normal abundance model, possibly zero-truncated, where animals might be seen multiple times during resighting surveys.

The key assumptions of these models are that (1) the marked animals are representative of unmarked animals, and (2) marked and unmarked animals are not misclassified. For example, if animals are marked via helicopter capture, and then resightings are conducted via helicopter surveys, the marked animals may behave differently during the resighting surveys because of their previous bad experience with helicopters swooping down and netting them during marking. Misclassification of marked and unmarked animals might occur with camera surveys if a mark is missed in a photo. The remedy to this issue is usually to use only those photos in which animals are clearly identifiable as either marked or not marked.

These models are implemented in Program MARK (White and Burnham 1999). In addition, Royle et al. (2014) provide a nice summary of past work with mark-resight models and provide extensions to spatially explicit mark-resight models, along with R code.

Spatially Explicit Capture-Recapture Models

Biologists have long recognized that home range size is an important component of detection probabilities, and, for example, Otis et al. (1978:76) recommended four traps per home range as a rule of thumb to obtain decent detection probabilities. The idea behind *spatially explicit capture-recapture* (SECR) models is that the spatial information on capture locations can be used to eliminate individual heterogeneity from exposure to traps, plus provide an estimate of density (D) of animals on the trapping grid (as opposed to an estimate of N, the "superpopulation" of animals that might use the grid). Trapping grids placed in a uniform habitat lack geographic closure (Fig. 5.1), and so you expect some animals at the edge or just off the edge of the grid to have far less exposure to traps, although still using the study area being trapped.

The most parsimonious model used to model exposure to traps is a Euclidian distance model, where animal home ranges are assumed to be a bivariate normal circle. The degree of exposure of an animal to a specific trap depends on the distance the animal's center of activity is from the trap, and a scale parameter, σ^2, that is the variance of a bivariate normal distribution centered on the animal's activity center. Mathematically, exposure to a trap is proportional to $e^{(-d^2/\sigma^2)}$, where d is the Euclidian distance between the trap and the animal's activity center. The exposure of the animal to the entire grid is the sum of this function for all traps. From this measure of exposure, we obtain the probability of capture, typically by assuming that the probability of capture is proportional to the measure of exposure. Home range centers, or centers of activity, are latent variables (unobserved), but the count of the centers divided by the area that they occur in provides an estimate of density.

Royle et al. (2014) provide a thorough treatment of spatially explicit models. They list some of the core assumptions for the simpler models (Royle et a. 2014:150): demographic closure, geographic closure (although the edge aspect described above is modeled), activity centers randomly distributed, detection is a function of distance (or some other measure of 'distance' from the center of an animal's home range, and independence of encounters among and within individuals. What they do not clearly state is that animals are assumed to have stable home ranges during the period of the trapping effort. Royle et al. (2014) discuss more complex models that model violations of several of these assumptions, although additional data in the form of covariates are usually required.

A violation of the basic assumptions of the SECR methods occurs when bait or other attractants are used, and animals are attracted to the grid. Stated differently, when the animal's home range is extended because of the detectors, the result will be an overestimate of density. The key assumption of the SECR method is that animal home ranges are fixed and *not* changing because of the detectors or during the survey. For example, Morton et al. (2016) could not use SECR methods for brown bears (*Ursus arctos*) with a grid of hair snares because the bears were moving from high elevation den sites to low elevation salmon streams during the survey; thus, the assumption of a fixed home range was violated during the time of the survey. Additional issues concerning bias of density estimates may arise if the size or the shape of home ranges varies widely by animal. The issue is that esti-

mates of the σ may be biased high because animals with small home ranges are captured only in a single trap and hence do not contribute information on estimation of σ, whereas animals with large home ranges or long, elliptical home ranges may be captured in traps far apart. Obviously, multiple captures of individuals are necessary to estimate σ.

A second major assumption of the SECR methods is that the density of animals is uniform across the area sampled (animals randomly distributed). That is, the habitat, as well as animal occupancy, must be uniform. The importance of this assumption has been recognized, and one approach to mitigate this assumption is to model density as a function of habitat covariates, implemented in the SECR R package and discussed by Royle et al. (2014).

A variety of software packages is available for SECR analyses, and all of the packages mentioned here are discussed in Royle et al. (2014). The CRAN package SECR for R is available from https://cran.r-project.org/web/packages/secr/index.html and provides frequentist estimation methods for estimating animal densities using closed model capture-recapture sampling. Alternatively, SPACECAP (http://www.mbr-pwrc.usgs.gov/software/spacecap.shtml) is a software package for estimating animal densities using closed model capture-recapture sampling based on photographic captures using Bayesian spatially-explicit capture-recapture models. The site https://sites.google.com/site/spatialcapture recapture/home contains supplemental materials, including ample R code, for the Royle et al. (2014) book, and provides information on oSCR, an R package for likelihood analysis of spatial capture recapture (SCR) models (oSCR means "open SCR"), the SCRbayes R package currently under development, and more information on SPACECAP.

Open Populations

The concept of a population estimate from an *open population* requires that there is a short interval of time over which the population is not changing during sampling (i.e., there is a fixed N value). Then, after some time, allowing emigration and deaths and in situ recruitment and immigration, another sample is taken. Jolly (1965) and Seber (1965) presented closed-form estimators for population size at each sampling occasion as well as the number of new animals entering the population (B) for a fully time-specific model. A more general approach using the concept of a "super population"—i.e., all the animals that are ever available for capture during the study (although with additional assumptions, all the animals that passed through the population between the first and last samples)—was presented by Schwarz and Arnason (1996). Arnason and Schwarz (1999) developed the package POPAN-5 to provide estimates using numerical methods. The estimator is also available in Program MARK (White and Burnham 1999). The POPAN estimator is numerically quite stable and is the recommended approach to estimating population size with open models.

The drawback of open population estimates is that the estimates are almost certainly biased low because only time-specific heterogeneity can be modeled. The effects of individual heterogeneity and behavioral response to capture cannot be accounted for with the open population estimates. Hence, when possible, robust designs are recommended over the traditional open population methods. More importantly, population rate of change can be obtained from the open population models, and the issues of bias with N are mitigated. Estimation of the population rate of change is discussed below. Royle et al. (2014) provide examples and code for open models, including spatially explicit open models.

Survival Rate Estimation

Fortunately, survival rates are generally much easier to estimate than population size, but with the caveat discussed in Preliminaries section of this chapter concerning the need to be cognizant of properly sam-

pling the population for which you are trying to estimate survival. The simplest finite survival estimate is what is commonly called *known fate*: $\hat{S} = \frac{y}{n}$, where y is the number of n animals that live over some interval. If we are willing to assume that the survival rate is identical for each of the n animals, i.e., assume the binomial distribution, we can then estimate the variance of the estimate as $\mathrm{Var}(\hat{S}) = \frac{\hat{S}(1-\hat{S})}{n}$, giving a standard error $\mathrm{SE}(\hat{S}) = \sqrt{\mathrm{Var}(\hat{S})}$. These estimators are the analytical maximum likelihood estimators when the survival process is assumed to follow a binomial distribution; i.e., to obtain the estimator of S above, the likelihood function is $L(S \mid y, n) = \binom{n}{y} S^y (1-S)^{n-y}$. Known fate estimates are typically computed from radio-tracking data, where each of the n animals has a radio-collar attached and is monitored for the entire interval. Extensions to the known fate approach include the Kaplan-Meier estimator (Kaplan and Meier 1958), which allows censoring of animals (e.g., radio-collar falls off part way through the interval—*right-censoring*), and staggered entry (e.g., animals are collared during the interval and inserted into the study—*left-censoring*; Pollock et al. 1989). Given no left or right censoring, the Kaplan-Meier estimator reduces to exactly the maximum likelihood known fate estimator.

In situations where the time of death is known on a fine scale, e.g., daily, the continuous time models from the human medical field can be used. The Cox proportional hazards model produces estimates equivalent to the known fate model (Cox 1972). However, the time of death must be known and not lumped into broad intervals.

Not all studies have the luxury of a big budget and radio telemetry equipment. More commonly, animals are marked and encounter histories are constructed with the goal of estimating survival. As a result, a suite of models has been developed depending on the encounter process.

Live Encounters

Survival can be estimated via the Cormack-Jolly-Seber model when marked animals are encountered alive (Cormack 1964; Jolly 1965; Seber 1965; Lebreton et al. 1992). Encounter histories are constructed based on recapture data after initial marking, and survival is estimated for the intervals between capture occasions. Traditionally, survival in these models is termed "apparent survival," because if an animal leaves the study area (area surveyed for captures and recaptures), i.e., emigrates permanently, the survival estimate reflects the emigration. More explicitly, apparent survival is $\varphi = S \times F$, where F is fidelity. Examples of encounter histories and the probability associated with each is given in Table 5.2. Note that when an animal is not marked until the second (or later) occasion, zeros are inserted as place-holders in the encounter history. Zeros in the encounter history after the initial mark indicate that the animal was not recaptured, not that it died or emigrated. The last row of Table 5.2 illustrates that the last two zeros can be either died or emigrated or remained on the study area and was not recaptured.

The encounter history probabilities (Table 5.2) are used to construct the likelihood function just as is done for the closed captures population estimators. The product of the probabilities raised to the power of the number of animals observed with that history is taken over the observed encounter histories

Note that animals need not be physically captured and marked if they are individually identifiable in

photos. That is, camera traps can be used for species such as tigers, which are individually identifiable from their stripes. The first photo that a tiger shows up in is when it is considered marked, and hence "released" into the population.

Another point to note for this model is that survival is being estimated for only the marked animals. That is, the inferences on survival pertain strictly to the marked population. To make inferences to the population from which the marked animals were sampled requires that the marked animals are representative of the entire population.

Dead Encounters

Marked animals encountered only as dead can also provide estimates of survival. Consider recoveries of hunter-harvested waterfowl. Birds are banded in the late summer before the hunting season, and bands are recovered throughout the hunting season. Two different parameterizations of this model are common. Seber (1970) presented the model in terms of S and r (probability that a marked animal that dies is reported, i.e., a conditional upon death recovery rate). Seber (1970) called this parameter λ, but I have changed the notation here to not cause confusion with the population rate of change later. Brownie et al. (1985) presented the model in terms of S and f (probability that a marked animal will be reported dead in the interval, not conditional upon death, although obviously only dead animals can be reported). Because a marked animal can be seen only

Table 5.2. Four examples of encounter history probabilities for a sample of encounter histories with $t = 4$ occasions from the Cormack-Jolly-Seber model. The terms $\Pr(x_i)$ are used to build the likelihood described in the text, where the definitions of the notation are also provided.

Encounter history x_i	$\Pr(x_i)$	Number observed
1101	$\varphi_1 p_2 \varphi_2 (1 - p_3) \varphi_3 p_4$	n_1
0101	$\varphi_2 (1 - p_3) \varphi_3 p_4$	n_2
1110	$\varphi_1 p_2 \varphi_2 p_3 [\varphi_3 (1 - p_4) + (1 - \varphi_3)]$	n_3
1100	$\varphi_1 p_2 [(1 - \varphi_2) + \varphi_2 (1 - p_3)\{\varphi_3 (1 - p_4) + (1 - \varphi_3)\}]$	n_4

Table 5.3. Expected recovery matrices for four cohorts over four years for the Seber (1970) and Brownie et al. (1985) parameterizations. The terms are used to build the likelihood described in the text, where the definitions of the notation are also provided.

	Released recoveries for the Seber (1970) parameterization			
	Year 1	Year 2	Year 3	Year 4
R_1	$R_1(1-S_1)\,r_1$	$R_1S_1\,(1-S_2)\,r_2$	$R_1S_1S_2\,(1-S_3)\,r_3$	$R_1S_1S_2S_3\,(1-S_4)\,r_4$
R_2		$R_2\,(1-S_2)\,r_2$	$R_2S_2\,(1-S_3)\,r_3$	$R_2S_2S_3\,(1-S_4)\,r_4$
R_3			$R_3\,(1-S_3)\,r_3$	$R_3S_3\,(1-S_4)\,r_4$
R_4				$R_4\,(1-S_4)\,r_4$
	Recoveries for the Brownie et al. (1985) parameterization			
R_1	R_1f_1	$R_1S_1f_2$	$R_1S_1S_2f_3$	$R_1S_1S_2S_3f_4$
R_2		R_2f_2	$R_2S_2f_3$	$R_2S_2S_3f_4$
R_3			R_3f_3	$R_3S_3f_4$
R_4				R_4f_4

once, the data are usually formatted into a recovery matrix, with a cohort of marked animals released and the number reported by occasion (year) thereafter. Expected recoveries (probability of being recovered in a particular year times the number of marked animals released) are demonstrated for the two parameterizations in Table 5.3. The likelihood is formed in the same way as described previously, by taking the product of the cell probabilities raised to the power of the number of animals observed with that history. Table 5.3 shows only the expected values for the recovery matrix, leaving out the important probability of bands never recovered, which for each cohort is just 1 minus the sum of probabilities of the bands recovered.

An important distinction between the live encounter model and the dead encounter model is in how the encounters of marked animals occur. For the live encounter model, the encounters are typically produced by the researcher on the same study area as the one where marking takes place. Thus the sampling process to apply marks and the reencounter process are similar if not identical. In contrast, with the dead encounter model, marking is done by biologists, but recoveries are provided by the public, i.e., hunters harvesting waterfowl. The sampling process for reencounters is separate from the marking process. As a result, there is no notion of emigration from the study area because birds can be recovered no matter where they go. Thus, true survival (S) is estimated in the dead encounters model, whereas apparent survival (φ) is estimated in the live encounters model.

The Seber (1970) and Brownie et al. (1985) parameterizations produce inferences from exactly the same data. The two parameterizations produce identical results when fully time-specific models are used because the Brownie et al. $f=(1-S)r$. However, with sparse data, the Brownie parameterization can produce estimates of survival >1, because $(1-S)$ is not constrained in this model to be $\leq f$ as is the case in the Seber parameterization. The Brownie et al. (1985) parameterization is used by waterfowl biologists because the f parameter is a measure of harvest when corrected for band reporting rates and wounding loss. The Seber (1970) parameterization is preferred when covariates are used in the model for S because the covariate is applied for both S and $(1-S)$.

Both Live and Dead Encounters plus Incidental Encounters

Burnham (1993) was the first to develop the full model for estimating survival with both live and dead encounter data. When parameterized with the Seber (1970) model for the dead encounters, four parameters result: S survival, F fidelity to the study area, p detection probability of live animals on the study area, and r probability of the mark being reported given that the animal has died. Emigration from the study area is assumed to be permanent; i.e., the animal does not return to the study area where marking took place.

Barker (1997) extended Burnham's (1993) model to include incidental sighting. He was modeling a brown trout population in New Zealand where tags were applied on the spawning area, with anglers providing dead recoveries as well as live resightings when trout were caught and then released after the angler read the tag. Hence both dead recoveries and incidental resightings took place over the interval, but live recaptures took place only on the study area where marking took place. The model has seven parameters:

- S_i = probability an animal alive at i is alive at $i+1$;
- p_i = probability an animal at risk of capture at i is captured at i;
- r_i = probability an animal that dies in i, $i+1$ is found dead, and the band is reported;
- R_i = probability an animal that survives from i to $i+1$ is resighted (alive) some time between i and $i+1$;
- R'_i = probability an animal that dies in i, $i+1$ without being found dead is resighted alive in i, $i+1$ before it died;
- F_i = probability an animal at risk of capture at i is at risk of capture at $i+1$;
- F'_i = the probability an animal not at risk of capture at i is at risk of capture at $i+1$.

The R and R' parameters handle the incidental sightings. Barker's model allows temporary emigration from the study area where marking is conducted, in that an animal can leave with probability $1-F$, and return with probability F'. The temporary emigration process can be Markovian (meaning that whether an

animal is present on the study area or not depends only on its status during the previous occasion), where both F and F' are estimated; random with $F = F'$ (i.e., presence on the study area does not depend on where the animal was during the previous occasion); and permanent as in Burnham's model with $F' = 0$ (i.e., animal is never allowed to immigrate back onto the study area). When no incidental sighting data are available, Barker's model is identical to Burnham's model with the constraints $R = 0$, $R' = 0$, and $F' = 0$.

Barker's model has been underutilized, probably because of the complexity of having seven parameters. On the other hand, the ability to get incidental sightings from the public provides an efficient sampling procedure. For example, neck-collared geese with identifiable codes on the collar could be used to provide incidental sightings. Incidental sightings of neck-collared mountain sheep in parks could also be useful. More researchers need to consider the possibilities of this model.

Robust Designs

The live encounters model (Cormack-Jolly-Seber) was "robustified" in what is known as *Pollock's robust design* (Pollock 1982, Kendall et al. 1995): instead of sampling a population just once, multiple samples are taken over a short period when population closure can be assumed. Then, any of the closed models can be used to estimate the probability of an animal being seen one or more times. So, for the closed models, the probability of being seen one or more times is $p* = 1 - \prod_{i=1}^{t}(1-p_i)$ as described above, and this quantity replaces p in the likelihood of the live encounters models providing the link between the closed model and the live encounters model. The advantages of this approach are several: increased precision because of better estimates of detection probabilities, ability to model individual heterogeneity in detection probabilities, and ability to evaluate the assumption of no temporary emigration in the live encounters model. That is, besides the φ and now reparameterized p of the live encounters model, and the parameters of the closed captures models for each primary occasion, additional parameters for movement off and back to the study area are included in the model (Fig. 5.3). The parameter γ'' is the probability of the animal's leaving the study area where live encounters are taking place given that the animal is currently on the study area, and $1-\gamma'$ is the probability of its returning to the study area given that the animal was not on the study area during the previous sampling period. Note that even with the robust design model, the survival rate is still apparent survival (death and emigration) because permanent emigration is still part of the survival parameter.

The same robust design approach has been applied to the Barker model described above (Kendall et al. 2013). Not only are the estimates for the temporary emigration much more precise from the robust model, but now a parameter for permanent emigration is also estimated. The information for estimating the permanent emigration parameter comes from the dead encounters and incidental sightings that take place off the study area, confirming that the animals was or is still alive, but apparently not on the study area because it was not detected.

Multistate Designs

Animals can move between different states that may affect their survival. Examples are breeding versus nonbreeding for a year, using different wintering areas in different years, or improving or declining body conditions. The *multistate model* was formulated by Brownie et al. (1993), with an application with three states representing three different wintering areas for an eastern U.S. population of Canada geese (Hestbeck et al. 1991), as shown in Fig. 5.4. This model is an extension of the live encounters Cormack-Jolly-Seber model, which has only one state.

Each of the three states in this example has an associated survival rate S and a detection probability p, with the state denoted in the superscript. In addition, there is a transition parameter Ψ that is the probability of transitioning from one state to another, again with the direction denoted in the double superscript, with all possible combinations between states shown, including not moving. The sum of the Ψ parameters for leaving or remaining in the state must sum to 1. Hence, when this model is fitted to data, one of the transition values is obtained by subtraction. For example, the probability of remaining in the current state could be obtained as $\psi_i^{rr} = 1 - \sum_{s \neq r} \psi_i^{rs}$.

Another key point that may not be obvious is that the transitions are assumed to take place at the end of the survival interval. Thus, going from A to B means surviving in A and then transitioning to B just before the next sampling occasion. When transitions take place at any time during the survival interval, thus violating this assumption, the resulting

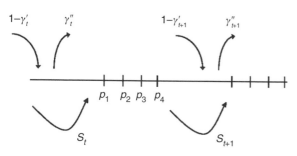

Fig. 5.3. Schematic of how temporary emigration is modeled in Pollock's robust design.

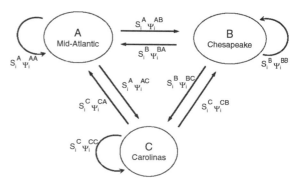

Fig. 5.4. Schematic of the Hestbeck et al. (1991) metapopulation of Canada geese wintering in three different wintering areas.

survival rates are a mixture of the state survival rates. The consequence is that state-specific survival rates cannot be estimated in that they all seem identical, or if not too badly violated, diminish the possibility of detecting differences. The multistate model can be data hungry because of the number of states. With g states, and all possible transitions estimated, there are $(g+1)*g$ parameters to be estimated.

The multistate model is a real chameleon, and with appropriate parameter fixing can emulate all the survival models discussed thus far except the robust design. However, the multistate model has also been extended to incorporate closed models on each sampling occasion ("robustified"). In addition, a multistate model assuming an open population during the sampling occasion has been developed by Kendall and Bjorkland (2001). Further, extensions for state uncertainty have been developed, for situations where detection of the animal may not clearly define its state (Kendall et al. 2012).

Survival Estimation Software

All of the survival models described here are available in Program MARK (White and Burnham 1999), including using MCMC with hyperdistributions across time (but not individuals). Program E-SURGE (Choquet 2008; Choquet et al. 2009; available at http://www.cefe.cnrs.fr/fr/ressources /films/34-french/recherche/bc/bbp/264-logiciels), provides a powerful tool for fitting complex multistate models, particularly hidden Markov models. I'm not aware of any comprehensive package that uses Bayesian methods, but obviously the models can be coded in WinBUGS or JAGS. Link and Barker (2010:252) provide some example code for the Cormack-Jolly-Seber (CJS) model, plus a parameterization of the Jolly-Seber model on page 257. R code from Royle and Dorazio (2008) are provided at http://www.mbr-pwrc.usgs.gov /pubanalysis/roylebook/chapters.htm. Royle et al. (2014) provide R code as well.

Population Growth Rates

Although the public nearly always wants to know "How many are there?" concerning a species of concern, a better question is "Is the population growing or declining?". The population growth rate, defined as $\lambda = \dfrac{N_{t+1}}{N_t}$, can also be estimated using marked animals with the Pradel (1996) or Link and Barker (2005) models, which are equivalent. These models can be parameterized in three ways: the seniority parameter γ that is the probability that an animal was present and not detected prior to being marked; the recruitment parameter f that is the number of new animals entering the population on the next sampling occasion divided by the number of animals in the population at the previous sampling occasion; and λ the population growth rate.

If the live encounter histories are reversed (turned end for end) and the data then analyzed with the live encounter model, the φ parameter is now the probability that the animal was in the population before it was marked (Pradel 1996). In other words, what were trailing zeros in the encounter history now become leading zeros in the encounter history, and with reversal of the history, φ is the probability the animal was present. Pradel (1996) developed the model to estimate this new seniority parameter γ as well as apparent survival φ and detection probability p. Pradel's key observation was that $N_t \varphi_t = N_{t+1} \gamma_{t+1}$, which when rearranged gives

$$\lambda_t = \frac{N_{t+1}}{N_t} = \frac{\varphi_t}{\gamma_{t+1}}.$$

Jolly's (1965) model had the relationship $N_t \varphi_t + B_t = N_{t+1}$, and with dividing through by N_t gives $\varphi_t + \dfrac{B_t}{N_t} = \dfrac{N_{t+1}}{N_t}$, or $\varphi_t + f_t = \lambda_t$. Therefore, whether the model is parameterized with γ, f, or λ, the results are the same. For assessing whether the population is changing because of changes in recruitment or survival, the φ and f parameterization provides the information to evaluate each component. To evaluate the trend or population growth rate, the φ and λ parameterization is used. The main advantage of the φ and γ parameterization is that it is numerically more stable than the others.

All three of these parameterizations are coded in Program MARK, with the λ parameter provided as a derived parameter for parameterizations where λ is not in the likelihood. Further, this model has been "robustified" by replacing p with $p^* = 1 - \prod_{i=1}^{t}(1-p_i)$, again with all three parameterizations.

Two major, long-term studies have used the Pradel models to assess population status. Reynolds et al. (2017) estimated rate of population change (λ) of the northern goshawk (*Accipiter gentilis atricapillis*) on the Kaibab Plateau, a forested sky island in northern Arizona. Dugger et al. (2016) estimated λ for northern spotted owls (*Strix occidentalis caurina*) in the Pacific Northwest, using data from 11 different study areas. Both of these studies also estimated sex- and age-specific survival rates using the Cormack-Jolly-Seber models.

Future

The world of estimating population parameters is changing rapidly because of the advancements in computing of these estimates through Bayesian methods, as well as the development of numerical methods to optimize likelihood functions. Furthermore, new and more complex models are constantly appearing in the literature. With the huge increase in computer processing speed we have observed over the last three decades, approaches only dreamed of previously are reality today. The trend will continue. Optimization of likelihoods with >1000 parameters has become possible. Previously, Bayesian methods just took too long to compute to be applied. Although taking much longer than frequentist methods, application of Bayesian methods is appearing more and more in the literature as improved software is made available and as users become more knowledgeable about computing, Markov Chain Monte Carlo (MCMC) sampling, and probability calculus. The level of statistical, mathematical, and programming skills needed for future quantitative analyses is well illustrated in Royle et al. (2014). Application of these methods really requires some background in mathematical statistics for the user to be comfortable with them.

Acknowledgments

This chapter has been written based on the knowledge gleaned from numerous colleagues with whom the author has worked over the past five decades—too numerous to mention.

LITERATURE CITED

Amstrup, S. C., T. L. McDonald, and B. F. J. Manly. 2005. Handbook of capture-recapture analysis. Princeton University Press, Princeton, NJ.

Arnason, A. N., and C. J. Schwarz. 1999. Using POPAN-5 to analyze banding data. Bird Study 46(Supplement): S157–S168.

Barker, R. J. 1997. Joint modeling of live-recapture, tag-resight, and tag-recovery data. Biometrics 53:666–677.

Brownie, C., D. R. Anderson, K. P. Burnham, and D. S. Robson. 1985. Statistical inference from band recovery data, a handbook, 2nd ed. U.S. Fish and Wildlife Service, Resource Publication 156, Washington, DC. 305 pp.

Brownie, C., J. E. Hines, J. D. Nichols, K. H. Pollock, and J. B. Hestbeck. 1993. Capture-recapture studies for multiple strata including non-Markovian transitions. Biometrics 49:1173–1187.

Borchers, D. L., S. T. Buckland, and W. Zucchini. 2002. Estimating animal abundance: Closed populations. Springer, London.

Burnham, K. P. 1993. A theory for combined analysis of ring recovery and recapture data. In Marked individuals in the study of bird population, edited by J.-D. Lebreton and P. M. North, pp. 199–213. Birkhauser Verlag, Basel, Switzerland.

Choquet, R. 2008. Automatic generation of multistate capture-recapture models. Canadian Journal of Statistics—Revue Canadienne de Statistique 36:43–57.

Choquet, R., L. Rouan, and R. Pradel. 2009. Program E-SURGE: A software application for fitting multievent models. In Modeling demographic processes in marked populations, Environmental and ecological statistics 3, edited by D. L. Thomson, E. G. Cooch, and M. J. Conroy, pp. 845–865. Springer, New York.

Cormack, R. M. 1964. Estimates of survival from the sighting of marked animals. Biometrika 51:429–438.

Coull, B. A., and A. Agresti. 1999. The use of mixed logit models to reflect heterogeneity in capture-recapture studies. Biometrics 55:294–301.

Cox, D. R. 1972. Regression models and life tables (with discussion). Journal of the Royal Statistical Society B 34:187–220.

Dinsmore, S. J., and D. H. Johnson. 2012. Population analysis in wildlife biology. In The wildlife techniques manual: Volume 1 Research, 7th ed. edited by N. J. Silvy, pp. 349–380. Johns Hopkins University Press, Baltimore, MD.

Dorazio, R. M., and J. A. Royle. 2003. Mixture models for estimating the size of a closed population when capture rates vary among individuals. Biometrics 59:351–364.

Dugger, K. M., E. D. Forsman, A. B. Franklin, R. J. Davis, G. C. White, C. J. Schwarz, K. P. Burnham, J. D. Nichols, J. E. Hines, C. B. Yackulic, J. Paul F. Doherty, L. L. Bailey, D. A. Clark, S. H. Ackers, L. S. Andrews, B. Augustine, B. L. Biswell, J. Blakesley, P. C. Carlson, M. J. Clement, L. V. Diller, E. M. Glenn, A. Green, S. A. Gremel, D. R. Herter, J. M. Higley, J. Hobson, R. B. Horn, K. P. Huyvaert, C. McCafferty, T. McDonald, K. McDonnell, G. S. Olson, J. A. Reid, J. Rockweit, V. Ruiz, J. Saenz, and S. G. Sovern 2016. The effects of habitat, climate, and Barred Owls on long-term demography of Northern Spotted Owls. The Condor: Ornithological Applications 118:57–116.

Fienberg, S. E. 1972. The multiple-recapture census for closed populations and incomplete 2^k contingency tables. Biometrika 59:591–603.

Gimenez, O., and R. Choquet. 2010. Individual heterogeneity in studies on marked animals using numerical integration: Capture-recapture mixed models. Ecology 91:951–957.

Hestbeck, J. B., J. D. Nichols, and R. A. Malecki. 1991. Estimates of movement and site fidelity using mark-resight data of wintering Canada geese. Ecology 72:523–533.

Huggins, R. M. 1989. On the statistical-analysis of capture experiments. Biometrika 76:133–140.

Huggins, R. M. 1991. Some practical aspects of a conditional likelihood approach to capture experiments. Biometrics 47:725–732.

Jolly, G. M. 1965. Explicit estimates from capture-recapture data with both death and immigration stochastic model. Biometrika 52:225–247.

Kaplan, E. L., and P. Meier. 1958. Nonparametric estimation from incomplete observations. Journal of American Statistical Association 53:457–481.

Kendall, K. C., J. B. Stetz, J. Boulanger, A. C. Macleod, D. Paetkau, and G. C. White. 2009. Demography and genetic structure of a recovering grizzly bear population. Journal of Wildlife Management 73:3–17.

Kendall, W. L., R. J. Barker, G. C. White, M. S. Lindberg, C. A. Langtimm, and C. L. Penaloza. 2013. Combining dead recovery, auxiliary observations and robust design data to estimate demographic parameters from marked individuals. Methods in Ecology and Evolution 4:828–835.

Kendall, W. L., and R. Bjorkland. 2001. Using open robust design models to estimate temporary emigration from capture-recapture data. Biometrics 57:1113–1122.

Kendall, W. L., K. H. Pollock, and C. Brownie. 1995. A likelihood-based approach to capture-recapture estimation of demographic parameters under the robust design. Biometrics 51:293–308.

Kendall, W. L., G. C. White, J. E. Hines, C. A. Langtimm, and J. Yoshizaki. 2012. Estimating parameters of hidden Markov models based on marked individuals: Use of robust design data. Ecology 93:913–920.

Kéry, M., and J. A. Royle. 2016. Applied hierarchical modeling in ecology: Analysis of distribution, abundance and species richness in R and BUGS Volume 1: Prelude and static models. Academic Press, San Diego, CA.

King, R., B. J. T. Morgan, O. Gimenez, and S. P. Brooks. 2010. Bayesian analysis for population ecology. Chapman & Hall/CRC, Boca Raton, FL.

Lebreton, J.-D., K. P. Burnham, J. Clobert, and D. R. Anderson. 1992. Modeling survival and testing biological hypotheses using marked animals: A unified approach with case studies. Ecological Monographs 62:67 118.

Link, W. A. 2003. Nonidentifiability of population size from capture-recapture data with heterogeneous detection probabilities. Biometrics 59:1123–1130.

Link, W. A., and R. J. Barker. 2005. Modeling association among demographic parameters in analysis of open population capture-recapture data. Biometrics 61:46–54.

Link, W. A., and R. J. Barker. 2010. Bayesian inference with ecological applications. Academic Press, San Diego, CA.

McClintock, B. T., and G. C. White. 2009. A less field-intensive robust design for estimating demographic parameters with mark-resight data. Ecology 90: 313–320.

McClintock, B. T., G. C. White, M. F. Antolin, and D. W. Tripp. 2009a. Estimating abundance using mark-resight when sampling is with replacement or the number of marked individuals is unknown. Biometrics 65:237–246.

McClintock, B. T., G. C. White, K. P. Burnham, and M. A. Pryde. 2009b. A generalized mixed effects model of abundance for mark-resight data when sampling is without replacement. In Modeling demographic processes in marked populations, Environmental and ecological statistics 3, edited by D. L. Thomson, E. G. Cooch, and M. J. Conroy pp. 271–289. Springer, New York.

Morton, J. M., G. C. White, G. D. Hayward, D. Paetkau, and M. P. Bray. 2016. Estimation of the brown bear population on the Kenai Peninsula, Alaska. Journal of Wildlife Management 80:332–346.

O'Connell, A. F., J. D. Nichols, and K. U. Karanth. 2011. Camera traps in animal ecology. Springer, New York.

Otis, D. L., K. P. Burnham, G. C. White, and D. R. Anderson. 1978. Statistical-inference from capture data on closed animal populations. Wildlife Monographs 62:1–135.

Pledger, S. 2000. Unified maximum likelihood estimates for closed capture-recapture models using mixtures. Biometrics 56:434–442.

Pledger, S. 2005. The performance of mixture models in heterogeneous closed population capture-recapture. Biometrics 61:868–873.

Pollock, K. H. 1982. A capture-recapture design robust to unequal probability of capture. Journal of Wildlife Management 46:757–760.

Pollock, K. H., S. R. Winterstein, C. M. Bunk, and P. D. Curtis. 1989. Survival analysis in telemetry studies: The staggered entry design. Journal of Wildlife Management 53:7–15.

Pradel, R. 1996. Utilization of capture-mark-recapture for the study of recruitment and population growth rate. Biometrics 52:703–709.

Reynolds, R. T., J. S. Lambert, C. H. Flather, G. C. White, B. J. Bird, L. S. Baggett, C. Lambert, and S. Bayard De Volo. 2017. Long-term demography of the Northern Goshawk in a variable environment. Wildlife Monographs 197:1–40.

Royle, J. A., R. B. Chandler, R. Sollmann, and B. Gardner. 2014. Spatial capture-recapture. Academic Press, Waltham, MA.

Royle, J. A., and R. M. Dorazio. 2008. Hierarchical modeling and inference in ecology. Academic Press, San Diego, CA.

Schwarz, C. J., and A. N. Arnason. 1996. A general methodology for the analysis of capture-recapture experiments in open populations. Biometrics 52:860–873.

Seber, G. A. F. 1965. A note on the multiple recapture census. Biometrika 52:249–259.

Seber, G. A. F. 1970. Estimating time-specific survival and reporting rates for adult birds from band returns. Biometrika 57:313–318.

Seber, G. A. F. 1982. The estimation of animal abundance and related parameters, 2nd ed. Charles Griffin, London.

White, G. C., D. R. Anderson, K. P. Burnham, and D. L. Otis. 1982. Capture-recapture and removal methods for sampling closed populations. LA-8787-NERP, Los Alamos National Laboratory, Los Alamos, NM. 235 pp.

White, G. C., and K. P. Burnham. 1999. Program MARK: Survival estimation from populations of marked animals. Bird Study 46(Supplement):S120–S139.

White, G. C., and E. G. Cooch. 2017. Population abundance estimation with heterogeneous encounter probabilities using numerical integration. Journal of Wildlife Management 81:322–336.

Williams, B. K., J. D. Nichols, and M. J. Conroy. 2001. Analysis and management of animal populations: Modeling, estimation, and decision making. Academic Press, San Diego, CA.

6 — Distance Sampling

STEPHEN T. BUCKLAND,
DAVID L. MILLER, AND
ERIC REXSTAD

Distance sampling relies on a combination of model-based and design-based methods.
— Buckland et al. (2004:vii)

Introduction

The term *distance sampling* refers to a suite of methods in which distances to detected objects are modeled to estimate object density and/or abundance in a survey region. The objects are typically animals or groups of animals (e.g., flocks, schools, herds), animal products such as dung or nests, cues to detection such as songs, calls, or whale blows, or plants. *Line transect sampling* is the most widely used technique. Its design usually involves a systematic grid of parallel lines placed at random over the survey region and an observer traveling along each line, recording distances from the line of detected objects. In some circumstances, it is more practical to record from a set of points rather than along lines, in which case the method is termed *point transect sampling*. For the standard methods, if an object is on the line or at the point, it is assumed to be detected with certainty. The probability of detection is assumed to fall smoothly with distance from the line or point. Modeling the *detection function* (the probability of detection as a function of distance from the line or point) is key to reliable estimation of abundance. Using the fitted detection function, we can estimate the proportion of objects detected within a strip centered on the line or a circle centered on the point, from which

it is straightforward to estimate object density on the sampled strips or circles. Given a randomized design, the density may be extrapolated to the wider survey region, and abundance can be estimated.

Thus, the standard methods represent a hybrid approach, in which model-based methods are used to estimate the detection function, and design-based methods are invoked, so that objects can be assumed to be located independently of the lines or points, and to allow extrapolation to the whole survey region. Increasingly, fully model-based methods are used in preference to this hybrid approach. For a summary of both the hybrid and the fully model-based approaches, together with details of extensions of the standard methods and of related methods, see Buckland et al. (2015). Windows-based software Distance (Thomas et al. 2010) and R packages (Fiske and Chandler 2011; Miller et al. in press) implement many of these methods.

Plot Sampling

Distance sampling is an extension of plot sampling, in which not all objects on a sampled plot are detected. We first consider plot sampling when all objects on each sampled plot are detected. The two forms of plot sampling that are of interest to us are (1) *circular plot sampling,* for which a sample of circular plots, each of radius w, is covered by an observer who records objects from the center point of each plot, and (2) *strip transect sampling,* in which an observer travels along the center line of each strip, recording all objects within the strip of half-width w.

Survey design is covered in detail by Buckland et al. (2001, 2004). For circular plot surveys, the design normally consists of a systematic grid of circles, positioned at random over the survey region, but sometimes, a simple random sample of plots is preferred. Survey regions are sometimes divided into geographic strata, with a systematic random or simple random sample selected from each stratum. Similar issues apply to the design of strip transect surveys, except that strips can be long, with each strip traversing the entire width of the survey region (or of the stratum), or they can be short, in which case the design is more closely comparable with a circular plot design.

Consider a circular plot survey comprised of K circular plots, each of radius w, and suppose that n objects are detected. Then assuming all objects on each sampled plot are counted, we can estimate object density D by

$$\hat{D} = \frac{n}{K\pi w^2},$$

and abundance in the survey region as

$$\hat{N} = \hat{D}A = \frac{nA}{K\pi w^2},$$

where A is the size of the survey region. For a strip transect survey with strips of half-width w and total length L, the equivalent expressions are

$$\hat{D} = \frac{n}{2wL},$$

and

$$\hat{N} = \hat{D}A = \frac{nA}{2wL}.$$

For both approaches, the density estimate can be expressed simply as $\hat{D} = \frac{n}{a}$, where a is the covered area; thus $a = K\pi w^2$ for circular plot sampling and $a = 2wL$ for strip transect sampling.

Measures of precision can be estimated for the estimated density by recognizing that uncertainty arises because sampled plots constitute only a portion of the entire study area for which inference is intended. The more variable the encounters per unit of effort between transects, the greater the uncertainty in the density estimate.

In standard plot sampling, no data are collected to test the assumption that all objects on a plot are detected. To cover the circle from the center point, or the strip from the center line, the radius or half-width w must be small, to ensure all objects on the plot are detected. This means that many objects are detected beyond w, and hence are not included in the analysis. Larger plots might be used if the observer is not restricted to the center point or line, but it may take some time to search each plot thoroughly. If objects are mobile, then it is difficult to track those objects that have already been counted, and additional objects may enter the plot during the count, resulting in overestimation of density. Distance sampling offers a solution to these difficulties, at the cost of having to measure the distance from the point or line of each detected object.

Estimating Detectability
Allowing for Imperfect Detectability

Given distances from the point or line to detected objects, we can readily assess whether all objects on a plot are detected. In the case of line transect sampling, if objects detected out to distance w either side of the line are recorded, then given random line placement, we expect roughly the same number of animals to be detected between say 0 and 0.1 w from the line, or any other strip of the same width, as say between 0.9 w and w, provided we detect all objects in the strip. Thus, the numbers of detections by strip should be realizations from a uniform distribution, and we can test the assumption of perfect detectability out to distance w by using, for example, a goodness-of-fit test. If fewer animals tend to be detected in strips farther from the line, then this suggests that not all objects have been detected. Point transect data can be used similarly, but in this case, the area of an "annulus" (a narrow, circular strip at a fixed radius from the point) of a given width increases linearly with distance from the point, and so given perfect detectability, we expect numbers of detections per annulus to likewise increase linearly (see example, in Buckland et al. 2015).

Given evidence that we do not have perfect detection out to distance w, we can go a step further and model the probability of detection as a function of distance y from the line or point. We can do this by selecting a smooth, nonincreasing function $g(y)$, termed the "detection function," that is unity at zero distance ($g(0) = 1$ corresponding to perfect detectability of objects at the line or point), and falls smoothly with distance.

Having fitted a detection function, we can estimate the probability of detection of an object at any given distance from the line or point. We can then estimate the probability P_a that an object is detected, given only that it is within the truncation distance w of a line or point, corresponding to the "covered area" a, where $a = 2wL$ for lines of total length L, or $a = K\pi w^2$ for K points. This probability corresponds to the proportion of objects that we expect to detect within the covered area. We illustrate estimation of P_a for line transect sampling in Fig. 6.1. If detectability remained perfect all the way to w, then the detection function would be a flat line, $g(y) = 1$, and all objects in the covered area would be detected. In this case, the assumptions of plot sampling would be satisfied.

This intuitive argument can be made rigorous via mathematics. The area of a rectangle with height

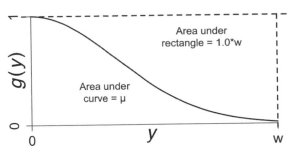

Fig. 6.1.   Detection probability estimation for line transect survey. Probability P_a that an object within distance w of a line is detected, represented by the proportion of the rectangle under the detection function curve $g(y)$.

$g(0) = 1$ and width w is equal to $1 \cdot w$. The area under the curve described by $g(y)$ to distance w is the integral $\mu = \int_0^w g(y)dy$. The probability that an object is detected if it is within w of the transect is the ratio of these two areas: the ratio of the area under the curve to the area under the rectangle, or $\int_0^w g(y)dy/w$. When the transects are points rather than lines, the calculation of P_a is more complex (Buckland et al. 2015).

Density estimation using either line or point transect sampling adjusts the density and abundance estimates shown for plot sampling by dividing the equations of the previous section by the estimated probability of detection $\hat{P}_a$.

There is one key design assumption (assumption 1 below) and three key model assumptions of standard distance sampling methods. We can guarantee that the design-based assumption holds by adopting an appropriate design. These four assumptions are:

1. Objects are distributed independently of the lines or points.
2. Objects on the line or at the point are detected with certainty ($g(0) = 1$).
3. Distance measurements are exact.
4. Objects are detected at their initial location.

For line transect sampling, the practical implication of the last assumption is that average speed of objects should be slow relative to the speed of the observer, and if objects move in response to the observer, their position before such movement should be recorded. These assumptions are discussed in detail by Buckland et al. (2001, 2015), and methods that allow the key assumptions to be relaxed or removed are described by Buckland et al. (2004, 2015).

Models of the Detection Function

A desirable property of detection functions is that they decline little if at all for small y. This is called the "shoulder property." At moderate values of y, the detection function should decrease smoothly, and for y large (approaching w), the detection function will typically level at a value near 0.

There are many functions that satisfy these criteria. The software Distance (Thomas et al. 2010) and the R package (Miller et al. in press) adopt an approach in which one of three key functions can be selected, and possibly augmented by series adjustments.

The key functions are the uniform

$$g(y) = 1 \quad \text{where} \ 0 \le y \le w,$$

which corresponds to perfect detection out to distance w (i.e., plot sampling) unless adjustment terms are added; the half-normal model

$$g(y) = \exp\left(\frac{-y^2}{2\sigma^2}\right) \quad \text{where} \ 0 \le y \le w,$$

with $g(0) = 1$ as desired and a scale parameter σ adjusting the rate at which the detection function declines; and finally, the hazard-rate model

$$g(y) = 1 - \exp\left[-\left(\frac{y}{\sigma}\right)^{-b}\right] \quad \text{where} \ 0 \le y \le w,$$

with not only a scale parameter σ but also an additional parameter b offering flexibility in the shapes the function can take on.

There are also three types of adjustment terms. They are needed only if the fit of the key function alone is poor. The user can select cosine terms, Hermite polynomial terms, or simple polynomial terms. Addition of adjustment terms can be done automatically by software using model selection criteria such as Akaike's Information Criterion (AIC).

Factors other than distance may influence probability of detection. These may relate to the individuals conducting the survey (experience, for example) or to the objects being surveyed (behavior, group size, or plumage/pelage), or to other factors such as weather conditions or habitat type. Measurements on these covariates recorded during the survey can be accommodated in the detection function by modeling the scale parameter σ in either the half-normal or hazard-rate key functions as a function of the covariates:

$$\sigma(\mathbf{z}_i) = \exp\left(\alpha + \sum_{q=1}^{Q} \beta_q z_{iq}\right),$$

where $\mathbf{z}_i = (z_{i1}, z_{i2}, \ldots, z_{iQ})^T$ is a vector of covariate values recorded for the ith detected animal, and α, $\beta_1, \ldots, \beta_Q$ are coefficients to be estimated.

We can now write the half-normal detection function for detection i as

$$g(y_i, \mathbf{z}_i) = \exp\left[\frac{-y_i^2}{2\sigma^2(\mathbf{z}_i)}\right] \quad \text{where} \ 0 \le y_i \le w,$$

and the hazard-rate detection function as

$$g(y_i, \mathbf{z}_i) = 1 - \exp[-(y_i / \sigma(\mathbf{z}_i))]^{-b} \quad \text{where} \ 0 \le y_i \le w.$$

It can be useful to model the effect of covariates other than distance on the detection function if we wish to understand what factors affect probability of detection, or if we wish to estimate object density at different levels of a covariate, such as different habitat types. If, however, we want just an overall estimate of abundance in the study area, there is no need to include additional covariates in the model, as pooling robustness ensures that provided $g(0)=1$ for all objects, heterogeneity among objects in detectability does not generate bias (Buckland et al. 2004, 2015).

A further advance in detection function modeling is relaxation of the assumption that all objects on the line or point are detected with certainty ($g(0)=1$). This assumption can fail for a variety of reasons, such as fossorial animals being underground, birds being high in the canopy above the head of the observer, or whales diving below the surface in a marine survey. As additional data are required to allow estimation of P_a, we again need additional data if we are to estimate $g(0)$. There are several ways in which these additional data may be gathered. We will focus on a simple method; other data collection protocols and modeling are described in Buckland et al. (2004).

If two observers independently search the same area, we can regard one observer as setting up trials for the other: given that observer 1 detects an object, whether observer 2 detects it gives us information on probability of detection, without having to assume $g(0)=1$. For some protocols, only one observer sets up trials; in others, each observer sets up trials for the other. Thus, for each detected object, it is detected by observer 1 only, by observer 2 only, or by both observers. Note the need for a field protocol that allows determination of whether an object was detected by both observers ("duplicate detections"). The method is like two-sample mark-recapture, in which one observer corresponds to one sample, and the other to the other. Hence the methods are described as *mark-recapture distance sampling* (MRDS).

When modeling the resulting data, it is convenient to scale the function $g(y, z)$ so that $g(0, z)=1$ by definition. We then denote the detection function by $p(y, z)$, which represents the probability that an animal at distance y from the line or point and with covariates z is detected. Thus $p(0, z) \leq 1$.

When there are two observers, we need to distinguish between their respective detection functions. We denote the detection function for observer 1 by $p_1(y, z)$ and that for observer 2 by $p_2(y, z)$. We will also need further detection functions: $p_{1|2}(y, z)$,

which is the probability that observer 1 detects an animal at distance y and covariates z *given that* observer 2 detects it, and the converse, $p_{2|1}(y, z)$. Finally, we need to define the probability that at least one observer detects the animal: $p.(y, z)$.

These different detection functions are related as follows:

$$p.(y, z) = p_1(y, z) + p_2(y, z) - p_2(y, z)p_{1|2}(y, z)$$
$$= p_1(y, z) + p_2(y, z) - p_1(y, z)p_{2|1}(y, z).$$

Early models for MRDS assumed *full independence*. Under this assumption, the previous equation becomes

$$p.(y, z) = p_1(y, z) + p_2(y, z) - p_1(y, z)p_2(y, z).$$

This provides a framework for the estimation of $g(0)$. However, as with traditional mark-recapture estimation, the existence of unexplained factors influencing detection probability may produce considerable bias in abundance estimation. This is the "curse of heterogeneity" described by Link (2003). Consequently, analysis of such MRDS data requires considerable modeling effort to incorporate sources of variation in detectability. Even then, more sophisticated models using the concepts of point and limiting independence (Buckland et al. 2015) may be needed to combat the bias induced by heterogeneity.

Standard Line and Point Transect Sampling Design and Field Methods

Valid inferences can be drawn only from carefully designed and executed data collection and distance sampling surveys.. When using design-based inference, if allocation of sampling effort fails to adhere to the principle that locations sampled are representative of locations over which inference is to be drawn, then resulting estimates will in general be biased. Random placement of lines or points, or random positioning of a grid of lines or points, ensures that objects are distributed independently of the lines or points.

Assumptions are made to make analysis of data simpler and easier. Trying to meet assumptions of distance sampling requires care during the design and conduct of distance sampling surveys. When assumptions cannot be met, they can be exchanged for the collection of additional data, which is likely to result in more complex analysis methods.

Both survey coordinators and observers should be mindful of the four key assumptions listed above and seek to collect data in a way that ensures the assumptions are met as far as possible. Detailed discussion of assumptions is given by Buckland et al. (2001, 2015). Here, we offer just a few illustrative thoughts.

Placement of line transects along roads or tracks violates assumption 1, because animals know the location of roads and tracks and may be either attracted or repelled. Because of the attraction or repulsion, the presumption that animals are as likely to be within a strip 2–4m from the transect as they are to be between 22m and 24m is untrue, resulting in biased estimates of density. Therefore, placement of transects should be representative of the study area.

Data gathering can often be refined to ensure that assumption 2 holds to a good approximation. For example, in shipboard surveys, multiple observers, perhaps with a mixture of naked eye and binocular search, can help. Assumption 3 can be met by use of technology. Items such as laser rangefinders involve a small cost in comparison to labor costs of nearly all surveys. Thus, the addition of uncertainty into the estimation process by guessing the distance to detections can be readily eliminated in most surveys for relatively low cost. Movement of animals prior to their detection by the observer (violation of assumption 4) will usually result in bias. For surveys on foot, observers should attempt to move quietly and unobtrusively, while trying to maintain a reasonable pace, because even if animals do not respond to the observer, if they are moving on average at around half the speed of the observer or faster, estimation will again be biased. If it is not possible to move appreciably faster than the study animals, then different field and/or analysis methods are required (Buckland

et al. 2001, 2015). It is generally more cost-effective to collect data that conform to the assumptions than to collect additional data to allow relaxation of those assumptions.

Estimation

When we estimate abundance in the survey region (of size A having sampled some smaller area a), we use the Horvitz-Thompson–like estimator:

$$\hat{N} = \frac{A}{a} \sum_{i=1}^{n} \frac{s_i}{\hat{P}_a(z_i)},$$

where s_i is the size of the ith group or cluster if the objects occur in groups (flocks, pods, etc.) but is simply 1 for all observations if the objects occur individually. If there are covariates in the detection function, $\hat{P}_a(z_i)$ is the estimated probability of detection for observation i with covariates z_i (without covariates this simplifies to $\hat{P}_a$). Note that if we set $\hat{P}_a(z_i) = 1$ and $s_i = 1$ for all i, we recover the strip transect estimators, above.

To estimate abundance, we need to estimate $\hat{P}_a(z_i)$, based on distance data. Both n and (if objects cluster) s_i are random variables, and when we calculate uncertainty in abundance, we need to include uncertainty about these quantities. For n we calculate the variance of the *encounter rate* (the number of objects detected divided by the effort—line length for line transect sampling, total number of visits to points for point transect sampling). The variance of group size is usually calculated empirically from the observed group sizes. Typically for line transect sampling, most of the uncertainty in our abundance estimate is due to uncertainty in estimating the encounter rate, while for point transect sampling, uncertainty in estimating the detection probability often exceeds that in estimating the encounter rate. See Buckland et al. (2015) for more information on variance and confidence interval estimation.

To demonstrate estimation of density when employing line and point transect sampling, we use data from Buckland (2006). It is a survey of woodland in Scotland, with a focus on a number of passerine species. For this example, we describe the analysis of detections of males of one species, the European robin (*Erithacus rubecula*). Data were collected from the same 32 ha site from two surveys: one using 19 line transects, walked twice, with a total length of 4.83 km, and a second using 32 points, each visited twice. In each case, as far as possible, exact distances from the line or point were measured to the individual robin, using a laser rangefinder.

Line Transect Estimates

Eighty-two detections were made during the line transect survey, with the maximum perpendicular distance of 100m. Truncating the most distant 2% of observations resulted in 80 detections with a trunca-

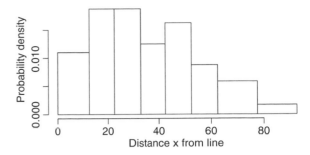

Fig. 6.2. Distribution of robin detections by distance from the line transect survey.

tion distance *w* of 95 m. A histogram of the perpendicular distances after truncation is shown in Fig. 6.2.

Fitting the three key functions with their preferred adjustments (Table 6.1) results in adequate fits to the data (Fig. 6.3). The *p*-values associated with a Kolomogorov-Smirnov goodness-of-fit test for exact distance data exceeded 0.2 for all three models. This

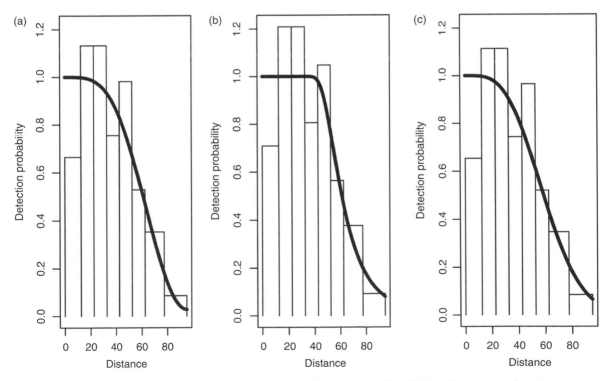

Fig. 6.3. Estimated detection functions for the line transect data for robins. Fits of the uniform key with cosine adjustments (left), hazard-rate (center), and Hermite polynomial (right) models are shown. Histograms are scaled such that area of the histogram is equal to the area under the detection functions; as a consequence, some of the histogram bars extend above 1.0.

Table 6.1. Analysis summary for three detection function models applied to the robin line transect data. ΔAIC gives the difference in AIC from the best model and CV indicates the coefficient of variation calculated as the ratio of the standard error to the point estimate (e.g., $se(\hat{D}) / \hat{D}$), a relative measure of uncertainty.

Key function	Adjustment type	No. adj. terms	ΔAIC	$\hat{P}_a$	$CV(\hat{P}_a)$	$\hat{D}$	$CV(\hat{D})$
Uniform	Cosine	2	0.00	0.636	0.162	68.5	0.193
Half-normal	Hermite poly	1	0.64	0.625	0.194	69.7	0.220
Hazard-rate	Simple poly	0	0.57	0.679	0.078	64.2	0.131

means that whichever model we choose, we base our inference upon models that fit the observed data well.

Note that the estimated probabilities of detection for the three models are close. Consequently, the density estimates $\hat{D}$ under the three models do not differ greatly. Furthermore, ΔAIC values (difference between the AIC of the model and that of the best model) for the three models are all small, suggesting that there is little to choose among the competing models. This is the luxury of collecting data that conform to distance sampling assumptions. The consequences of decisions from the analysis, such as which model to select, are minor for well-behaved line transect data. If we use the model with the smallest AIC, the uniform with two cosine adjustment terms, we obtain a density (number of males per km²) of 68.5.

Point Transect Estimates

Fifty-two detections were made when surveying the point transects; distances ranged from 20 m to 120 m. Truncation to $w = 110$ m resulted in the elimination of the two outermost detections. A histogram of the radial distances after truncation is shown in Fig. 6.4.

Fitting the three key functions with their preferred adjustments (Table 6.2) results in adequate fits to the data (Fig. 6.5). The p-values associated with a Kolomogorov-Smirnov goodness-of-fit test for exact distance data exceeded 0.2 for all three models. As with the line transect data, we are basing our in-

ference upon models that fit the observed data well, regardless of which model we choose.

There is a wider disparity between models in AIC scores for the point transect survey. AIC shows preference for the hazard-rate model, which has a wider shoulder that accommodates the dearth of detections at small distances better than other models. Despite the difference in shapes of the fitted detection functions (Fig. 6.5), there remains close agreement between the density estimates. The largest density estimate (produced by the uniform-cosine model) is only 9% larger than the density estimate produced by the preferred hazard-rate model, with an estimate of 59.7 males per km².

Model-Based Distance Sampling

Conventional distance sampling uses model-based methods for the detection process and design-based methods to estimate animal abundance in the study

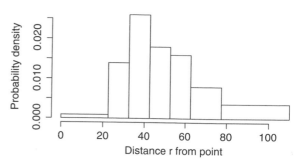

Fig. 6.4. Distribution of robin detections by distance (snapshot point transect sampling).

Table 6.2. Analysis summary for three detection function models applied to the robin snapshot point transect data from the Montrave case study (Buckland 2006).

Key function	Adjustment type	No. adj. terms	ΔAIC	$\hat{P}_a$	CV($\hat{P}_a$)	$\hat{D}$	CV($\hat{D}$)
Uniform	Cosine	1	2.78	0.316	0.063	65.1	0.164
Half-normal	Hermite poly	1	2.68	0.329	0.573	62.5	0.593
Hazard-rate	Simple poly	0	0.00	0.344	0.171	59.7	0.229

region, given estimated probabilities of detection. Increasingly, fully model-based methods are being used. Although they tend to be less robust for estimating animal abundance than conventional methods, they offer several advantages: they allow the analyst to explore how animal density varies by habitat or topography and to estimate abundance for any subregion of interest; and they provide tools for analyzing data from designed distance sampling experiments to assess treatment effects. Buckland et al. (2016) summarize model-based methods and provide a general framework.

Designed distance sampling experiments can be used to assess the success of different management strategies, or for environmental impact assessment. The data are plot counts (where the plot is either a strip centered on the line or a circle centered on the point), and count models can be used, with a correction for detectability. The correction can be estimated in a first step, followed by a count model with the correction as an offset fitted in a second step, or the model can be fitted in a single step, which is more easily done using a Bayesian framework. Examples of these two approaches are provided by Buckland et al. (2015).

We focus here on the use of model-based methods to estimate how density of objects varies through the study region.

Spatial Modeling

In this section, we seek to build a fully spatially explicit model of the population—that is, a *species distribution model*. Such models can be useful for exploring how object density varies as a function of spatially referenced covariate data such as altitude, bathymetry, habitat, or climate.

Spatial models for distance sampling data are an active area of research, and many alternative approaches have been suggested. Until recently, models that treat distance sampling data as point processes have been computationally expensive and difficult to estimate. Recent work by Yuan et al. (2017) addresses this directly, modeling object locations as a realization of a spatial point process, with the locations of detected objects resulting from a thinned process.

Here we focus on *density surface models* (DSMs), a two-stage approach to modeling abundance, which models detectability at the first stage, and spatial variation in density at the second. Once a detection function has been fitted to the distance data, estimated detection probabilities are used to adjust the observed counts, and a spatial model can then be constructed to describe variations in density in both geographical and covariate space. Second-stage models can be purely spatial, considering only location, or can include (temporally varying) environmental covariates. We refer to the models as "spatial models" here, even though density surface models do not necessarily incorporate explicitly spatial terms.

To build a spatial model, we need to know something about the location of each observation. For line transect sampling, given that line locations are known, recording distance along the line as well as distance from the line of each detection is sufficient, while for point transect sampling, distance and bear-

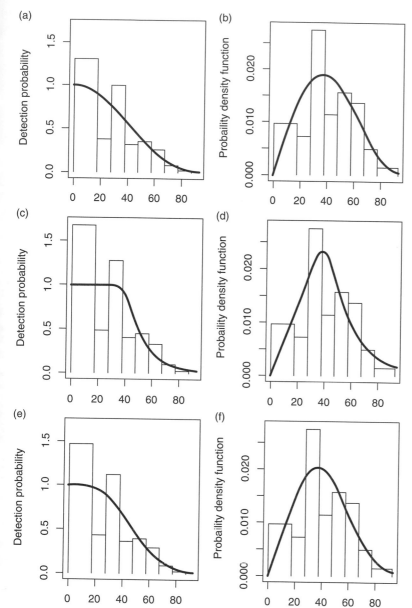

Fig. 6.5. Estimated detection functions (left) and probability density functions (right) for the snapshot point transect data for robins. Fits of the uniform key with cosine adjustments (a, b), the hazard-rate model (c, d), and the half-normal key with Hermite adjustments (e, f) are shown.

ing of each detection from the point at which it was detected are sufficient.

The response for a density surface model is the number of (estimated) objects for a given sample. In the case of point transects, the samples are the points, and we simply need counts or estimates for each point. For line transect data, the transects are usually long compared to the potential changes in ani-

mal density, so we cut the transects into short segments. A guide is to aim for segments that are approximately square (so they are about $2w \times 2w$ units in dimension), although environmental covariate resolution is relevant in deciding on segment length. We allocate each observation to a segment, and these are the sample units for the spatial model. Fig. 6.6 shows a segmentation scheme. We use the

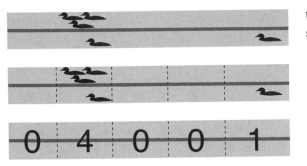

Fig. 6.6. Segmentation process for a line transect. Top: the line transect (dark grey horizontal line) with observations of *Gavia immer*; middle: the segments (indicated by dashed vertical lines) are overlaid; bottom: each segment's response is the count in that segment (the n_j in Equation (6.1)). *Source:* Gavia immer *graphic from http://phylopic.org by Unknown (photo), John E. McCormack, Michael G. Harvey, Brant C. Faircloth, Nicholas G. Crawford, Travis C. Glenn, Robb T. Brumfield & T. Michael Keesey, Creative Commons BY license.*

word "segment" below to refer to the sample unit of a DSM, whether line or point transects are used.

Density Surface Models

We denote the area of segment j by a_j, and the number of objects seen in a segment (or plot) j by n_j. To model these counts, we use smooth functions f_k of one or more *environmental covariates*, taking the value z_{jk} for segment j. To consider detectability, we multiply the estimated probability of detection in the segment, $\hat{p}_j$, by the segment area (a_j) to calculate an effective area. We write the model as:

$$E(n_j) = a_j \hat{p}_j \exp\left(\beta_0 + \sum_k f_k(z_{jk}) \right) \quad (6.1)$$

This model falls into the *generalized additive modeling* (GAM) framework, which includes many models as subcases.

Equation 6.1 allows for detection functions only where a covariate is recorded at the level of the segment. To include covariates that are recorded at the level of the observation, we use the Horvitz-Thompson–like estimator to summarize the counts at

the segment level, using the estimated abundance per segment as the response. The model is formulated as:

$$E(\hat{N}_j) = a_j \exp\left(\beta_0 + \sum_k f_k(z_{jk}) \right)$$
$$\text{where } \hat{N}_j = \sum_i \frac{s_{ij}}{\hat{P}_a(z_{ij})}, \quad (6.2)$$

where $\hat{P}_a(z_{ij})$ is the estimated probability of detection of individual i detected in segment j. The response is estimated with the same type of Horvitz-Thompson–like estimator used previously, but summing over only the observed groups in that segment (hence the j subscript); as previously, if objects occur individually, we set s_{ij} to 1 for all observations.

For models of either type (Equation 6.1 or Equation 6.2), we use count distributions to model the response. Although quasi-Poisson distributions have been used historically, Tweedie or negative binomial distributions provide more flexible mean-variance relationships. Discussions of these response distributions can be found in Shono (2008) and Ver Hoef and Boveng (2007).

Example

To illustrate density surface modeling, we use data collected by the University of Rhode Island as part of the Ocean Special Area Management Plan. The survey consisted of 41 flights off the coast of Rhode Island between October 2010 and July 2012. In each flight, a subset of 24 line transects perpendicular to the coastline was surveyed (see Fig. 6.7). There was one observer on each side of the aircraft, recording seabirds in three bins: 44–163 m, 164–432 m, and 433–1000 m, with cut points marked on the wing struts. We restrict ourselves to observations of common loons (or great northern divers, *Gavia immer*) on the water.

A hazard-rate function with flock size as a covariate had a lower AIC than without the covariate (ΔAIC = 3.49). The detection function fitted with

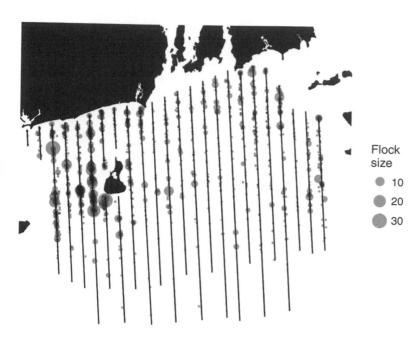

We fitted density surface models using the following spatial covariates: space (as centroids of the segments, projected from latitude and longitude; 2-dimensional smooth); a long-term geometric mean of chlorophyll-A (as a proxy for primary productivity); bottom roughness and bathymetry (both as proxies for prey availability). We can select between terms using *p*-values or use shrinkage-type methods to remove terms during model fitting (in which case we can view the GAM as an empirical Bayes model); here we show results using a shrinkage smoother. Both Tweedie and negative binomial response distributions were fitted to the data, with the negative binomial providing better fit according to standard GAM diagnostics performed on the (deviance and randomized quantile) residuals. Estimated smooth relationships are shown in Fig. 6.9.

Once we have fitted models, we can perform inference: we can think about the biological interpretation of the smooth effects (Fig. 6.9), or we can build maps of the distribution of the population. Such a map is shown in Fig. 6.10 alongside a map of uncertainty. The latter is essential to understanding where to believe the former; overlaying the effort on the uncertainty map allows us to keep in mind where we are extrapolating and where we should be cautious in

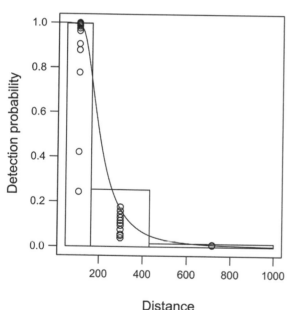

Fig. 6.8. Hazard-rate detection function with flock size as a covariate. Dots indicate the fitted values of the detection function at the observed values of flock size, while the line indicates the average detection function. Estimated parameters are very similar to that of a model without the size covariate, though AIC favors the model with flock size included.

software Distance (Miller et al., in press) is shown in Fig. 6.8.

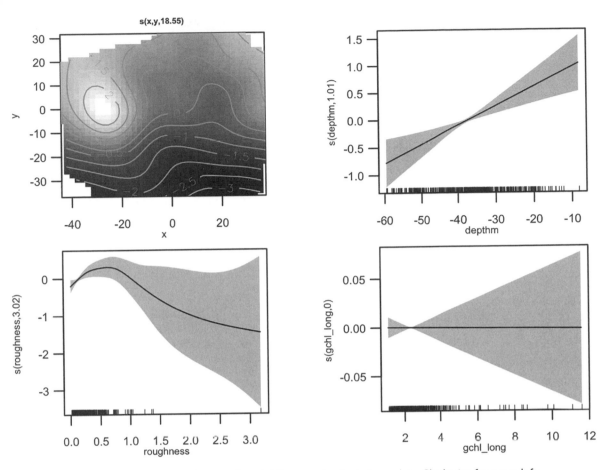

Fig. 6.9. Smooth components from the spatial model fitted to the *Gavia immer* data. Clockwise from top left: bivariate smooth of projected coordinates (x, y), depth in meters (depth m), geometric mean of 2002–2012 chlorophyll A (gchl_long), and bottom roughness (roughness). In the top left plot, with contours giving values; black dots indicate the centroids of the segments. In the remaining three plots, black lines indicate smooth functions, grey bands show +/–2 standard errors from the mean, and the rug plot (lines above the horizontal axis) indicates covariate values at the segments. Note that none of these is directly interpretable, as effects are on the *log* (link) scale and do not include effort. We see that the chlorophyll A measure has been removed from the model during fitting, leading to a zero effect. We should also be suspicious of the roughness term in the model as there are no data in the range 1.5-3; the wide band should be a warning also. We may also need to be cautious in any extrapolation using depths outside the modeled range as the linear effect may lead to extreme predictions.

our interpretation. Miller et al. (2013) give a worked example of the DSM approach in an appendix, and we refer readers to this for a more detailed analysis.

Variations and New Technologies

There are several variations on standard distance sampling methods. Buckland et al. (2015) give further details.

Cue Counting

In *cue counting*, an observer travels along lines, or visits a sample of points, and records cues, together with estimates of their distances from the observer. The cues are generally ephemeral, such as bird song bursts or calls or whale blows. Essentially standard point transect methods are used to analyze the distances. For standard point transects, effort is defined

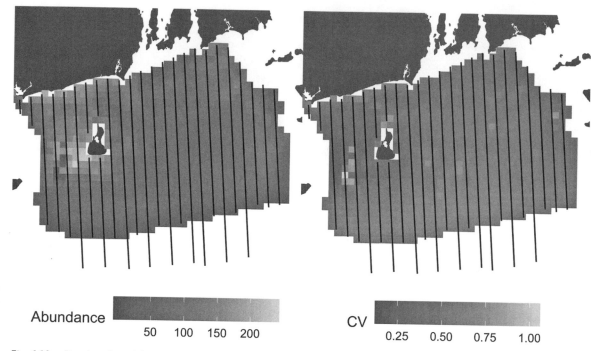

Fig. 6.10.   Results of modeling the *Gavia immer* data. Left: map of abundance corresponding to the fitted model; right: uncertainty for the same model, plotted as coefficient of variation per cell (calculated as standard error divided by abundance estimate for that cell). Note the very high uncertainty corresponds to a cell with a roughness value of 3.23, which lies where the confidence band shown in the lower left plot of Fig. 6.9 is very wide.

to be the number of visits to points, whereas for cue counting, it is the total time "on effort"—that is, total time spent detecting cues. In contrast to line and point transect methods, the analysis methods are identical whether the observer travels along lines or visits a sample of points. The analysis yields estimates of the number of cues per unit time per unit area. To convert these estimates to estimates of object density, we require an estimate of the cue rate—the average number of cues per unit time per object. Then object density is estimated as

$$\hat{D} = \frac{2n}{T\phi w^2 \hat{P}_a \hat{\eta}},$$

where $\hat{\eta}$ is the estimated cue rate, $\hat{P}_a$ is the estimate of the proportion of cues in the covered area a that is detected, n is the number of cues detected in time T, and ϕ is the angle in radians of the sector surveyed.

If the survey is conducted from a sample of points, typically the whole circle is surveyed, so that $\phi = 2\pi$. If the survey is conducted along lines, a sector ahead of the observer is usually surveyed, perhaps with ϕ equal to $\pi/2$ or $\pi/3$ radians.

Indirect Surveys

For some animal populations, it is easier to survey signs produced by the animals rather than the animals themselves. Examples include surveys of ape nests, as well as of dung of many species, including deer and elephants.

There are two main approaches. For the clearance plot method, all sign is removed from the plots, or is marked. After say d days, a survey is conducted of new sign appearing since the plots were cleared. We assume that d is chosen so that no new sign has time to decay. New sign density, say R, is then estimated using standard distance sampling methods, and animal density is then estimated as

$$\hat{D} = \frac{\hat{R}}{d\hat{\eta}},$$

where $\hat{\eta}$ is an estimate of the sign production rate (that is, the average number of sign produced per animal per day).

For the standing crop method, in which plots are not cleared initially, we also need an estimate $\hat{\xi}$ of the mean time (in days) to decay. Then animal density is estimated as

$$\hat{D} = \frac{\hat{R}}{\hat{\xi}\hat{\eta}}.$$

Trapping and Lure Point Transects

The design considerations for a trapping or lure point transect survey are the same as for point transect sampling. A trap or lure is placed at each point. For those animals recorded at the trap or lure, we are unable to observe its initial location when the trap or lure was set. Thus, in addition to the main survey, a trial survey must be carried out, in which a trap or lure is set at a known distance from an animal, and we note whether it is subsequently recorded at the trap or lure. Our data are thus binary, and we expect the proportion of zeros (corresponding to the animal not being recorded at the trap or lure) to increase with distance from the point. We can model the binary data, to estimate the detection function as a function of distance, and possibly of additional covariates. For these trials, we observe distances for the undetected animals, as well as the detected animals, so we need not assume that detection is certain if the initial distance is zero. Buckland et al. (2015) illustrate lure point transect sampling using a survey of the endemic Scottish crossbill (*Loxia scotica*) and trapping point transect sampling with a study of the Key Largo woodrat (*Neotoma floridana smalli*).

New Technologies

New technologies allow distance sampling methods to be applied to more populations. Observers are increasingly being replaced by acoustic sensors, which allow cue count methods to be applied to a wide range of species of mammals (both marine and terrestrial), birds, amphibians, and insects. Similarly, aerial observers are being replaced by high-resolution imagery. In both cases, software that can search the large volumes of data generated is being developed, so that human experts need to examine only a subset of recordings or images, to identify species, and to check automated classifications.

Radar surveys of migrating birds and sonar surveys of marine resources require three-dimensional distance sampling methods, in which the distribution of objects with depth cannot be assumed uniform (Buckland et al. 2015).

As aerial observers are replaced by high-resolution digital image recorders, and as drone and automated vehicle technology improves, we can expect that most line transect surveys of seabirds and marine mammals and of large terrestrial animals in open habitats will be carried out using autonomous platforms. Drones might also be used to place acoustic detectors in inaccessible locations. Camera traps provide another means of surveying populations, and methods of analysis using both distance sampling and spatial capture-recapture are under development (Howe et al. 2017).

Conclusions

Distance sampling methods are widely used for estimating the abundance of animal populations. Various adaptations and advances ensure that the methods can provide reliable estimates for a very wide range of populations. We note in passing that spatial capture-recapture (SECR) methods, widely used in the analysis of camera trap data, reside on a spectrum between traditional mark-recapture methods (without spatial information) and distance sampling methods, where animal locations are known (Borchers et al. 2015). The concept of a detection function, central to distance sampling, also lies at the heart of estimation using SECR methods.

Current research areas are driven by advances in technology (such as camera traps, acoustic detectors, and drones), software development (such as software that can detect objects of interest in large quantities of high-resolution images), and statistical methods (such as methods for efficient fitting of point process models). While distance sampling methods were originally developed solely for estimating the size of populations of interest, tools are now available to explore a range of other issues, such as how detectability is affected by various covariates, how density varies as a function of spatial covariates, how different management regimes or developments affect animal densities, or how biodiversity changes in response to climate change, land use change, or habitat loss. We expect the methods to continue to evolve in response to new technologies and methodologies and changing demands of policymakers.

Acknowledgments

The authors thank Kris Winiarski, Peter Paton, and Scott McWilliams for their permission to use the Rhode Island line transect survey data. Data were collected by K. Winiarski, J. Veale, and E. Jedrey with pilots J. Chronic and D. McGowan of New England Specialized Aviation Services. Their study was supported by grants from the State of Rhode Island for the Ocean Special Area Management Plan.

LITERATURE CITED

Borchers, D. L., B. C. Stevenson, D. Kidney, L. Thomas, and T. A. Marques. 2015. A unifying model for capture-recapture and distance sampling surveys of wildlife populations. Journal of the American Statistical Association 110:195–204.

Buckland, S. T. 2006. Point transect surveys for songbirds: Robust methodologies. The Auk 123:345–357.

Buckland, S. T., D. R. Anderson, K. P. Burnham, J. L. Laake, D. L. Borchers, and L. Thomas. 2001. Introduction to distance sampling. Oxford University Press, Oxford.

Buckland, S. T., D. R. Anderson, K. P. Burnham, J. L. Laake, D. L. Borchers, and L. Thomas (eds). 2004. Advanced distance sampling. Oxford University Press, Oxford.

Buckland, S. T., C. S. Oedekoven, and D. L. Borchers. 2016. Model-based distance sampling. Journal of Agricultural, Biological and Environmental Statistics 21:58–75.

Buckland, S. T., E. A. Rexstad, T. A. Marques, and C. S. Oedekoven. 2015. Distance sampling: Methods and applications. Springer, New York.

Fiske, I., and R. Chandler. 2011. Unmarked: An R package for fitting hierarchical models of wildlife occurrence and abundance. Journal of Statistical Software 43:1–23.

Howe, E. J., Buckland, S. T., Després-Einspenner, M.-L., and Kühl, H. S. 2017. Distance sampling with camera traps. Methods in Ecology and Evolution 8: 1558–1565.

Link, W. A. 2003. Nonidentifiability of population size from capture-recapture data with heterogeneous detection probabilities. Biometrics 59:1123–1130.

Miller, D. L., M. L. Burt, E. A. Rexstad, and L. Thomas. 2013. Spatial models for distance sampling data: Recent developments and future directions. Methods in Ecology and Evolution 4:1001–1010.

Miller, D. L., E. Rexstad, L. Thomas, L. Marshall, and J. Laake. In press. Distance sampling in R. Journal of Statistical Software. https://doi.org/10.1101/063891.

Shono, H. 2008. Application of the Tweedie distribution to zero-catch data in CPUE analysis. Fisheries Research 93:154–162.

Thomas, L., S. T. Buckland, E. A. Rexstad, J. L. Laake, S. Strindberg, S. L. Hedley, J. R. B. Bishop, T. A. Marques, and K. P. Burnham. 2010. Distance software: Design and analysis of distance sampling surveys for estimating population size. Journal of Applied Ecology 47:5–14.

Ver Hoef, J. M. and P.L. Boveng. 2007. Quasi-Poisson vs. negative binomial regression: How should we model overdispersed count data? Ecology 88:2766–2772.

Yuan, Y., F. E. Bachl, F. Lindgren, D. L. Borchers, J. B. Illian, S. T. Buckland, H. Rue, and T. Gerrodette. 2017. Point process models for spatio-temporal distance sampling data from a large-scale survey of blue whales. Annals of Applied Statistics 11:2270–2297.

7 — Occupancy Modeling Applications

CHRIS SUTHERLAND AND
DANIEL W. LINDEN

A bird in the hand is worth two in the bush . . . if $p = 0.5$.

Introduction

Occupancy modeling has proliferated with development and application across the ecological literature since its formalization (MacKenzie et al. 2002, 2003), and its place among the most widely applied ecological models is testament to its tremendous utility and flexibility (Bailey et al. 2014). In this chapter, we will develop an appreciation for how seemingly simple detection-nondetection data offer powerful insights into interesting spatiotemporal distribution patterns across large spatial scales with relatively little cost.

Arguably, the ultimate goal of many modern ecological monitoring programs is to make inferences about an ecological state variable using statistical models, which can in turn be used to inform management or conservation (Williams et al. 2002). The gold standard of ecological state variables is *density*, the number of individuals per unit area (Krebs 1994). In addition to static properties such as abundance and density, there is often a keen interest in estimating dynamic properties of a system, including survival and recruitment of individuals (Pollock et al. 1990: chap. 5), or colonization and extinction of populations (Hanski 1998: chap. 8). In practice, however, estimating parameters of spatially and temporally dynamic processes usually requires collecting the types of individual-level data that impose some level of logistical constraint. For example, the estimation of population density typically requires that individuals be captured (directly or indirectly), uniquely identified (by natural or artificial markings, but see Chapter 6 of this volume), and subsequently captured again within a relatively short timeframe (Otis et al. 1978; Williams et al. 2002). To take another example, the estimation of meaningful demographic parameters such as survival and/or recruitment requires that sufficient numbers of known individuals be observed repeatedly over time (Amstrup et al. 2010; McCrea and Morgan 2014). The difficulty in collecting such data is exacerbated for rare, low-density species for which robust statistical inference is arguably most valuable. In short, obtaining sufficient individual-level data to estimate population size and/or demographic rates, while feasible at smaller spatial scales, is often prohibitively expensive to carry out at the landscape scale, and so other approaches are required.

The issue of making statistical inferences efficiently across large spatial scales can be somewhat resolved by the appreciation that abundance, the number of individuals, and occurrence, the presence of at least one individual at a site, are directly related

(Kéry and Royle 2015), and that knowledge about how species are distributed in space, i.e., species occurrence, can be as useful and informative as knowing *how many* individuals there are. Moreover, in addition to direct observation of the focal species, determining whether a site is occupied or not can be achieved using indirect observations, e.g., tracks or sign, and does not require having to capture, or even count, individuals. Therefore, occurrence surveys are logistically convenient, making if far more efficient to sample larger areas with fewer resources. Thus, species occurrence—whether or not a species occupies a particular part of the landscape—has become a state variable of great interest in ecological studies, as have the associated spatiotemporal occurrence dynamics, i.e., the colonization and extinction patterns of occurrence through time.

Like almost every other ecological sampling technique, however, the ability to detect the presence of a species is imperfect (Kellner and Swihart 2014), and failure to account for observation error has the potential to introduce substantial bias in the resulting inference (Guillera-Arroita et al. 2014). For example, even when a breeding amphibian population occupies a pond, there is a chance that on any single visit to that pond you will not observe evidence that a population is present, which could cause one to naively conclude that fewer ponds are occupied than is actually the case. Again, this is particularly true for rare, or low-density, species, but it can also be related to the environmental conditions during a survey, the experience of the surveyor, or even the ecology of the focal species. Capture-recapture methods were developed with exactly this issue in mind and are built on the premise that, by subjecting a population to repeated sampling in a short space of time, capture rates can be estimated and thus population size inferred, or by subjecting individuals to repeated sampling over time, vital rates such as survival and recruitment can be estimated. The same is true for occupancy: repeatedly sampling a location in search for evidence that a species is present will generate detection-nondetection data that can be used to estimate detection probability and, as a result, error-corrected estimates of the proportion of the landscape that is occupied. As is the case with capture-recapture, the notion of repeated sampling is the cornerstone of occupancy modeling.

The occupancy modeling framework (MacKenzie et al. 2002, 2003) has been developed around the ideas that patterns of occupancy in space and time can be of tremendous use in understanding ecological processes (e.g., Sutherland et al. 2014; Linden et al. 2017), guiding management actions (e.g., Fuller et al. 2016), and informing conservation practices (Hayward et al. 2015), that occurrence data are straightforward and efficient to collect (MacKenzie and Royle 2005) and that detection is imperfect and observation error must be accounted for (Guillera-Arroita et al. 2014). For the rest of this chapter we will focus on the development and application of occupancy models, including descriptions of sampling designs required to collect the necessary data and the analytical approaches used to fit models to those data.

The Occupancy Model

The primary inference objective when applying occupancy models is to estimate the proportion of an area occupied by a focal species at some point in time while accounting for imperfect detection (MacKenzie et al. 2002). Given that most areas of interest in wildlife science are too large to sample entirely, a subset of sampling units or sites are selected in a probabilistic manner (e.g., simple random selection) to accommodate these logistical constraints while allowing model inferences to extend to the whole area. Sites may be defined by naturally occurring discrete units (e.g., ponds or islands) or by an arbitrary delineation of continuous space (e.g., grid cells or quadrats), depending on the biology of the target species. Regardless of how sites are defined, the occupancy model assumes that sites are independent units such that detection of the target species in one site does not influence detection in a nearby site. To satisfy this *independence* assumption, sites should be spaced far enough apart to eliminate the possibility of movement between them by the target species. If independence is not, or cannot, be achieved by design, then non-independence must be accounted for statistically.

Once defined and selected for sampling, sites are surveyed multiple times using a standardized protocol to detect the target species. The protocol can involve some type of direct (e.g., visual, audible) or indirect (e.g., tracks, latrines) confirmation that the species is present and the site is occupied. Failure to detect the species is not necessarily indicative of true absence, given that imperfect detection may result in false absences. Repeat surveys at a site should occur within a short enough period that the occupancy status of the site is closed (i.e., occupancy state does not change between survey occasions), another important assumption in occupancy modeling. This *closure* assumption is critical because it means that the pattern of detections across survey occasions is informative about the probability of detecting a species, given that it occurs. For any site where the species is observed on at least one occasion, the occupancy status for the site is certain and occasions at the site that do not yield detections are known to be false absences. Only those sites with no detections on all occasions are potentially true species absences, and, with information about detection probability, it is possible to estimate the probability that such sites are occupied despite having never observed evidence of the species.

The period during which occupancy states are assumed to be closed and during which repeated surveys occur is often called a *season*, or a *primary sampling period*, and the repeat surveys within a season are called *occasions*, or *secondary sampling periods*. Seasons will be defined by species biology, specifically a temporal scale at which population processes such as births and deaths or immigration and emigration can be assumed to have little impact on inferences regarding species distribution. For example, the breeding season for temperate songbirds would be a period when adult males could be easily detected

by song once territories are established and the pattern of occupancy in a given landscape could be considered a reasonably static and representative snapshot of the system.

The collection of detections (1) and nondetections (0) across repeated survey occasions for a given site is called a *detection history*, and the detection histories for all sampled sites comprise the core data in an occupancy model. For example, a detection history $y_i = 010$ indicates that during the $K = 3$ occasions for site i, the target species was detected only during the second survey. Note the bold notation to indicate that y_i is a vector, whereas the individual observations for site i and survey j might be indicated by $y_{i,j}$. The distinction between site-specific detection histories (y_i) and site- and survey-specific detections ($y_{i,j}$) is a matter of notational and computational convenience, which may become apparent later. Regardless, the data are represented by binary values for each of N sites and K occasions (Fig. 7.1).

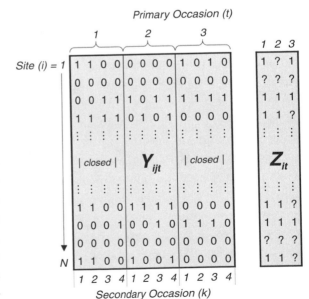

Fig. 7.1. Illustration of detection history data and the partially observed stated for an occupancy model. The figure illustrates the primary (*t*) and secondary (*j*) sampling structure.

An occupancy model formulates the data likelihood as a hierarchical function of two processes: the *state process* = whether a site is occupied by the species; and the *observation process* = whether the species is detected at a site given that the site is occupied. The state process is partially latent, which means sometimes the occupancy state is known (i.e., for sites with ≥ 1 detection) and sometimes it is unknown (i.e., for sites with "all-zero" detection histories). When a site is truly occupied, the occupancy state has a value $z_i = 1$, whereas unoccupied sites have $z_i = 0$ (Fig. 7.1). The occupancy states are described as Bernoulli random variables such that $z_i \sim$ Bernoulli(ψ), where ψ represents the probability of occurrence. If the values of z_i were known for all sites (i.e., if detection was perfect), estimation of this probability could be achieved with a simple logistic regression. Given that the occupancy states are partially latent, however, we must combine the Bernoulli regression of occurrence with another regression model for the observation process such that the detection data are conditional on the occupancy state: $y_{i,j} \mid z_i \sim$ Bernoulli($z_i p$). Here, $\Pr(y_{i,j} = 0) = 1$ when $z_i = 0$, recognizing that it is impossible to detect the species if it does not occupy a site. If the site is occupied ($z_i = 1$), then detections are a function of p, the detection probability. Thus, the state process and the observation process are described by conditionally related generalized linear models (GLMs), which are used to jointly estimate the probabilities of occurrence (ψ) and detection (p).

The Bernoulli regressions shown above are analogous to the intercept-only models discussed in previous chapters of this volume. They assume that the probabilities are constant both across sites for occupancy and across sites and surveys for detection. Of course, this is rarely the case in reality. For example, environmental features that influence habitat quality for the species might cause spatial variation in occupancy probability among sites, while factors such as vegetation cover (site specific) and weather (both site and survey specific) may influence detection probability among sites and/or across survey occasions.

Variation in the expected occupancy and detection probabilities can be easily accommodated by extending the intercept-only GLM to include some number of covariates using a link function; for example:

$$\text{logit}(\psi_i) = \beta_0 + \beta_1 \times \text{habitat}_i$$
$$\text{logit}(p_{i,j}) = \alpha_0 + \alpha_1 \times \text{weather}_{i,j}.$$

Here, the probabilities of occupancy and detection now vary according to linear models on the logit scale, similar to a logistic regression. As evidenced by the indices for *habitat* and *weather*, the covariates for occupancy can vary only by site (i), while those for detection can vary by site (i), survey (j), or both (i,j). Another important assumption for occupancy models is that there is *no unexplained heterogeneity* in detection probability, which could otherwise bias the estimation of occupancy probability. This means that detection probability should either be constant across sites and surveys (again, this is rarely possible) or be adequately explained by the covariates included in the GLM.

Parameter estimation for an occupancy model is typically achieved with maximum likelihood methods or with Bayesian approaches involving Markov chain Monte Carlo (MCMC). A full description of these estimation techniques is beyond the scope of this chapter (see Kéry and Royle 2015), though we later present software options for model fitting and assessment. One difference to note in these approaches is that with maximum likelihood the latent occupancy states are marginalized out of the model, while Bayesian approaches directly estimate the possible latent states for each site. Parameter estimates for the linear models (e.g., α and β in the model above) will typically match to within some negligible error between the approaches, given reasonable sample sizes. While maximum likelihood estimation is computationally faster and allows for straightforward model selection options (e.g., using Akaike's Information Criterea (AIC)), Bayesian estimation accommodates greater flexibility in model structures and can better accommodate sparse data.

Finite-sample inferences can be made with Bayesian approaches, if the modeling is intended for understanding only the selection of sampled sites (Royle and Dorazio 2008); empirical Bayes methods can actually generate finite-sample estimates from models fit by maximum likelihood. We direct interested readers to Kéry and Royle (2015) for in-depth yet accessible and pragmatic illustrations of hierarchical modeling approaches, including those for occupancy.

Woodpecker Occupancy in California

We illustrate an application of single-season occupancy modeling with data from a study by Linden and Roloff (2015) on white-headed woodpeckers (*Picoides albolarvatus*) in recently harvested forests of northern California. This species of woodpecker historically nested in open pine savannahs that have mostly been replaced with dense mature forest, and the regional management practices now result in "islands" of open nesting habitat. The authors surveyed 66 recent clear-cuts for the presence of nesting woodpeckers with the main objective of examining how nest-site availability, as approximated by snag density, influenced the probability of nest presence. Similar to the initial case study provided by Mackenzie et al. (2002), this application involved discrete sites where the interpretation of species occupancy is relatively straightforward (Efford and Dawson 2012). Linden and Roloff (2015) used more complex occupancy models (i.e., multi-season, multi-state) for their inferences, so here we restrict our analysis to data from a single season and focus on site occupancy by nesting woodpeckers.

Sites were surveyed on 2–3 occasions >1 week apart between May and July in 2010. The surveys involved 10-min point counts conducted in the morning, during which an observer recorded white-headed woodpecker detections from a central location at each site. When evidence of nesting behavior was detected (e.g., bird entering a tree cavity), the observer attempted to locate and confirm the presence of a nest. Detection probability was hypoth-

esized to increase over time as nests developed from eggs to fledglings, and adult woodpeckers visited to feed offspring more frequently (thus, they would be more detectable). It was also hypothesized that clear-cuts with greater snag densities would potentially provide more opportunities for male woodpeckers to construct new cavities and increase the probability that the clear-cut would be occupied by a cavity nest. Therefore, we modeled variation in detection according to the ordinal date (number of days since January 1) of the survey and variation in occupancy according to snag density (# ha^{-1}) in the clear-cut. To facilitate model fitting and interpretation, we scaled both covariates to have a mean = 0 and variance = 1 (after log-transforming snag density). We used the unmarked package (Fiske and Chandler 2011) in R for model fitting (see the section below on software for more details) and employed model selection with AIC to rank four combinations of models that either included or excluded the "date" and "snags" covariates for detection and occupancy, respectively.

One important difference between our illustrated example and the methods of Linden and Roloff (2015) is that they did not model detections of nests after the first detection at a site. This is known as a *removal design* (Farnsworth et al. 2002) and was intended to accommodate a potential violation of the independence assumption. When a cavity nest is identified at a given site, the observer would likely remember the location of that nest during future surveys and be able to detect its presence more easily. Rotating multiple observers would be another design-based approach to solving this problem.

The surveys in 2010 observed a total of 43 out of the 66 sites occupied with a nest for a naive occupancy proportion of 0.65. Model selection results indicated that the fully parameterized model with a date effect on detection and a snag effect on occupancy was best supported (Table 7.1), since it had the lowest AIC score and a high model weight. Given the overwhelming support for the top model, we can probably justify using it to make inferences without needing model averaging. The logit-scale intercept for detection (Table 7.2) translated to an average detection probability of 0.86 (95% credible interval [CI]: 0.74, 0.93), and the date effect indicated that detection increased during the course of the sampling period (Fig. 7.2). This supports the hypothesis that woodpecker nests are more detectable when offspring are older and require more frequent attention from the adults. The logit-scale intercept for occupancy (Table 7.2) translated to an average occupancy probability of 0.71 (95% CI: 0.57, 0.82), and the snag effect indicated that nest occupancy increased with snag density in clear-cuts (Fig. 7.2). The finite-sample proportion of sites occupied was 0.67 (95% CI: 0.65, 0.85). Note that the lower confidence interval of the mean occupancy estimate was actually lower than the observed proportion of occupied sites (0.57 < 0.65)—this reflects the fact that the occupancy estimate is a population average potentially relevant to other clear-cuts in the region, and that the true mean occupancy could be lower than what was observed. Conversely, the finite-sample proportion is relevant only to the 66 sites that were sampled and cannot be lower than what was observed. Linden and Roloff (2015) used the es-timated snag effect to argue for forest management practices that retained high densities of snags during harvest to provide increased nesting opportunities for white-headed woodpeckers.

The Dynamic Occupancy Model

The occupancy models described so far are called *single season* models because they quantify a static distribution of species occurrence across sites during a single primary sampling period. *Multi-season*, or *dynamic*, occupancy models extend the single season model to accommodate changes in site occupancy states across multiple seasons, or primary sampling periods, due to dynamic local population processes including colonization and extinction. *Colonization* refers to the process whereby a site that is unoccupied by the target species at time t becomes occupied at time $t + 1$, while *extinction* refers to the opposite process, where an occupied site becomes unoccupied. Colonization is analogous to recruitment, while extinction is the complement of survival. That these population processes can occur implies a demographically open system between primary sampling

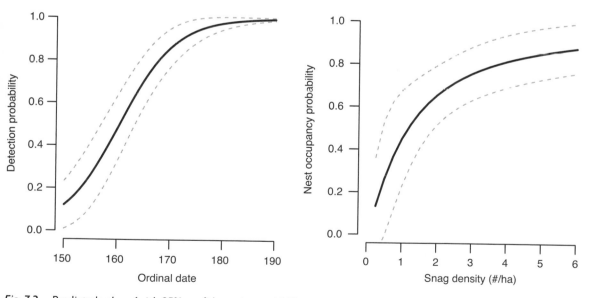

Fig. 7.2.   Predicted values (with 95% confidence intervals) illustrating covariate relationships from the top occupancy model for white-headed woodpeckers (*Picoides albolarvatus*) in northern California, USA. Covariates include the effects of ordinal date on detection probability and snag density on nest occupancy probability.

Table 7.1. Model selection results for nesting white-headed woodpeckers during 2010 in northern California. Model structures are listed with number of parameters (K), the AIC score, relative difference (ΔAIC), model weights (AICwt), and the negative log-likelihood (LL).

Model	K	AIC	ΔAIC	AICwt	LL
$p(date)\ \psi(snags)$	4	154.82	0.00	0.97	−73.41
$p(date)\ \psi(.)$	3	162.05	7.24	0.03	−78.03
$p(.)\ \psi(snags)$	3	214.19	59.37	0.00	−104.09
$p(.)\ \psi(.)$	2	223.93	69.11	0.00	−109.96

Table 7.2. Parameter estimate means (SEs) from the occupancy models for nesting white-headed woodpeckers during 2010 in northern California. Parameters include logit-scale intercepts for probabilities of detection (α_0) and occupancy (β_0), as well as the effects of relevant covariates.

Model	Detection parameters		Occupancy parameters	
	α_0	date	β_0	snags
$p(date)\ \psi(snags)$	1.809 (0.379)	2.268 (0.411)	0.884 (0.316)	1.002 (0.368)
$p(date)\ \psi(.)$	1.706 (0.370)	2.253 (0.396)	0.870 (0.299)	–
$p(.)\ \psi(snags)$	1.037 (0.223)	–	0.963 (0.357)	1.162 (0.417)
$p(.)\ \psi(.)$	1.027 (0.224)	–	0.830 (0.301)	–

periods. This is in contrast to the assumption of closure between secondary sampling periods within a season.

The data required for a multi-season dynamic occupancy model are similar to those for a single season model with N sites and K secondary sampling periods, but with additional replication across T primary sampling periods. The additional replication extends to the state process model, which now defines multiple states for each site over time such that $z_{i,t} = 1$ when site i is occupied during primary period t, and is 0 otherwise. The probability of occurrence for a given primary period is then simply $\psi_t = \Pr(z_{i,t} = 1)$, as in the single season case when $T = 1$. The standard parameterization for dynamic occupancy models, as illustrated in Mackenzie et al. (2003), defines an initial probability for the occupancy states in primary period $t = 1$, with subsequent occupancy states (for

$t = 2, 3, \ldots, T$) controlled by probabilities of extinction (ε) and colonization (γ). Site occupancy probability in the initial period is

$$z_{i,1} \sim \text{Bernoulli}(\psi_1),$$

and site occupancy in period $t = 2, 3, \ldots, T$ is

$$z_{i,t} \mid z_{i,t-1} \sim \text{Bernoulli}(\pi_{i,t}),$$

where

$$\pi_{i,t} = z_{i,t-1}(1 - \varepsilon_{t-1}) + (1 - z_{i,t-1})\gamma_{t-1}.$$

In this way, the probability that a site is occupied in time t is dependent on the occupancy state in time $t-1$ for all $t \geq 2$. When a site is occupied at $t-1$, this probability is determined by the inverse of the extinction probability, $1-\varepsilon_{t-1}$ (also known as "survival" and often denoted by φ). On the other hand, if a site is unoccupied at $t-1$, the state probability is determined by the colonization probability, γ_{t-1}. Note that with T primary periods there are only $T-1$ possible transitions.

The observation process model links the data to the season-specific occupancy states in the same manner as a single season model. The only difference is that the parameters are now specific to the season, t, during which the observations were collected:

$$y_{i,j,t} \mid z_{i,t} \sim \text{Bernoulli}(z_{i,t}p_t).$$

Variation in the parameters can be accommodated, as before, with logit-link functions for each of the transition probabilities. The initial occupancy probability can be modeled as a function of covariates that vary by site (i), while colonization and extinction probabilities can vary by site (i), season (t), or both (i,t). For example:

$$\text{logit}(\gamma_{i,t}) = \beta_1 + \beta_2 \times \text{mastcrop}_{i,t};$$
$$\text{logit}(\varepsilon_{i,t}) = \beta_3 + \beta_4 \times \text{winter}_{i,t}.$$

Here the colonization probability is a function of the local mast crop the previous fall, while extinction probability is a function of the severity of the preceding winter. Both of these covariates are specific to the site and season in which they were measured. Note that since the colonization and extinction probabilities describe changes over time, the covariates should reflect conditions that would be hypothesized to influence population processes *between* the primary sampling periods. This does not mean the conditions must vary over time (i.e., they may vary only by site); rather, it means that the effects are relevant to *changes* in the occupancy states. Detection probability can also be modeled with a GLM as before, with covariates that vary by site (i), survey (j), season (t), or any combination thereof.

A variety of biologically interesting derived parameters can be calculated to provide summaries of the dynamic system. For example, the occupancy

probabilities in each primary period, ψ_t, from $t=2, 3, \ldots, T$, can be derived as follows:

$$\psi_t = \psi_{t-1}(1-\varepsilon_{t-1}) + (1-\psi_{t-1})\gamma_{t-1}.$$

It should be clear that this equation is similar to that for determining the site-specific probabilities ($\pi_{i,t}$) of the occupancy states. The metapopulation growth rate (i.e., the change in the proportion of area occupied from one time period to the next), λ_t, can be calculated as the change in occupancy probability from time t to time $t+1$:

$$\lambda_t = \frac{\psi_{t+1}}{\psi_t}.$$

Other potentially useful summaries include season-specific turnover probabilities and the overall equilibrium occupancy probability (Royle and Dorazio 2008).

Water Vole Occupancy Dynamics in Scotland

A classic problem for which the dynamic occupancy model is ideally suited is testing predictions from metapopulation theory (MacKenzie et al. 2003). Metapopulation theory predicts that patch occupancy dynamics emerge according to the *area-isolation paradigm* (Hanski 1998); that is, extinction probabilities are related to patch size, a proxy for local population size, and therefore demographic stochasticity (the "area"), and it also predicts that colonization probabilities are related to how connected a patch is (the "isolation"). Metapopulation theory is therefore motivated in terms of conditional-on-state transition dynamics (colonization and extinction), in exactly the same way that the dynamic occupancy model has been developed. Classical metapopulation models were, however, developed under the much-criticized assumption that occupancy states are observed without error (Moilanen 2002), which the development of hierarchical occupancy models explicitly addresses

To illustrate the application of the dynamic occupancy model, we use a subset of the data described in Sutherland et al. (2014), a study of a population of water voles *Arvicola amphibius*, a riparian specialist species, in northwest Scotland. Water vole colonies occupy discrete patches of lush riparian habitat embedded within a matrix of unsuitable habitat and exhibit high levels of patch turnover, and as such, represent a rare example of a classically functioning metapopulation (Lambin et al. 2012; Sutherland et al. 2012, 2014). The network studied consisted of 114 discrete habitat patches, each of which were surveyed 2–4 times between July and August, the breeding season, from 2009 to 2012. Water voles use conspicuously placed latrines as territory marking, and, therefore, water vole occupancy was determined through a series of latrine surveys whereby each linear riparian patch was surveyed, and the detection of any latrines recorded. Given the suggestion that this system appears to function as a classical metapopulation, it is natural to test whether there is empirical support for the area-isolation paradigm—we do this using the dynamic occupancy model.

Our primary focus is testing whether ideas from metapopulation theory can explain variation in transition probabilities. To test the "area" prediction of the area-isolation paradigm, we consider two extinction models: a model in which extinction probability is related to patch size, which, for the semi-aquatic riparian habitat specialists, is the length of the waterway defined as suitable habitat, and a second model that assumes extinction is constant across all patches regardless of size (the "null" model). To test the "isolation" prediction, we consider two colonization models: first, a model in which colonization is related to a measure of connectivity, and a second model that assumes extinction is constant across all patches regardless of connectivity (the "null" model). Here we define the connectivity of a patch as the sum of the weighted contribution of all surrounding patches (Moilanen and Nieminen 2002):

$$\text{connectivity}_i = \sum_{i \neq j} \exp(-0.33 \times d_{i,j}),$$

where the term $\exp(-0.33 \times d_{i,j})$ is the estimated dispersal kernel in Sutherland et al. (2014) where contributions to connectivity reduce as pairwise distances, $d_{i,j}$, increase, and -0.33 is the estimated spatial scale parameter that translates to a typical dispersal distance of 2.1 km.

In addition to the transition probabilities, we were particularly interested in investigating year-to-year variation in latrine detectability, so again, we compared two detection models, a model with year-specific detection probability, and a model in which detection probability is constant across years. And finally, we estimated ψ_1, the proportion of the 114 patches that were occupied in the first year (2009).

There was no a priori reason to suspect that any particular combination of these models is more likely than any other, so we compared all combinations of the model components described above, resulting in a total of eight competing models (Table 7.3). Again, we used unmarked (Fiske and Chandler 2011) for model fitting and employed model selection with AIC to compare and select among the candidate model set.

Using AIC to compare models, we found that there was overwhelming support for the colonization-by-connectivity hypothesis, for year-to-year variation in detectability, and, interestingly, for constant extinction rates across the patch network (Table 7.3). As before, the full support of the top model allows us to make inferences without having to do any model averaging. Under the most supported model, extinction probability is assumed to be constant across all patches, which is in contrast to metapopulation theory predictions; estimated extinction probability was 0.16 (95% CI: 0.09, 0.26). The results did, however, support the hypothesized relationship between colonization and connectivity: colonization probability was highest for more connected patches and lowest for more isolated patches (Fig. 7.3). The estimated intercept and slope of the colonization-connectivity relationship on the linear predictor scale was -1.07 (95% CI: -1.09, -0.24) and 0.15 (95% CI: 0.06, 0.25), respectively. This translates to

Table 7.3. Model selection results for the Scottish water vole metapopulation. Model structures are listed with number of parameters (K), the AIC score, relative difference (ΔAIC), model weights (AICwt), and the negative log-likelihood (LL).

Model	K	AIC	ΔAIC	AICwt	LL
ψ(.) γ(connectivity) ε(.) p(yr)	8	1330.69	0.00	1	−657.34
ψ(.) γ(.) ε(.) p(yr)	7	1344.44	13.75	0	−665.22
ψ(.) γ(connectivity) ε(.) p(.)	5	1349.03	18.35	0	−669.52
ψ(.) γ(connectivity) ε(size) p(yr)	9	1355.87	25.18	0	−668.93
ψ(.) γ(.) ε(.) p(.)	4	1363.17	32.49	0	−677.59
ψ(.) γ(.) ε(size) p(yr)	8	1372.35	41.66	0	−678.18
ψ(.) γ(connectivity) ε(size) p(.)	6	1433.62	102.93	0	−710.81
ψ(.) γ(.) ε(size) p(.)	5	1451.42	120.73	0	−720.71

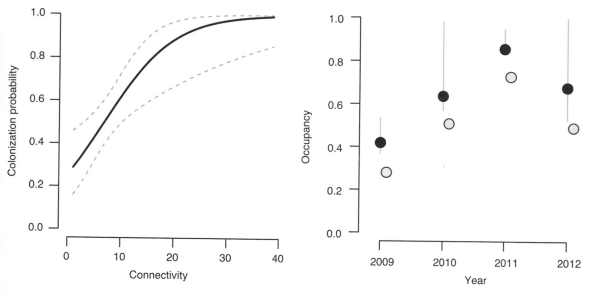

Fig. 7.3. Predicted values (with 95% confidence intervals) illustrating covariate relationships from the top dynamic occupancy model for the multi-season analysis of the Scottish water vole (*Arvicola amphibus*) data. On the left is the predicted relationship between colonization and connectivity. On the right is the estimated yearly occupancy (black circles) and the observed (naive) occupancy, i.e., the number of sites where latrines were detected at least once in the year.

colonization probabilities of 0.29, 0.55, and 0.99 for the least (1.04), average (8.29), and most (38.90) connected patches, respectively (Fig. 7.3). Detection probability was markedly lower in 2012 (0.50, 95% CI: 0.40, 0.60) than in the three earlier years (2009: 0.70 [0.60, 0.79]; 2010: 0.72 [0.64, 0.79]; 2011: 0.78 [0.72, 0.83]). The fact that detection probability in each year was <1 results in a discrepancy between observed and estimated occupancy rates, the largest

of these discrepancies being in 2012 when detectability is lowest (Fig. 7.3).

Advanced Topics

The two formulations of the occupancy model described above, i.e., the single season and the dynamic multi-season models, represent arguably *the fundamental modeling framework for analyzing* detection-nondetection data for inference about species occurrence. The framework has proven to be extremely flexible and, as such, has given rise to many useful extensions developed to address important issues ranging from a failure to meet basic model assumptions to a desire to model more complex systems with increasingly biologically realistic models (Bailey et al. 2014). Although we have focused in this chapter on developing an intuitive understanding of the basic occupancy models and the tools used to apply them, with this knowledge these more complex model extensions should become more readily accessible. Therefore, in anticipation, we briefly describe some of the most widely used of the more complex occupancy models. We note also that all of the extensions described have been developed for both the single and the multi-season cases.

The first two noteworthy extensions, the multi-state occupancy model and the multi-species occupancy model, relate to interesting variations in the state variable occupancy. Multi-state occupancy models extend the modeling framework to inference about multiple occupancy states and state-specific detectability (Nichols et al. 2007; MacKenzie et al. 2009). The popular application of this model has involved three states: (1) unoccupied, (2) occupied without reproduction, and (3) occupied with reproduction. This model formulation assumes a hierarchy of state uncertainty, such that sites observed in state 3 are known with certainty (i.e., observed state 3 = true state 3), while those observed in state 2 may in truth be state 2 or 3 and those observed in state 1 may actually be in any one of the three states. Dynamic versions of these models have been used to examine changes in occupancy and reproduction and mechanisms behind state transitions over time (Martin et al. 2009; MacKenzie et al. 2012). The case study of woodpecker occupancy in California presented earlier is a simplified version of a dynamic multi-state model used by Linden and Roloff (2015) to understand white-headed woodpecker nesting; they found that snag density had a stronger relationship with nest persistence (i.e., the probability of nesting in year t given nesting in year $t-1$) than with the initial nesting probability.

Multi-species models extend inference to more than one species and include several alternative approaches that are question dependent. MacKenzie et al. (2004) developed an occupancy model for investigating co-occurrence patterns between pairs of species, which involves explicit estimation of species interaction factors in both detection and occurrence, although the estimation of these parameters requires more data than single species occupancy models. A less data-intensive model is a restricted version of MacKenzie et al. (2004), which involves assuming a unidirectional "dominant-subordinate" relationship between species pairs (Waddle et al. 2010). Oftentimes it is not the co-occurrence of species pairs that is of interest, but rather the number, species richness, and composition of a collection of sites (i.e., a metacommunity). This can be addressed by combing several single species occupancy models and allowing parameters to be shared across species, although Dorazio et al. (2010) went one step further and developed a model that acknowledges that some (rare) species may go undetected. Assuming that species detection probabilities come from a common distribution (i.e., a random effect), Bayesian multi-species formulation of the occupancy model uses data augmentation to estimate the latent species richness by accounting for imperfect detection (see also Iknayan et al. 2014). These models have been further extended to quantify drivers of geographic variation in community structure (Sutherland et al. 2016) and composition (Tenan et al. 2017).

The repeated visit design allows for estimation of the false-negative observation rate but assumes that false positives, recording a species detection when a site is *not* occupied, do not occur or are negligible (MacKenzie et al. 2006). However, Royle and Link (2006) demonstrated that even at very low rates, false positives can introduce substantial bias in estimates of occupancy and developed a generalized occupancy model that accounts for both error rates. Subsequently, Miller et al. (2011) improved on the

Royle and Link (2006) model by incorporating information about detection state uncertainty and the ability to utilize information from multiple methods that independently detect species but with varying efficacy. These "false-positive" models have proven useful for situations in which species can be easily misidentified, such as avian or anuran call surveys (McClintock et al. 2010), where there is variation in observer ability (Molinari-Jobin and Kéry 2012) and transience (Sutherland et al. 2013).

Finally, multi-scale occupancy models (Nichols et al. 2008) were developed in response to the increasing use of multiple data collection methods in wildlife monitoring studies. A precursor to the multi-method false-positive model of Miller et al. (2011), the multi-scale model makes efficient use of multi-method detection data with the added advantages of explicitly dealing with nonindependence of sampling locations in close proximity while simultaneously allowing inference to be made about local scale (detector-level) and large scale (site-level) occupancy. Another approach to multi-scale occupancy involves a single sampling method with site locations clustered in space and/or time to make similar inferences about occupancy at both the local scale (i.e., availability) and the large scale (Mordecai et al. 2011; Pavlacky et al. 2012).

Software

The popularity of occupancy modeling in ecology has led to the subsequent development of several user-friendly options for analyzing detection/nondetection data for inference about species occurrence. The most widely used software for fitting occupancy models are the program PRESENCE (Hines 2006), and the R package 'unmarked' (Fiske and Chandler 2011), although other software such as MARK (White and Burnham 1999) and E-SURGE (Choquet et al. 2009), both traditionally capture-recapture programs, include occupancy models. Because of their prevalence, we focus on PRESENCE and unmarked.

PRESENCE is a stand-alone software program developed specifically for occupancy estimation and was the result of model developments by MacKenzie et al. (2002) and MacKenzie et al. (2003). PRESENCE can be used to fit single season and dynamic occupancy models and can be used to fit the broadest range of important variants including, among others, multiple detection methods, protocols with spatial dependence, the presence of false positives (in addition to false negatives), two-species interaction models, and multi-state models. GenPRES, a sister program, is a convenient simulation-based program that can be used to explore the effects of sampling effort, sampling design, and detection heterogeneity on the bias and precision of parameter estimates and specifically to explore the trade-offs in the number of site visits versus the number of sites visited. As a result of the proliferation of R in ecological analysis, the developers have recently created RPresence, an R package that allows analysis of all the models in PRESENCE completely within R.

Another R package, 'unmarked', contains a suite of functions to fit a wide range of hierarchical models, including occupancy models (Fiske and Chandler 2011). The major advantage of 'unmarked', and one of the reasons it has gained a lot of traction, is the availability of a variety of hierarchical models that can be fit using a single common workflow. The package is, however, limited in terms of occupancy modeling capabilities when compared to the occupancy-specific program PRESENCE. That said, the most widely applied occupancy models, i.e., the single season and dynamic occupancy models, are available as are models for accommodating false-positive detection errors.

In addition to detailed documentation, PRESENCE and unmarked have very active and supportive web-based forums, and both platforms are responsive to the needs of the users in terms of prioritizing development and extensions. As noted above, we have focused exclusively on classical (frequentist) likelihood methods. Oftentimes, however, "nonstandard" occupancy models are of interest, e.g., the community (Royle and Dorazio 2008; Sutherland et al. 2016) or spatial models (Sutherland et al. 2014; Chandler et al. 2015). The application of more sophisticated occupancy models typically requires the use of Bayesian methods, which involve more investment and the development of bespoke, analysis-specific model code. The recent books by Kéry and Schaub (2011) and Kéry and Royle (2015) provide extensive examples with code for most occupancy models. In addition to above resources, the data and R code to conduct both analyses are available here: https://github.com/chrissuthy/Occupancy_Chapter.

Acknowledgments

The water vole data used in this chapter is the product of many years of terrific work by the many volers and volettes who have worked on the project, the hospitality shown to us by Chris Rix at Inchnadamph Lodge, and all of the Assynt landowners who gave us permission to access and survey on their property.

LITERATURE CITED

Amstrup, S. C. S., T. T. L. McDonald, and B. F. J. B. Manly. 2010. Handbook of capture-recapture analysis. Princeton University Press, Princeton, NJ.

Bailey, L. L., D. I. Mackenzie, and J. D. Nichols. 2014. Advances and applications of occupancy models. Methods in Ecology and Evolution 5: 1269–1279.

Chandler, R. B., E. Muths, B. H. Sigafus, C. R. Schwalbe, C. J. Jarchow, and B. R. Hossack. 2015. Spatial occupancy models for predicting metapopulation dynamics and viability following reintroduction. Journal of Applied Ecology 52:1325–1333.

Choquet, R., L. Rouan, and R. Pradel. 2009. Program E-SURGE: A software application for fitting multi-event models. In Modeling demographic processes in marked populations, edited by D. L. Thomson, E. G. Cooch, and M. J. Conroy. Springer, Boston, MA, 2009. 845–865.

Dorazio, R., M. Kéry, J. A. Royle, and M. Plattner. 2010. Models for inference in dynamic metacommunity systems. Ecology 91:2466–75.

Efford, M. G., and D. K. Dawson. 2012. Occupancy in continuous habitat. Ecosphere 3:art32.

Farnsworth, G. L., K. H. Pollock, J. D. Nichols, T. R. Simons, J. E. Hines, and J. R. Sauer. 2002. A removal model for estimating detection probabilities from point-count surveys. Auk 119:414–425.

Fiske, I. J., and R. B. Chandler. 2011. unmarked: An R package for fitting hierarchical models of wildlife occurrence and abundance. Journal of Statistical Software 43:1–23.

Fuller, A. K., D. W. Linden, and J. A. Royle. 2016. Management decision making for fisher populations informed by occupancy modeling. Journal of Wildlife Management 80:1–9.

Guillera-Arroita, G., J. J. Lahoz-Monfort, D. I. MacKenzie, B. A. Wintle, and M. A. McCarthy. 2014. Ignoring imperfect detection in biological surveys is dangerous: A response to "fitting and interpreting occupancy models." PLoS ONE 9(7):e99571.

Hanski, I. 1998. Metapopulation dynamics. Nature 396:41–49.

Hayward, M. W., L. Boitani, N. D. Burrows, P. J. Funston, K. U. Karanth, D. I. Mackenzie, K. H. Pollock, and R. W. Yarnell. 2015. Ecologists need robust survey designs, sampling and analytical methods. Journal of Applied Ecology 52:286–290.

Iknayan, K. J., M. W. Tingley, B. J. Furnas, and S. R. Beissinger. 2014. Detecting diversity: Emerging methods to estimate species diversity. Trends in Ecology and Evolution 29:97–106.

Kellner, K. F., and R. K. Swihart. 2014. Accounting for imperfect detection in ecology: A quantitative review. PLoS ONE 9(10):e111436.

Kéry, M., and J. A. Royle. 2015. Applied hierarchical modeling in ecology: Analysis of distribution, abundance and species richness in R and BUGS: Volume 1: Prelude and Static Models.

Kéry, M., and M. Schaub. 2011. Bayesian population analysis using WinBUGS: A hierarchical perspective. Academic Press.

Krebs, C. J. 1994. Ecology: The experimental analysis of distribution and abundance, 6th ed. HarperCollins, New York.

Lambin, X., D. Le Bouille, M. K. Oliver, C. S. Sutherland, E. Tedesco, and A. Douglas. 2012. High connectivity despite high fragmentation: Iterated dispersal in a vertebrate metapopulation. In Dispersal ecology and evolution, edited by J. Clobert, M. Baguette, T. G. Benton, and J. M. Bullock, pp. 405–412. Oxford University Press, Oxford.

Linden, D. W., A. K. Fuller, J. A. Royle, and M. P. Hare. 2017. Examining the occupancy-density relationship for a low-density carnivore. Journal of Applied Ecology 54:2043–2052.

Linden, D. W., and G. J. Roloff. 2015. Improving inferences from short-term ecological studies with Bayesian hierarchical modeling: White-headed woodpeckers in managed forests. Ecology and Evolution 5:3378–3388.

Mackenzie, D. I., L. Bailey, and J. D. Nichols. 2004. Investigating species co-occurrence patterns when species are detected imperfectly. Journal of Applied Ecology 73:546–555.

MacKenzie, D. I., J. D. Nichols, J. E. Hines, M. G. Knutson, and A. B. Franklin. 2003. Estimating site occupancy, colonization, and local extinction when a species is detected imperfectly. Ecology 84:2200–2207.

MacKenzie, D. I., J. D. Nichols, G. Lachman, S. Droege, J. A. Royle, and C. A. Langtimm. 2002. Estimating site occupancy rates when detection probabilities are less than one. Ecology 83:2248–2255.

MacKenzie, D. I., J. D. Nichols, J. A. Royle, K. H. Pollock, L. L. Bailey, and J. E. Hines. 2006. Occupancy estimation and modeling: Inferring patterns and dynamics of species occurrence. Academic Press, London.

MacKenzie, D. I., J. D. Nichols, M. E. Seamans, and R. J. Gutiérrez. 2009. Modeling species occurrence dynamics with multiple states and imperfect detection. Ecology 90:823–835.

MacKenzie, D. I., and J. A. Royle. 2005. Designing occupancy studies: General advice and allocating survey effort. Journal of Applied Ecology 42:1105–1114.

MacKenzie, D. I., M. E. Seamans, R. J. Gutiérrez, and J. D. Nichols. 2012. Investigating the population dynamics of California spotted owls without marked individuals. Journal of Ornithology 152: 597–604.

Martin, J., C. L. Mcintyre, J. E. Hines, J. D. Nichols, J. A. Schmutz, and M. C. MacCluskie. 2009. Dynamic multistate site occupancy models to evaluate hypotheses relevant to conservation of Golden Eagles in Denali National Park, Alaska. Biological Conservation 142:2726–2731.

McClintock, B. T., L. L. Bailey, K. H. Pollock, and T. R. Simons. 2010. Experimental investigation of observation error in anuran call surveys. Journal of Wildlife Management 74:1882–1893.

McCrea, R., and B. Morgan. 2014. Analysis of capture-recapture data. Chapman and Hall/CRC.

Miller, D. A. W., J. D. Nichols, B. T. McClintock, E. H. C. Grant, L. L. Bailey, and L. A. Weir. 2011. Improving occupancy estimation when two types of observational error occur: Non-detection and species misidentification. Ecology 92:1422–1428.

Moilanen, A. 2002. Implications of empirical data quality to metapopulation model parameter estimation and application. Oikos 96:516–530.

Moilanen, A., and M. Nieminen. 2002. Simple connectivity measures in spatial ecology. Ecology 83:1131–1145.

Molinari-Jobin, A., and M. Kéry. 2012. Monitoring in the presence of species misidentification: The case of the Eurasian lynx in the Alps. Animal Conservation 15:266–273.

Mordecai, R. S., B. J. Mattsson, C. J. Tzilkowski, and R. J. Cooper. 2011. Addressing challenges when studying mobile or episodic species: Hierarchical Bayes estimation of occupancy and use. Journal of Applied Ecology 48:56–66.

Nichols, J. D., L. L. Bailey, A. F. O'Connell Jr., N. W. Talancy, E. H. C. Grant, A. T. Gilbert, E. M. Annand, T. P. Husband, and J. E. Hines. 2008. Multi-scale occupancy estimation and modelling using multiple detection methods. Journal of Applied Ecology 45:1321–1329.

Nichols, J. D., J. E. Hines, D. I. MacKenzie, M. E. Seamans, and R. J. Gutiérrez. 2007. Occupancy estimation and modeling with multiple states and state uncertainty. Ecology 88:1395–1400.

Otis, D. L., K. P. Burnham, G. C. White, and D. R. Anderson. 1978. Statistical inference from capture data on closed animal populations. Wildlife Monographs 62:3–135.

Pavlacky, D. C., J. A. Blakesley, G. C. White, D. J. Hanni, and P. M. Lukacs. 2012. Hierarchical multi-scale occupancy estimation for monitoring wildlife populations. Journal of Wildlife Management 76:154–162.

Pollock, K. H., J. D. Nichols, C. Brownie, and J. E. Hines. 1990. Statistical inference for capture-recapture experiments. Wildlife Monographs 107:3–97.

Royle, J. A., and R. M. Dorazio. 2008. Hierarchical modeling and inference in ecology: The analysis of data from populations, metapopulations and communities. Academic Press, Oxford.

Royle, J. A., and W. A. Link. 2006. Generalized site occupancy models allowing for false positive and false negative errors. Ecology 87:835–841.

Sutherland, C. S., M. Brambilla, P. Pedrini, and S. Tenan. 2016. A multiregion community model for inference about geographic variation in species richness. Methods in Ecology and Evolution:783–791.

Sutherland, C. S., D. A. Elston, and X. Lambin. 2012. Multi-scale processes in metapopulations: Contributions of stage structure, rescue effect, and correlated extinctions. Ecology 93:2465–2473.

Sutherland, C. S., D. A. Elston, and X. Lambin. 2013. Accounting for false positive detection error induced by transient individuals. Wildlife Research 40:490–498.

Sutherland, C. S., D. Elston, and X. Lambin. 2014. A demographic, spatially explicit patch occupancy model of metapopulation dynamics and persistence. Ecology 95:3149–3160.

Tenan, S., M. Brambilla, P. Pedrini, and C. S. Sutherland. 2017. Quantifying spatial variation in the size and structure of ecologically stratified communities.8: 976–984.

Waddle, J. H., R. M. Dorazio, S. C. Walls, K. G. Rice, J. Beauchamp, M. J. Schuman, and F. J. Mazzotti. 2010. A new parameterization for estimating co-occurrence of interacting species. Ecological Applications 20:1467–75.

White, G. C., and K. P. Burnham. 1999. Program MARK: Survival estimation from populations of marked animals. Bird Study 46:S120–S139.

Williams, B. K., J. D. Nichols, and M. J. Conroy. 2002. Analysis and management of animal populations: Modeling, estimation and decision making. Academic Press.

PART III DYNAMIC MODELING OF PROCESSES

8

JAMIE S. SANDERLIN,
MICHAEL L. MORRISON, AND
WILLIAM M. BLOCK

Analysis of Population Monitoring Data

In numerous areas that we call science, we have come to like our habitual ways, and our studies that can be continued indefinitely. We measure, we define, we compute, we analyze, but we do not exclude. And this is not the way to use our minds most effectively or to make the fastest progress in solving scientific questions.
—Platt (1964:352)

What Is Monitoring?

Monitoring is a system for detecting trends over a period of time in areas such as habitat, water, and more. Herein we focus on animal populations and evaluate common quantitative methods for monitoring wildlife species. Animal distribution, abundance, survival, and productivity are examples of important biological parameters that can show trends based on ecosystem condition. Although there are different monitoring types (Morrison and Marcot 1995), monitoring is at its best with *targeted* (or *focused*) *monitoring* (Nichols and Williams 2006), when specific goals and objectives are identified and evaluated under rigorous sampling designs. Herein we focus on targeted monitoring. The process of defining goals and objectives, sampling, evaluating population response, and directing further monitoring and/or management actions should be classified as scientific research (Morrison et al. 2008).

We first distinguish between the biological versus sampled population, as these are seldom the same entity. Since it is rare to census all individuals in a population, the sampled population usually represents a subset of the biological population, but processes in the former may not fully reflect those in the latter (Morrison 2012). The related biological and sampled populations are a result of a hierarchical framework of processes (see Chapter 10): the true underlying ecological process (state process) and the observation process (sampling process) (Fig. 8.1). A biological population consists of all individuals of the same species that occur in the same area at a specific time and is characterized by births, deaths, emigration, and immigration. A community (or assemblage) of species consists of multiple populations of different species that occur in the same area at a specific time. The current state of a community is characterized by number of species (species richness), species composition, number of species and relative abundance of each species (diversity), and local species extinction and colonization.

A robust monitoring data set requires careful planning and adequate support to ensure durability. Researchers and managers should

- formulate clear questions and objectives;
- select appropriate state variables (variables that describe system state, such as abundance) and vital rates to assess study questions;
- determine if there is sufficient power to answer questions within limitations (optimization);
- develop sampling design that fulfills four main questions of what, where, when, and how;

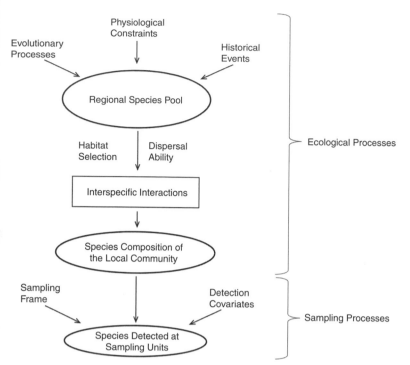

Fig. 8.1. Example of ecological and sampling processes with a study of species in a community. The sampling frame incudes the study design and target population. *Source: Adapted from Morin (1999).*

- ensure data quality controls are in place before collecting data;
- conduct study, and simultaneously conduct quality assurance and quality control (QA/QC), as well as complete QA/QC after study is completed;
- use appropriate models for questions to analyze data;
- modify sampling design or models based on current status.

Qualities of a Robust Monitoring Approach

Collecting robust monitoring data is challenging, but not impossible. A critical first step is to clearly articulate monitoring questions and objectives (Yoccoz et al. 2001). In conjunction with clear objectives, selection of appropriate parameters to assess change over time is paramount. For example, abundance may not change over the course of a monitoring study, but vital rates such as survival or reproduction could change and thus represent more informative variables. If abundance changes, monitoring of addi-

tional vital rates provides understanding of the causes underlying those changes. The reason for change, however, may not be the monitoring objective. Therefore, the appropriate parameters ultimately depend on what will provide the strongest inference for the monitoring objective(s).

Given appropriate state variables and vital rates, the next crucial step is determining if the study design provides sufficient power (probability of correctly rejecting the null hypothesis) to answer questions with a given effect size (magnitude of change with a given precision during a certain time frame). Effort level (sampling frequency and intensity) required to meet study objectives is an important part of determining if there is sufficient power. Expected sampling time frame for state variables and vital rates depends on the rate that population processes function; they could be short (i.e., monitoring small mammal abundance change after a disturbance event) or long (i.e., monitoring adaptation). Ideally, pilot studies can help identify if sufficient power is available with a given effect size for study parameters. We recognize, however, that pilot studies are not always feasible. Information from previous studies (i.e., variance of important parameter[s]) on similar systems could be useful in sampling design development. The pilot study can also pinpoint potential sampling efficiencies for a monitoring study (see "How" section below). Inventories, which quantify metrics of species at one point in time, may be used as pilot studies. Many monitoring schemes consist of repeat inventories, which allow us to estimate state variables at discrete points in time and then extract trends; an alternate approach involves estimating change metrics through demographic modeling, which does not require repeat sampling, per se. Multiple inventories require more effort, but usually provide more accurate trend depictions. Monitoring is pragmatic; thus, whatever approach provides more accurate trend estimates at reduced effort should be adopted.

With an appropriate study design, we assume the sampled population is an unbiased random sample representative of the biological population. This is a critical assumption for all estimation methods we discuss below. Given limited resources, it is also important to optimize study designs for maximum parameter accuracy at reduced effort. Additional robust monitoring data-set qualities are well-designed data input options (see "Data Quality Controls for Monitoring Data" section below) and use of appropriate models (see "Analytical Methods" below, and Table 8.1). Finally, a robust monitoring data set results from a process of feedback loops used to modify sample design and/or models based on current monitoring objective status in an adaptive management framework. All of these steps increase likelihood of success with a monitoring scheme, but monitoring may still not work, even with well-designed efforts, due to reasons beyond the researcher's or manager's control (i.e., reduced budgets and/or commitment to collect data half-way through a study).

Study Design

Given state variables and vital rates that correspond to monitoring objectives, study design then dictates choice of analysis method. Study designs for monitoring data sets can include (from strong to weak inference): experimental, quasi-experimental, and observational approaches (Table 8.2). Experimental approaches test biological relationships, and can be classified as true or quasi-experimental approaches. True experimental approaches have randomly assigned and replicated experimental units to treatments and controls, whereas quasi-experimental approaches, which often occur with ecological studies, rarely have randomly assigned treatments and controls with experimental units or treatment replication. Experimental approaches directly measure cause and effect, while observational approaches are correlational and evaluate patterns. Most monitoring can be classified as observational (although see Russell et al. 2009).

The basic principles (what, where, when, how) of establishing a monitoring study are the same, regardless of temporal or spatial scales. Monitoring objectives dictate whether data sets are massive (i.e., monitoring animal population responses to global climate change or forest succession) or specific to a targeted area (i.e., animal population response to a localized disturbance event). There are challenges, however, with long-term data sets. Often, researchers or managers are limited by number of sampling locations due to budgetary and time constraints. Limits on sampling locations may also restrict abilities to address study objectives. Thus, innovative approaches (i.e., citizen science [Dickinson et al. 2010]) may be needed. Regardless of monitoring study scale or duration, a valid study design fulfills four main questions of what, where, when, and how.

1. *What?* What parameter(s) are most sensitive to index change? The most essential study design aspect is determining the most informative parameter(s) and precision needed for monitoring objectives. *A priori* knowledge and literature review of species' biology should assist with this aspect. Since these preconditions are not always possible with rare species,

Table 8.1. Animal population state variables and vital rates for monitoring, analysis methods, analysis programs, and sampling considerations. Sampling considerations were condensed from Williams et al. (2002), unless otherwise noted. Although there are several more, we list the most common analysis methods and programs used for monitoring state variables and vital rates. Analysis approaches could be frequentist or Bayesian (Gelman et al. 2004; see Chapter 14 in this volume). With Bayesian approaches, state-space models (see Chapter 10 in this volume) and programs OpenBUGS (Speigelhalter et al. 2007; Lunn et al. 2000) and JAGS (Plummer 2003) are typically used with these state variables and vital rates, with the exceptions of genetic measures.

Monitoring state variable(s) and vital rates	Analysis method	Analysis program(s)	Sampling considerations
Abundance, density, λ (finite rate of increase)	Closed (Otis et al. 1978) and open (Jolly-Seber model [Jolly 1965; Seber 1965]) CMR models (or both closed and open models, *sensu* robust design models [Pollock 1982]), distance-based methods (Buckland et al. 2001; see Chapter 6, this volume), spatially-explicit capture recapture (SECR; Royle and Young 2008), matrix models[2] (Caswell 2001)	MARK,[1] CAPTURE (Otis et al. 1978; Rexstad and Burnham 1991), DISTANCE (Laake et al. 1996), SPACECAP (Gopalaswamy et al. 2012)	• Requires extensive effort to capture, mark, and recapture animals (CMR), although physical marks not necessary (i.e., noninvasive genetic CMR, camera traps). • Population closure, tag loss (or with genetic CMR, genotyping error [Taberlet et al. 1999]), and sampling nonindependence (from mates, siblings, or animal clusters) should be minimized or incorporated into population models (i.e., Wright et al. 2009). Assumptions include tags are read correctly and mark does not influence animal behavior. Population closure is minimized by limiting number of repeat sample days and avoiding sampling during migration or high rates of mortality/recruitment. For animals that naturally cluster and show high site fidelity, multiple independent groupings need to be sampled. • Multiple sampling methods (Abadi et al. 2010) may improve abundance or density estimates. Evaluation of costs and precision gain is needed with joint data models (Sanderlin et 2019). • For distance-based methods: accurate estimation of distance/angles from observer to animal; animals located on the line/point have a detection probability of one; animals are not influenced by observation (see Buckland et al. 2001 for specific considerations). • For density methods, the effective trapping area may not be known, and Spatially Explicit Capture-Recapture (SECR) methods are preferable (Royle and Young 2008).
Survival, nest success	Known-fate analyses (i.e., Kaplan-Meier method [Kaplan and Meier 1958; Pollock et al. 1989a,b]), band-recovery analyses (Brownie et al. 1978; Brownie and Pollock 1985, Barker 1997), Cormack-Jolly-Seber model (CJS; Cormack 1964; Jolly 1965; Seber 1965), nest survival (Dinsmore et al. 2002)	MARK[1]	• Effort for nest survival can be allocated toward increased time searching for nests or more visits. • For known-fate analyses, survival probability should be restricted to days when the fate could be observed and disturbance minimized to nests and surrounding habitat during visits. • Accurate estimation of death times and well-specified origin times are needed with telemetry studies. • Censoring is random and independent of survival. • Sampling periods are instantaneous, and recaptured animals are released immediately. • Emigration from the sampled area is permanent.
Reproduction, recruitment	Jolly-Seber model (Jolly 1965; Seber 1965), reverse-time models (Pollock et al. 1974: Nichols et al. 2000; Nichols 2016)	MARK[1]	• Sampling periods are instantaneous, and recaptured animals are released immediately. • Emigration from the sampled area is permanent.

Monitoring state variable(s) and vital rates	Analysis method	Analysis program(s)	Sampling considerations
Movement/dispersal	Multi-strata (multistate) CMR (Arnason 1972, 1973; Brownie et al. 1993; Schwarz et al. 1993), least cost models (Adriaensen et al. 2003), geneflow via population substructure (Hedrick 2005), genetic assignment tests (Waser and Strobeck 1998)	MARK,[1] GIS, STRUCTURE (Pritchard et al. 2000), GENELAND (Guillot et al. 2005)	• Simultaneous sampling of multiple sites is preferable. • CMR sampling is often costly, while telemetry may provide more information. • Indirect methods (e.g., genetic monitoring) may increase sample sizes and power to detect changes.
Occupancy, local extinction and colonization probability	occupancy models (MacKenzie et al. 2006; see Chapter 7, this volume)	PRESENCE (Hines 2006), MARK,[1] R package *unmarked* (Fiske and Chandler 2011)	• Require repeat surveys (visits and/or sampling points) during a period of closure. • Sampling unit and definition of season appropriate for species. • See MacKenzie and Royle (2005) for more details on occupancy study design.
Range distribution	Maximum entropy modeling with presence-only data (Phillips et al. 2006), occupancy models with presence-absence data (MacKenzie et al. 2006; see Chapter 7, this volume), abundance/density models (see above), abundance indices	MaxEnt (Phillips et al. 2005), PRESENCE (Hines 2006), R package *unmarked* (Fiske and Chandler 2011)	• Yackulic et al. (2013) provide recommendations based on detection probability (equal to one, less than one, and constant varies) and sampling probability (varies, constant). • Many assumptions (e.g., constant sampling and detection probability with respect to environmental covariates that determine occupancy) of presence-only data are difficult to meet in practice (Yackulic et al. 2013).
Genetic measures	Hybridization, geographic range, genetic variation, effective population size, movement, population substructure	STRUCTURE (Pritchard et al. 2000), GENEPOP (Raymond and Rousset 1995), ARLEQUIN (Schneider et al. 2000), GENELAND (Guillot et al. 2005)	• Methods rely on reduced genotyping errors (Bonin et al. 2004). Quality and quantity of DNA is important. • Genetic changes with population changes are easier to detect after severe disturbances (Schwartz et al. 2007).
Diversity, species richness, species turnover, species evenness	Multi-species occupancy models (Dorazio et al. 2006; MacKenzie et al. 2006), hierarchical Bayesian analyses and MCMC (see Chapter 14, this volume)	JAGS (Plummer 2003), OpenBUGS (Speigelhalter et al. 2007; Lunn et al. 2000), R (R Development Core Team 2016)	• Require repeat surveys (visits and/or sampling points) during period of closure. • Sampling unit and definition of season appropriate for all species in the community. • Optimal methods for species richness include focusing on number of sampling occasions, while sampled area percentage more important for rare species occupancy probability in a community (Sanderlin et al. 2014). • Sampling design recommendations are needed with multi-season, multi-species occupancy models.

Program MARK (White and Burnham 1999).

This analysis method results in a change metric, not a metric derived from repeated inventories.

Table 8.2. Differences in the design aspects of true experiments and quasi-experiments

Design aspect	Type of experimental design	
	True experiment	Quasi-experiment
Treatments	Randomly allocated to experimental units	Self-assigned to experimental units
Controls	Randomly allocated to experimental units	Randomly allocated to experimental units, self-assigned, or lacking
Confounding factors	Controlled by design	Not controlled by design
Cause and effect	Directly inferable	Not directly inferable
Inference	Strong	Weak
Potential design	Randomized complete block; completely randomized; factorial treatments; split plot	Nonequivalent controls; interrupted time series; before-after-control-impact

Table from Block et al. (2001).

studies with similar species may contribute to identifying important parameters with rare species.

2. *Where?* What is the geographic scale for sampling? Researchers and managers should first identify, as closely as possible, the extent of the biological population of interest. For example, there may be interest in monitoring population trend of black bears (*Ursus americanus*) on a wildlife management area (WMA) to make harvest management decisions specific to the WMA, even though black bears have a larger distribution across the landscape. In this case, the target population, and thus geographic scale, would be the WMA only, even though this is a subset of the biological population. This is a cautionary example, however, since black bears that do not reside on the WMA may be subject to different pressures than those that influence the WMA population subset. Any valid study design should extend beyond the WMA because population processes are unlikely to be captured fully on the WMA.

3. *When?* Length of time for monitoring depends on the study system and monitoring questions. For example, results of a study of fire effects on birds would be different for one year, five years, and 20 years after the fire, respectively. The time frame for detecting change will also affect power to detect change and monitoring feasibility.

4. *How?* The sampling method will depend on objectives and biology of the species. For example, mist-netting and banding with capture-mark-resight might be more advantageous for estimating passerine bird survival than point counts, which are more suitable for estimating occupancy (MacKenzie et al. 2006) or density (Buckland et al. 2001). Potential sampling designs include: simple random sampling, stratified random sampling, systematic sampling, adaptive sampling, cluster sampling, and two-stage cluster sampling (Thompson 2004; Morrison et al. 2008). Several sampling designs can implemented based on species rarity and elusiveness (Thompson 2004) and animal distribution across space (clumped versus randomly distributed). Rare or elusive species may be easier to detect via non-invasive methods with DNA (Waits 2004) or camera traps (Karanth et al. 2004), as opposed to traditional capture-mark-recapture (CMR) animal trapping methods.

Timing (i.e., season, day or night) and sampling length depend on the species and monitoring goals. Is the species available for detection (i.e., birds during the breeding season will sing more often)? Is it better to sample before or after breeding? Should sampling occur over different seasons (i.e., birds could be sampled during the spring migratory season versus winter for residents)? It is important to sample during the same season so comparisons can be made between and among years.

Lifespan of the species should also be considered: short life spans may not require as many monitoring years as long-lived individuals. For example, for a small mammal, such as the white-footed mouse (*Peromyscus leucopus*), the full life cycle could be monitored over the course of a year, but the full life cycle of a northern goshawk (*Accipiter gentilis*) could require a decade to monitor. Species with shorter life spans could add more variance to abundance estimate trends than long-lived species of a population.

Without an adequate budget and commitment to collect data by investigators, well-planned studies can fail. Monitoring should not be done if there is inadequate support, since objectives will not be met with available resources (Legg and Nagy 2006). Resources are better allocated to projects that could meet study objectives. Optimal (increased parameter accuracy with reduced effort) study designs (i.e., Field et al. 2005; Sanderlin et al. 2014; Williams et al. 2018) for monitoring are critical for allocating limited resources to key monitoring parameters.

Depending on analysis method, there are important study design sampling considerations (Table 8.1). Methods of increasing capture probabilities reduce bias in monitoring parameter estimates for all analyses (Williams et al. 2002). CMR methods tend to be more labor intensive than occupancy methods, although advances in technologies such as noninvasive genetic sampling and camera traps have improved sampling capacity and observer safety, especially with rare or elusive species. Noninvasive methods are usually more cost efficient for efforts to detect rare or elusive species (compared to physically capturing animals), although genotyping error (Bonin et al. 2004) and other potential biases (Foster and Harmsen 2012) should be considered when designing a monitoring program. For example, De Barba et al. (2010) demonstrated that it was more cost-effective to sample and monitor a small brown bear (*Ursus arctos*) population using noninvasively collected hair and fecal genetic samples than radio-telemetry or physical trapping. They further showed that noninvasive genetic sampling designs (hair traps and opportunistic sampling) have different costs and efficiencies.

Value Added

Most sampling methods have ancillary data that are collected in addition to the main objectives. These additional data could be valuable to project objectives, and may sway decisions about which sampling methods to use. For example, known-fate methods could be selected to estimate survival changes, but the use of Global Positioning System (GPS) or Very High Frequency (VHF) collars could also provide information to develop habitat maps. It is important, however, that ancillary data collection does not increase sampling effort substantially because that would detract from the main study objective(s).

Adaptive Management

Adaptive management, or adaptive decision-making, is an iterative process for simultaneously managing and gaining knowledge about natural resources. Adaptive management is important to ensuring a robust monitoring data set, as it allows sample design and/or model modification based on current status or monitoring objective. The monitoring phase in adaptive management contributes to knowledge of progress toward objectives, resource status and dynamics, and decisions on modifications to resource dynamics models (Williams 2011). Monitoring under adaptive management, if done correctly via valid study design, will have thresholds that trigger specific actions (Nie and Schultz 2012; Schultz et al. 2013). Biological or management relevance will determine thresholds and triggers, which are important for ensuring efficient monitoring efforts (Schwartz et al. 2015). Adaptive management is distinctly different from a "trial and error" approach, wherein one watches what is happening and then decides on a management option. Moir and Block (2001) identify three monitoring feedback loops that address the following questions and prompt specific actions: (1) Are the right things being monitored? (2) Is the monitoring design being implemented correctly? (3) Has the project met stated goals and objectives, or are modifications in order?

Data Quality Controls for Monitoring Data

Without foresight and planning, consistent monitoring data collection is unmanageable. Monitoring data usually include population data from multiple sampling years, locations, and observers, as well as several environmental and biological covariates. Careful attention to detail when training observers to collect data, develop a data management plan, and archive data is essential for ensuring data integrity with data analyses. Data quality steps also ensure that appropriate data are collected to answer study questions and meet project goals and that covariates are collected at correct scale(s).

Relational Databases

Relational database programs (i.e., MySQL, Microsoft Access) offer project flexibility in electronically storing and retrieving data for quality control and analysis steps. A relational database is a collection of linked tables that contain records (e.g., rows), with each record having fields (e.g., columns). Relationships between entities can be classified as one-to-one, one-to-many, or many-to-many (Figs. 8.2 and 8.3), linked by pairs of values from a primary key (e.g., unique identifier) and foreign key (e.g., a field in a second table that is a primary key in a first table).

Designing a database and import forms before data collection ensures consistency and allows data to be imported and checked for quality as data are collected. Computer and field data logger-data import forms allow users to restrict import fields to valid data input values (i.e., only integer values that are valid with study year field).

Relational databases can also be used to sort and clean data fields for errors by using queries. For example, a common error in multiple-species avian

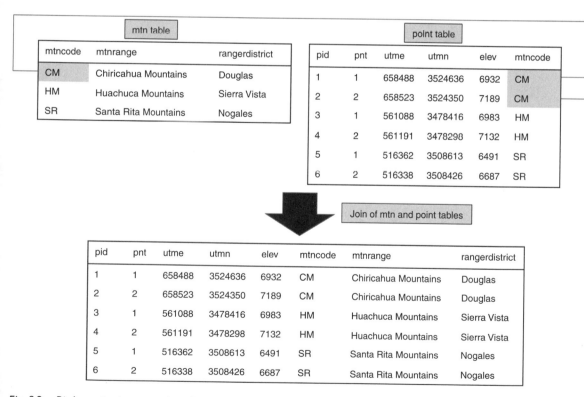

Fig. 8.2. Bird monitoring examples of one-to-one (*mtn* table) and one-to-many (*mtn* table and *point* table) relationships. Each "*mtncode*" in the *mtn* table is unique to each *mtnrange*. The one-to-many relationship is illustrated with multiple points having the same *mtncode*. Data are stored more efficiently in two tables than in the bottom "join" table.

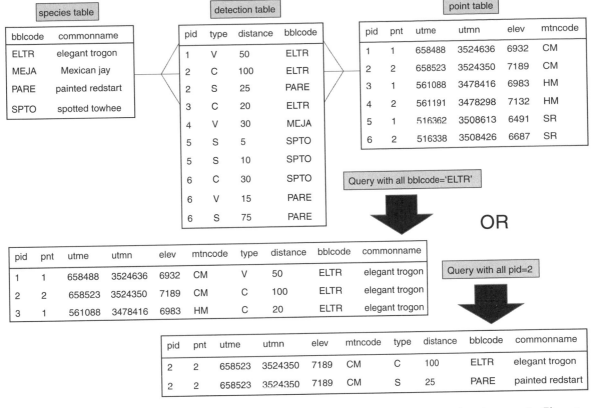

Fig. 8.3. Bird monitoring example of many to many relationship. We illustrate two queries whereby a species Elegant trogon (*Trogon elegans*) can be detected at many point count stations, and a point count station (point count station identity [pid] 2) can have many species detected there. Data are stored more efficiently for access and queries in three tables than in one large table.

point count surveys is recording a wrong letter with a species' Breeding Bird Laboratory mnemonic code; therefore, one can query all codes that are not in the species' code table for efficient corrections. Quick report summaries are also possible. For example, with a bird monitoring data set one could be interested in number of different species detected by year and sampling location. Accurate and efficient summaries are accomplished with queries (Watson 2004 provides query examples).

Data Archive

Another critical aspect of data integrity is archiving cleaned data, both physically and electronically. Essential considerations are security and preservation; hence issues ranging from tampering to future availability and back-up procedures need to be addressed. The primary data archive is the personal data archive, ensuring consistency in future data use. The secondary data archive could include several choices depending on data types (i.e., Dryad [http://datadryad.org] or GenBank [https://www.ncbi.nlm.nih.gov/genbank]). Journals are starting to require (except with funding agency restrictions and private lands access restrictions) data availability with published studies and are providing archiving capabilities. A data archive must ensure that complete metadata (what, where, why, and how data were collected; data reliability; known data issues) are included with data sets so that information necessary for future analyses will be available.

Analytical Methods

Analysis method choice depends on study design, state variables, and vital rates, which correspond to monitoring objectives. Sometimes the analysis method may depend on the species of interest (Thompson et al. 1998). A few examples include band-recovery methods for survival (Brownie et al. 1978; Brownie and Pollock 1985; Barker 1997), which are more common with birds; removal estimators (e.g., White et al. 1982) for abundance, which tend to be used more with fisheries and small mammal trapping; and nest survival methods (i.e., Dinsmore et al. 2002), which are frequently used with birds. We describe several state variables and vital rates and their most commonly associated analysis methods, analysis programs, and sampling considerations (Table 8.1). In Table 8.1 we also distinguish between two approaches to monitoring—namely, repeated inventories versus change metrics. Our list is not meant to be exhaustive, but it does provide an overview of the methods and programs for state variables and vital rates most commonly associated with monitoring animal populations. Depending on monitoring data availability, novel methods with integrated population models (i.e., Schaub and Abadi 2011) allow multiple state variables and vital rates to be monitored simultaneously (i.e., Tempel et al. 2014).

State Variables and Vital Rates

ABUNDANCE, DENSITY, AND FINITE RATE OF INCREASE

The most common monitoring state variables are abundance (number of individuals in the sampled population), density (number of individuals in the population per unit area), and finite rate of increase (λ; change in population abundance or density over a unit of time). The most common abundance and density sampling methods include: capture-mark-recapture (CMR), distance-based methods (Buckland et al. 2001), and spatially-explicit capture recapture methods (SECR; Royle and Young 2008). The most common λ methods and monitoring approaches include: metrics based on multiple inventories to estimate abundance or density, which are then used to extract trends, and change metrics based on matrix models for λ (Caswell 2001). Capture-mark-recapture can be categorized into three types, depending on whether the population is closed to population changes from births, deaths, immigration, and/or emigration (Table 8.1).

In general, abundance estimates are more precise with closed versus open models (Thompson et al. 1998; Williams et al. 2002). From a sampling perspective, however, it is more difficult to maintain closure from movement or demographics. Open models (Jolly 1965; Seber 1965; Schwarz and Arnason 1996) and temporal symmetry models (Nichols 2016) permit survival, reproduction, and movement estimation, which is more informative regarding population trajectory than abundance. Robust design models (models based on sampling that occurs during primary sessions [population is open] and secondary sessions [population is closed]) are more advantageous than open models because abundance estimates are less biased (Pollock et al. 1993). However, a robust design requires additional sampling effort (compared to open models) and hence more time and money for secondary sampling sessions.

Whereas abundance and density are important, they are not necessarily good surrogates for λ, survival, or reproduction. There is insufficient information to address possible monitoring objectives using only abundance estimates. A population size of 100 with high survival and reproduction rates will have a very different trajectory than another population of size 100 with low survival and reproduction rates. It is important to note, however, that abundance could be the primary monitoring objective and thus survival or reproduction may not be of interest. In these cases, the most efficient way to measure change would be to monitor abundance changes. An advantageous monitoring objective would assess the "why" of abundance changes (assuming minimal additional costs), rather than just monitoring abundance. A good example of monitoring using λ, in combination with apparent survival and fecundity, is with the northern spotted owl (*Strix occidentalis caurina*) in the Pacific Northwest, a species with which contentious conservation issues are associated because of its association with old forests (Anthony et al. 2006). Monitoring studies often use several vital rates, as in this example, to understand why a population has a certain trajectory. The main study objective of Anthony at al. (2006) was to determine if owl populations were stationary ($\lambda = 1$), increasing ($\lambda > 1$), or declining ($\lambda < 1$). They concluded that populations were declining in nine study areas and stationary in four.

OCCUPANCY

Occupancy probability (MacKenzie et al. 2006), or probability of species presence, is becoming more prevalent in monitoring endeavors (i.e., MacKenzie and Nichols 2004), given a perceived benefit of reduced effort. Occupancy provides more opportunity to monitor changes in rare species compared to CMR or distance-based methods for which it is difficult to obtain adequate sample sizes. But whereas less effort may be needed for occupancy than abundance monitoring, it is not as effective for gaining information about population status.

Occupancy is based on species presence, and does not indicate abundance, unless assumptions about the relationship between detection and population size have been made (Royle and Nichols 2003). If a species is present, occupancy models do not distinguish between 2 individuals versus 100, which may be critical with monitoring. Although occupancy provides less potential information than abundance (or other state variables), it is better to quantify occupancy well than to do a poor job of estimating abundance. As with the abundance and λ relationship for detecting monitoring trends, the change metrics of interest with occupancy would be local extinction and colonization probabilities.

Ganey et al. (2004) used a pilot study of Mexican spotted owls (*S. o. lucida*) to assess abundance monitoring feasibility, but determined it was cost-prohibitive, and thus recommended occupancy monitoring as a population size index because plot size (1 km^2) was assumed to have one individual if present. The U.S. Fish and Wildlife Service Mexican Spotted Owl Recovery Team (2012) then used this information about detection probabilities, occupancy rates, and habitat variables to design an occupancy monitoring program with sufficient sampling effort (number of call stations per survey plot, number of visits per plot, numbers of plots).

DIVERSITY, SPECIES RICHNESS, SPECIES TURNOVER, SPECIES EVENNESS

Multi-species monitoring is also becoming more frequent (i.e., DeWan and Zipkin 2010; Noon et al. 2012) using multi-species occupancy models (Dorazio et al. 2006; MacKenzie et al. 2006). The multi-species occupancy model extends the single-species occupancy model (MacKenzie et al. 2006) and allows estimation of unknown species richness (number of species in an area) and species turnover (local extinction and colonization), while accounting for imperfect detection probabilities (Royle and Dorazio 2008). This is an area of rapid statistical method growth, and holds vast potential for including detection probability in diversity measures (number of species and abundance of each species) and community dynamics (Iknayan et al. 2014).

There are, however, limitations to adopting a multiple-species approach. An optimal sampling design will differ for rare and for common or abundant species (Sanderlin et al. 2014), and maximizing the number and range of species that are adequately detected can be challenging (Manly et al. 2004). Still, monitoring multiple species simultaneously, may be preferable to a single-species approach, especially for a study with limited resources. A single-species approach to monitoring as representative of the community runs risks of not representing other species in the community, having weak environmental relationships with management actions, being poor future community indicators, or having so much uncertainty in single-species responses that community response interpretations are not possible (Manly et al. 2004).

Similar to abundance, occupancy probability and species richness are valuable to monitoring objectives, but do not give species' trajectories, such as local colonization and extinction probabilities. Thus, from a monitoring perspective, more information about persistence is available from local colonization and extinction probabilities compared to occupancy probability. Russell et al. (2009) used a BACI (before-after-control-impact) design to evaluate effects of prescribed fire on avian communities in a ponderosa pine (*Pinus ponderosa*) forest using multi-species occupancy models to estimate species richness, species turnover, and extinction. One advantage of using these methods is that in addition to obtaining information about fire impacts on common species, it was possible to obtain this information for rare or elusive species, which is important for land management agencies with limited budgets. Although Russell et al. (2009) concluded that prescribed fire treatments had little to no short-term effect on the avian community, this did not preclude the possibility that longer-term effects were present. Multi-species occupancy models allow long-term monitoring of turnover and extinction trends.

SURVIVAL AND REPRODUCTION

Many analysis methods for detecting vital rate trends of survival and reproduction or recruitment are based on CMR sampling methodology or radio-telemetry (Table 8.1), although recent advances in noninvasive genetic sampling have elevated the potential in monitoring survival and reproduction change, (Schwartz et al. 2007). Sampling effort is similar to that of abundance methods and requires multiple sample occasions across years (CMR) or days to years (radio-telemetry).

Returning to the northern spotted owl example, the other study objectives were to identify reasons for owl population changes by quantifying temporal and age-specific survival and fecundity rate trends (An-thony et al. 2006). The researchers found that 6 of the 14 study areas had declining fecundity rates, while 5 study areas had declining and 9 had stable survival rates. Study results suggested fluctuations in fecundity and survival related to weather or prey abundance, which helped guide future monitoring recommendations and identified possible reasons for population trajectories.

MOVEMENT

There are different sampling methods (and thus analysis approaches) for estimating movement trends over time: multi-strata CMR (Arnason 1972, 1973; Brownie et al. 1993; Schwarz et al. 1993), gene flow via population substructure (Hedrick 2005), genetic assignment tests (Waser and Strobeck 1998), and radio and global positioning systems (GPS) telemetry with geographic information systems (GIS) analyses (White and Garrott 1990; Adriaensen et al. 2003). CMR methods, which involve sampling at discrete points in time, are often costly and yield low recapture or resighting probabilities, especially for migratory animals such as birds (Webster et al. 2002). Satellite telemetry can provide valuable movement path information; however, sample sizes (thus power to detect change) are often limited due to high costs, and telemetry methods are often more suitable for larger species due to technological limitations (Hebblewhite and Haydon 2010).

Indirect methods with genetic monitoring and stable isotopes may be more cost-effective and provide more movement information with higher sample sizes or information from individuals that dispersed and reproduced. For example, although Coulon et al. (2008) concluded that dispersal and gene flow between Florida scrub-jay (*Aphelocoma coerulescens*) metapopulations were similar on the basis of separate analyses of demographics and genetics, dispersal and genetic grouping results would have more closely matched if all jays were sampled with the demographic analysis. In this example, jays that dispersed longer distances were missing from the demographic analysis.

RANGE DISTRIBUTION

Species distribution models (SDMs) rely on correlations between species occurrence data and spatial environmental data, resulting from multiple modeling approaches: presence-only data (i.e., maximum entropy modeling; Phillips et al. 2006), presence-absence data (occupancy models; MacKenzie et al. 2006), abundance or density models (detailed above), or abundance indices. In a monitoring context, range distribution trends are quantified by expansions or contractions in species' ranges. Many assumptions (e.g., constant sampling and detection probability with respect to environmental covariates that determine occupancy) of presence-only data are difficult to meet in practice (Yackulic et al. 2013). In an eight-year Swiss breeding bird study monitoring European crossbill (*Loxia curvirostra*), Kéry et al. (2013) illustrated how detection probability was an important source of variation in space and time; had they not included it in their dynamic occupancy models, SDMs would have been biased.

GENETIC MEASURES

There are several analysis methods available to detect trends in genetic measures of hybridization, geographic range, genetic variation, effective population size, and population substructure (i.e., Raymond and Rousset 1995; Pritchard et al. 2000; Schneider et al. 2000; Guillot et al. 2005). Genetic monitoring applications are relatively new, compared to the other methods described above, and show great promise (Schwartz et al. 2007), especially for monitoring adaptive responses to environmental changes such as climate change (Hansen et al. 2012) and for monitoring rare or elusive species (Waits 2004). Genetic monitoring can be more cost-effective than traditional methods, although costs of accounting for and reducing genotyping errors should be considered (Bonin et al. 2004). For example, Beacham et al. (2008) demonstrated that population genetic assignment of Chinook salmon (*Oncorhynchus tshawytscha*) from the Yukon River was both accurate and more cost-efficient than traditional methods of de-

termining population origin. This approach also allowed managers to determine harvest limits or quotas at specific locations more efficiently—an important consideration in a management context. Data preprocessing and quality control steps sometimes differ based on the genotyping technologies employed due to the amount of data generated (i.e., approaches with microsatellites or mitochondrial DNA have fewer loci than next-generation technologies, which could produce millions of single nucleotide polymorphisms or the whole genome of an organism). Massive data sets can create bioinformatics challenges, but also have potentially more information and power to detect population changes.

Environmental DNA (eDNA) monitoring using trace amounts of DNA from many different species in water or soil, for example, is also a novel approach to genetic monitoring (Bohmann et al. 2014). The data collection phase is efficient and cost-effective, yet still poses challenges in the laboratory (e.g., false positives or negatives) and for data analysis (e.g., efficiently handle large amounts of data) challenges. There are questions of how accurate the genetic information is with eDNA technologies (i.e., representative sample of target populations, detection probability biases, false positives due to study design) and whether it could be useful for determining not only presence-absence, but also relative abundance, effective population size, and in combination with other traditional sampling approaches (Bohmann et al. 2014). In a monitoring context, eDNA can be used to detect trends in occupancy (e.g., local extinction and colonization), and possibly changes in relative abundance and effective population size over time.

Analysis Programs

There is a suite of analysis tools available to evaluate monitoring data sets (Table 8.1). Many of these tools can be used with simulation studies (i.e., Field et al. 2005) to evaluate whether there is sufficient power to answer monitoring questions with a given effect size by using estimated quantities (i.e., detection probability, sample size) from pilot studies, previous studies on or near the study area, or simulated values based on similar studies. It is important to consider the analysis method, animal distribution and spatial scale, and costs when evaluating monitoring sampling designs (Thompson et al. 1998). There are also programs that allow users to evaluate different study designs (sample size), effect size, and associated power for abundance (Program Monitor [Gibbs and Ene 2010]) or occupancy (GENPRES [Bailey et al. 2007]; SPACE [Ellis et al. 2014]).

We illustrate the interactions of sample size, effect size, and power using a fictitious small mammal data set (Fig. 8.4). As sampling effort increases, power to detect a non-zero trend increases. In this example, ability to detect a trend (percentage change in abundance) is different with effect size: there is more power with a bigger effect size. Although this example is not exhaustive, one can see potential for sampling efficiencies with monitoring. The population may not need to be monitored all years and/or at the same sampling intensity to achieve a desirable power, and thus design optimization is valuable. Additionally, it is valuable to evaluate sensitivity (Williams et al. 2002) of state variables and vital rates for persistence (e.g., identify which state variables and vital rates are the greatest contributors to persistence; persistence would show the most variation with perturbations of these state variables and vital rates) and to determine appropriate state variables and vital rates for monitoring.

Different packages and programming languages may be needed, depending on state variables and vital rates, as well as problem complexity. Some monitoring questions may be answered with analysis packages (Table 8.1), while other questions are more complex, have a broader scope, or occur over larger scales, and require tailored analyses (i.e., Bayesian approaches) using programming languages like R (R Development Core Team 2016) or Python (van Rossum et al. Python Software Foundation, https://www.python.org/). For example, single or two-species occupancy models can be analyzed with PRESENCE (Hines 2006), but Bayesian approaches are necessary with multi-species occupancy models. Although custom-designs can be advantageous, analysis and output time is typically longer due to complexity and to ensuring accurate parameter estimation. Computational time may be critical in time-sensitive monitoring analyses (i.e., setting harvest regulations).

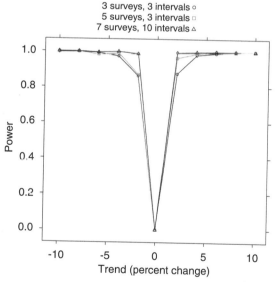

Fig. 8.4. Monitoring data example of the power to detect a trend with different sampling effort amounts. Data were created to be similar to a small mammal abundance study consisting of five plots using program Monitor (Gibbs and Ene 2010). Data were simulated for 10 years, 1,000 iterations, with 2 tails, with rounding, and truncation with counts < 0. The survey design consisted of monitoring for 10 intervals (all years) or 3 intervals (years 0, 4, 9) with route regression and the log-normal measure type.

Conclusions

Advances in statistical methods, software, and computational ability have provided some improvement in our ability to collect and analyze high-quality monitoring data sets, yet important challenges remain. We do not always follow the steps described above, especially those concerned with gathering preliminary data and conducting power analyses to determine if objectives can be met with available budgets.

Critical evaluation of cost-effectiveness and efficiencies is needed when deciding on sampling and analysis methods for detecting population changes. Analysis and method choices depend on study design, state variables and vital rates corresponding to monitoring objectives, and species of interest. Selecting appropriate methods in a cost-efficient manner is essential to effective monitoring.

Acknowledgments

The authors appreciate comments from Bret Collier, Kevin McKelvey, and the editors of this volume on earlier drafts. The views presented here are those of the authors and do not necessarily reflect the view of the USDA Forest Service or Texas A&M University.

LITERATURE CITED

Abadi, F., O. Gimenez, R. Arlettaz, and M. Schaub. 2010. An assessment of integrated population models: Bias, accuracy, and violation of the assumption of independence. Ecology 91:7–14.

Adriaensen, F., J. P. Chardon, G. De Blust, E. Swinnen, S. Villalba, H. Gulinck, and E. Matthysen. 2003. The application of "least-cost" modelling as a functional landscape model. Landscape and Urban Planning 64:233–247.

Anthony, R. G. , E. D. Forsman, A. B. Franklin, D. R. Anderson, K. P. Burnham, G. C. White, C. J. Schwarz, J. D. Nichols, J. E. Hines, G. S. Olson, S. H. Ackers, L. S. Andrews, B. L. Biswell, P. C. Carlson, L. V. Diller, K. M. Dugger, K. E. Fehring, T. L. Fleming, R. P. Gerhardt, S. A. Gremel, R. J. Gutiérrez, P. J. Happe, D. R. Herter, J. M. Higley, R. B. Horn, L. L. Irwin, P. J. Loschl, J. A. Reid, and S. G. Sovern. 2006. Status and trends in demography of northern spotted owls, 1985–2003. Wildlife Monographs 163:1–48.

Arnason, A. N. 1972. Parameter estimates from mark-recapture experiments on two populations subject to migration and death. Researches on Population Ecology 13:97–113.

Arnason, A. N. 1973. The estimation of population size, migration rates, and survival in a stratified population. Researches on Population Ecology 15:1–8.

Bailey, L. L., J. E. Hines, J. D. Nichols, and D. I. MacKenzie. 2007. Sampling design trade-offs in occupancy studies with imperfect detection: Examples and software. Ecological Applications 17:281–290.

Barker, R. J. 1997. Joint modeling of live-recapture, tag-resight, and tag-recovery data. Biometrics 53:666–677.

Beacham, T. D., M. Wetklo, C. Wallace, J. B. Olsen, B. G. Flannery, J. K. Wenburg , W. D. Templin, A. Antonovich, and L. W. Seeb. 2008. The application of microsatellites for stock identification of Yukon River Chinook salmon. North American Journal of Fisheries Management 28:283–295.

Block, W. M., A. B. Franklin, J. P. Ward Jr., J. L. Ganey, and G. C. White. 2001. Design and implementation of monitoring studies to evaluate the success of ecological restoration on wildlife. Restoration Ecology 9: 293–303.

Bohmann, K., A. Evans, M. T. P. Gilbert, G. R. Carvalho, S. Creer, M. Knapp, D. W. Yu, and M. de Bruyn. 2014. Environmental DNA for wildlife biology and biodiversity monitoring. Trends in Ecology and Evolution 29:358–367.

Bonin, A., E. Bellemain, P. Bronken Eidesen, F. Pompanon, C. Brochmann, and P. Taberlet. 2004. How to track and assess genotyping errors in population genetics studies. Molecular Ecology 13:3261–3273.

Brownie, C., D. R. Anderson, K. P. Burnham, and D. R. Robson. 1978. Statistical inference from band recovery data—A handbook. U.S. Fish and Wildlife Service, Resource Publication 131, U.S. Fish and Wildlife Service, U.S. Department of the Interior, Washington, DC.

Brownie, C., J. E. Hines, J. D. Nichols, K. H. Pollock, and J. B. Hestbeck. 1993. Capture-recapture studies for multiple strata including non-Markovian transitions. Biometrics 49:1173–1187.

Brownie, C., and K. H. Pollock. 1985. Analysis of multiple capture-recapture data using band-recovery methods. Biometrics 41:411–420.

Buckland, S. T., D. R. Anderson, K. P. Burnham, J. L. Laake, D. L. Borchers, and L. Thomas. 2001. Introduction to distance sampling: Estimating abundance of biological populations. Oxford University Press, New York.

Caswell, H. 2001. Matrix population models: Construction, analysis, and interpretation, 2nd ed. Sinauer Associates, Sunderland, MA.

Cormack, R. M. 1964. Estimates of survival from the sighting of marked animals. Biometrika 51:429–438.

Coulon, A., J. W. Fitzpatrick, R. Bowman, B. M. Stith, C. A. Makarewich, L. M. Stenzler, and I. J. Lovette. 2008. Congruent population structure inferred from dispersal behaviour and intensive genetic surveys of the threatened Florida scrub-jay (*Aphelocoma coerulescens*). Molecular Ecology 17:1685–1701.

De Barba, M., L. P. Waits, P. Genovesi, E., Randi, R. Chirichella, and E. Cetto. 2010. Comparing opportunistic and systematic sampling methods for non-invasive genetic monitoring of a small translocated brown bear population. Journal of Applied Ecology 47:172–181.

DeWan, A. A., and E. F. Zipkin. 2010. An integrated sampling and analysis approach for improved biodiversity monitoring. Environmental Management 45:1223–1230.

Dickinson, J. L., B. Zuckerberg, and D. N. Bonter. 2010. Citizen science as an ecological research tool: Challenges and benefits. Annual Review of Ecology, Evolution, and Systematics 41:149–172.

Dinsmore, S. J., G. C. White, and F. L. Knopf. 2002. Advanced techniques for modeling avian nest survival. Ecology 83:3476–3488.

Dorazio, R. M., J. A. Royle, B. Söderström, and A. Glimskär. 2006. Estimating species richness and accumulation by modeling species occurrence and detectability. Ecology 87:842–854.

Ellis, M. M., J. S. Ivan, and M. K. Schwartz. 2014. Spatially explicit power analyses for occupancy-based monitoring of wolverine in the U.S. Rocky Mountains. Conservation Biology 28:52–62.

Field, S. A., A. J. Tyre, and H. P. Possingham. 2005. Optimizing allocation of monitoring effort under economic and observational constraints. Journal of Wildlife Management 69:473–482.

Fiske, I. J., and R. B. Chandler. 2011. unmarked: An R package for fitting hierarchical models of wildlife occurrence and abundance. Journal of Statistical Software 43:1–23.

Foster, R. J., and B. J. Harmsen. 2012. A critique of density estimation from camera-trap data. Journal of Wildlife Management 76:224–236.

Ganey, J. L., G. C. White, D. C. Bowden, and A. B. Franklin. 2004. Evaluating methods for monitoring populations of Mexican spotted owls: A case study. In Sampling rare or elusive species: Concepts, design, and techniques for estimating population parameters, edited by W. L. Thompson, pp. 337–385. Island Press, Washington, DC.

Gelman, A., J. B. Carlin, H. S. Stern, and D. B. Rubin. 2004. Bayesian data analysis, 2nd edition. Chapman and Hall/CRC, New York.

Gibbs, J. P., and E. Ene. 2010. Program Monitor: Estimating the statistical power of ecological monitoring programs. Version 11.0.0. www.esf.edu/efg/gibbs/monitor/.

Gopalaswamy, A. M., J. A. Royle, J. E. Hines, P. Singh, D. Jathanna, N. S. Kumar, and K. U. Karanth. 2012. Program SPACECAP: Software for estimating animal density using spatially explicit capture-recapture models. Methods in Ecology and Evolution 3:1067–1072.

Guillot, G., F. Mortier, and A. Estoup. 2005. GENELAND: A computer package for landscape genetics. Molecular Ecology Notes 5:712–715.

Hansen, M. M., I. Olivieri, D. M. Waller, E. E. Nielsen, and The GeM Working Group. 2012. Monitoring adaptive genetic responses to environmental change. Molecular Ecology 21:1311–1329.

Hebblewhite, M., and D. T. Haydon. 2010. Distinguishing technology from biology: A critical review of the use of GPS telemetry data in ecology. Philosophical Transactions of the Royal Society B 365:2303–2312.

Hedrick, P. W. 2005. A standardized genetic differentiation measure. Evolution 59:1633–1638.

Hines, J. E. 2006. PRESENCE3: Software to estimate patch occupancy and related parameters. USGS, Patuxent Wildlife Research Center, Laurel, MD. https://www.mbr-pwrc.usgs.gov/software/presence.html.

Iknayan, K. J., M. W. Tingley, B. J. Furnas, and S. R. Beissinger. 2014. Detecting diversity: Emerging methods to estimate species diversity. Trends in Ecology and Evolution 29:97–106.

Jolly, G. M. 1965. Explicit estimates from capture-recapture data with both death and immigration-stochastic model. Biometrika 52:225–247.

Kaplan, E. L. and P. Meier. 1958. Nonparametric estimation from incomplete observations. Journal of the American Statistical Association 53:457–481.

Karanth, K. U., J. D. Nichols, and N. S. Kumar. 2004. Photographic sampling of elusive mammals in tropical forests. In Sampling rare or elusive species: Concepts, design, and techniques for estimating population parameters, edited by W. L. Thompson, pp. 229–247. Island Press, Washington, DC.

Kéry, M., G. Guillera-Arroita, and J. J. Lahoz-Monfort. 2013. Analysing and mapping species range dynamics using occupancy models. Journal of Biogeography 40:1463–1474.

Laake, J. L., S. T. Buckland, D. R. Anderson, and K. P. Burnham. 1996. DISTANCE user's guide V2.2. Colorado Cooperative Fish & Wildlife Research Unit, Colorado State University, Fort Collins.

Legg, C. J. and L. Nagy. 2006. Why most conservation monitoring is, but need not be, a waste of time. Journal of Environmental Management 78:194–199.

Lunn, D. J., A. Thomas, N. Best, and D. Spiegelhalter. 2000. WinBUGS—A Bayesian modelling framework: Concepts, structure, and extensibility. Statistics and Computing 10:325–337.

MacKenzie, D. I., and J. D. Nichols. 2004. Occupancy as a surrogate for abundance estimation. Animal Biodiversity and Conservation 27.1:461–467.

MacKenzie, D. I., J. D. Nichols, J. A. Royle, K. H. Pollock, L. L. Bailey, and J. E. Hines. 2006. Occupancy estimation and modeling. Academic Press, New York.

MacKenzie, D. I., and J. A. Royle. 2005. Designing occupancy studies: General advice and allocating survey effort. Journal of Applied Ecology 42:1105–1114.

Manly, P. N., W. J. Zielinski, M. D. Schlesinger, and S. R. Mori. 2004. Evaluation of a multiple-species approach to monitoring species at the ecoregional scale. Ecological Applications 14:296–310.

Moir, W. H., and W. M. Block. 2001. Adaptive management on public lands in the United States: Commitment or rhetoric? Environmental Management 28:141–148.

Morin, P. J. 1999. Community ecology. Blackwell Science, Malden, MA.

Morrison, M. L. 2012. The habitat sampling and analysis paradigm has limited value in animal conservation: A prequel. Journal of Wildlife Management 76:438–450.

Morrison, M. L., W. M. Block, M. D. Strickland, B. A. Collier, and M. J. Peterson. 2008. Wildlife study design, 2nd ed. Springer-Verlag, New York.

Morrison, M. L., and B. G. Marcot. 1995. An evaluation of resource inventory and monitoring program used in National Forest planning. Environmental Management 19:147–156.

Nichols, J. D. 2016. And the first one now will later be last: Time-reversal in Cormack-Jolly-Seber models. Statistical Science 31:175–190.

Nichols, J. D., J. E. Hines, J. -D. Lebreton, and R. Pradel. 2000. The relative contributions of demographic components to population growth: A direct estimation approach based on reverse-time capture-recapture. Ecology 81:3362–3376.

Nichols, J. D., and B. K. Williams. 2006. Monitoring for conservation. Trends in Ecology and Evolution 21: 668–673.

Nie, M. A., and C. A. Schultz. 2012. Decision-making triggers in adaptive management. Conservation Biology 26:1137–1144.

Noon, B. R., L. L. Bailey, T. D. Sisk, and K. S. McKelvey. 2012. Efficient species-level monitoring at the landscape scale. Conservation Biology 26:432–441.

Otis, D. L., K. P. Burnham, G. C. White, and D. R. Anderson. 1978. Statistical inference from capture data on closed animal populations. Wildlife Monographs 62: 1–135.

Phillips, S. J., R. P. Anderson, and R. E. Schapire. 2006. Maximum entropy modeling of species geographic distributions. Ecological Modeling 190:231–259.

Phillips, S. J., M. Dudik, and R. E. Schapire. 2005. Maxent software for species distribution modeling. https://www.cs.princeton.edu/~schapire/maxent/.

Platt, J. R. 1964. Strong inference. Science 146:347–353.

Plummer, M. 2003. JAGS: A program for analysis of Bayesian graphical models using Gibbs sampling. Proceedings of the Third International Workshop on Distributed Statistical Computing (DSC 2003), March 20–22, Vienna, Austria.

Pollock, K. H. 1982. A capture-recapture design robust to unequal probability of capture. Journal of Wildlife Management 46:752–757.

Pollock, K. H., W. L. Kendall, and J. D. Nichols. 1993. The "robust" capture-recapture design allows components of recruitment to be estimated. In Marked individuals in the study of bird populations, edited by J.-D. Lebreton, and P. M. North, pp. 245–252. Brikhäuser Verlag, Basel, Switzerland.

Pollock, K. H., D. L. Solomon, and D. S. Robson. 1974. Tests for mortality and recruitment in a K-sample tag-recapture experiment. Biometrics 30:77–87.

Pollock, K. H., S. R. Winterstein, C. M. Bunck, and P. D. Curtis. 1989a. Survival analysis in telemetry studies: The staggered entry design. Journal of Wildlife Management 53:7–15.

Pollock, K. H., S. R. Winterstein, and M. J. Conroy. 1989b. Estimation and analysis of survival distributions for radio-tagged animals. Biometrics 45:99–109.

Pritchard, J. K., M. Stephens, and P. Donnelly. 2000. Inference of population structure using multilocus genotype data. Genetics 155:945–959.

R Development Core Team. 2016. R: A language and environment for statistical computing. R Foundation for Statistical Computing, Vienna, Austria.

Raymond, M., and F. Rousset. 1995. GENEPOP (version 1.2): Population genetics software for exact tests and ecumenicism. Journal of Heredity 86:248–249.

Rexstad, E. A., and K. P. Burnham. 1991. User's guide for interactive program CAPTURE. Abundance estimation of closed animal populations. Colorado Cooperative Fish and Wildlife Research Unit, Colorado State University, Fort Collins.

Royle, J. A., and R. M. Dorazio. 2008. Hierarchical modeling and inference in ecology. Academic Press, New York.

Royle, J. A., and J. D. Nichols. 2003. Estimating abundance from repeated presence-absence data or point counts. Ecology 84:777–790.

Royle, J. A., and K. V. Young. 2008. A hierarchical model for spatial capture-recapture data. Ecology 89:2281–2289.

Russell, R. E., J. A. Royle, V. A. Saab, J. F. Lehmkuhl, W. M. Block, and J. R. Sauer. 2009. Modeling the effects of environmental disturbance on wildlife communities: Avian responses to prescribed fire. Ecological Applications 19:1253–1263.

Sanderlin, J. S., W. M. Block, and J. L. Ganey. 2014. Optimizing study design for multi-species avian monitoring programmes. Journal of Applied Ecology 51:860–870.

Sanderlin, J. S., W. M. Block, B. E. Strohmeyer, V. A. Saab, and J. L. Ganey. 2019. Precision gain versus effort with joint models using detection/non-detection and banding data. Ecology and Evolution 9:804–817. https://doi.org/10.1002/ece3.4825.

Schaub, M., and F. Abadi. 2011. Integrated population models: A novel analysis framework for deeper insights into population dynamics. Journal of Ornithology 152:227–237.

Schneider, S., D. Roessli, and L. Excoffier. 2000. Arlequin: A software for population genetics data analysis. User manual Version 2.000. Genetics and Biometry Lab, Department of Anthropology, University of Geneva, Switzerland.

Schultz, C.A., T. D. Sisk, B. R. Noon, and M. A. Nie. 2013. Wildlife conservation planning under the United States Forest Service's 2012 planning rule. Journal of Wildlife Management 77:428–444.

Schwartz, M. K., G. Luikart, and R. S. Waples. 2007. Genetic monitoring as a promising tool for conservation and management. Trends in Ecology and Evolution 22:25–33.

Schwartz, M. K., J. S. Sanderlin, and W. M. Block. 2015. Manage habitat, monitor species. In Wildlife habitat conservation: Concepts, challenges, and solutions, edited by M. L. Morrison and H.A. Mathewson pp. 128–142. Johns Hopkins University Press, Baltimore, MD.

Schwarz, C. J., and A. N. Arnason. 1996. A general methodology for the analysis of capture-recapture experiments in open populations. Biometrics 52:860–873.

Schwarz, C. J., J. F. Schweigert, and A. N. Arnason. 1993. Estimating migration rates using tag-recovery data. Biometrics 49:177–193.

Seber, G. A. F. 1965. A note on the multiple-recapture census. Biometrika 52:249–259.

Spiegelhalter, D. J., A. Thomas, N. G. Best, and D. Lunn. 2007. OpenBUGS user manual. Version 3.0.2. MRC Biostatistics Unit, Cambridge, UK.

Taberlet, P., L. P. Waits, and G. Luikart. 1999. Noninvasive genetic sampling: Look before you leap. Trends in Ecology and Evolution 14:323–327.

Tempel, D. J., M. Z. Peery, and R. J. Gutiérrez. 2014. Using integrated population models to improve conservation monitoring: California spotted owls as a case study. Ecological Modelling 289:86–95.

Thompson, W. L., ed. 2004. Sampling rare or elusive species: Concepts, design, and techniques for estimating population parameters. Island Press, Washington, DC.

Thompson, W. L., G. C. White, and C. Gowan. 1998. Monitoring vertebrate populations. Academic Press, San Diego, CA.

U.S. Fish and Wildlife Service, Mexican Spotted Owl Recovery Team. 2012. Mexican spotted owl (Strix occidentalis lucida) recovery plan, first revision. U.S. Fish and Wildlife Service, Southwest Region, Albuquerque, NM. 413 pp.

Van Rossum, G., et al. The Python Language Reference, Python Software Foundation. https://docs.python.org/3/reference/index.html.

Waits, L. P. 2004. Using noninvasive genetic sampling to detect and estimate abundance of rare wildlife species. In Sampling rare or elusive species: Concepts, design, and techniques for estimating population parameters, edited by W. L. Thompson, pp. 211–228. Island Press, Washington, DC.

Waser, P. M., and C. Strobeck. 1998. Genetic signatures of interpopulation dispersal. Trends in Ecology and Evolution 13:43–44.

Watson, R. T. 2004. Data management: Databases and organizations, 4th ed. John Wiley & Sons, New York.

Webster, M. S., P. P. Marra, S. M. Haig, S. Bensch, and R. T. Holmes. 2002. Links between worlds: Unraveling migratory connectivity. Trends in Ecology and Evolution 17:76–83.

White, G. C., D. R. Anderson, K. P. Burnham, and D. L. Otis. 1982. Capture-recapture and removal methods for sampling closed populations. Los Alamos National Laboratory, Los Alamos, NM. LA-8787-NERP.

White, G. C., and K. P. Burnham. 1999. Program MARK: survival estimation from populations of marked animals. Bird Study 46(Supplement 1):S120–S139.

White, G. C., and R. A. Garrott. 1990. Analysis of wildlife radio-tracking data. Academic Press, San Diego, CA.

Williams, B. K. 2011. Adaptive management of natural resources—Framework and issues. Journal of Environmental Management 92:1346–1353.

Williams, B. K., J. D. Nichols, and M. J. Conroy. 2002. Analysis and management of animal populations. Academic Press, San Diego, CA.

Williams, P. J., M. B. Hooten, J. N. Womble, G. G. Esslinger, and M. R. Bower. 2018. Monitoring dynamic spatio-temporal ecological processes optimally. Ecology 99:524–535.

Wright, J. A., R. J. Barker, M. R. Schofield, A. C. Frantz, A. E. Byrom, and D. M. Gleeson. 2009. Incorporating genotype uncertainty into mark-recapture-type models for estimating abundance using DNA samples. Biometrics 65:833–840.

Yackulic, C. B., R. Chandler, E. F. Zipkin, J. A. Royle, J. D. Nichols, E. H. Campbell Grant, and S. Veran. 2013. Presence-only modelling using MAXENT: When can we trust the inferences? Methods in Ecology and Evolution 4:236–243.

Yoccoz, N. G., J. D. Nichols, and T. Boulinier. 2001. Monitoring of biological diversity in space and time. Trends in Ecology and Evolution 16:446–453.

9 Systems Analysis and Simulation

Stephen J. DeMaso
and Joseph P. Sands

A system isn't just any old collection of things. A system is an interconnected set of elements that is coherently organized in a way that achieves something.
—Meadows (2008:11)

Systems Approach to Solving Problems

There are many useful qualitative and quantitative approaches to decision-making and problem-solving. For instance, trial and error is the most widespread and useful approach to solving many problems on a daily basis (Grant ct al. 1997). However, some problems are too complex (e.g., too many treatments, experiments take extensive time, and/or are costly) to rely on trial and error for a solution. Fortunately, a systems approach integrates relevant information gained through trial and error or expert opinions and the scientific method to facilitate a formal description of the structure and dynamics of complex systems. A good example of this is found in DeMaso et al. (2012) who used a systems approach to simulate various functional forms of density dependence in a South Texas northern bobwhite (*Colinus virginianus*) population. Some functional forms appeared to work, while others gave clearly erroneous population projections.

Some Foundations of the Systems Approach

Lindeman's (1942) insightful paper "The Trophic Dynamic Aspect of Ecology" is considered by many to be the cornerstone of ecosystem ecology, the integrated study of living (biotic) and non-living (abiotic) components of ecosystems and their interactions within an ecosystem framework. These systems are often very complex with many parts. Aldo Leopold (1949:190) stated, "To keep every cog and wheel is the first precaution of intelligent tinkering." What Leopold was getting at was how important the interconnectedness of plants and animals are to a healthy environment (i.e., the system [environment] is more important than some of the individual parts [plants and animals]). These are some of the earliest examples of natural resource professionals thinking in terms of systems and viewing nature as a system. Van Dyne (1969) provided an overview of the meaning, origin, and importance of ecosystem concepts and discusses field research that addresses the ecosystem concept and how it can be useful in both research and management in natural resource sciences.

Odum (1969) viewed ecosystems as highly organized systems with strong parallels to organisms and human societies. He argued that all ecosystems follow a common strategy leading to a steady state equilibrium characteristic of mature systems. Odum claimed that maturing ecosystems increasingly develop self-regulation or homeostasis in the same way that organisms do. This ecosystem homeostasis is mediated by such characteristics as high species diversity, niche specialization, complex food webs, high ratios of biomass to production, and increased interdependence among species in the form of mutualism.

Patten (1971) covered the transition of ecology from synecology (the branch of ecology dealing with the relations between natural communities and their environments) to systems ecology. Patten (1972) also outlined for ecologists the concepts upon which systems science as a discipline is built and presented examples of applications of systems analysis methods to ecosystems. This approach also included new ecological theory, including an investigation into the feasibility of several nonlinear formulations for use in compartment modeling of ecosystems, and the important topic of connectivity in systems.

Patten (1975, 1976) published the proceedings of a landmark symposium entitled "Modeling and Analysis of Ecosystems," held at the University of Georgia in 1973. The goals of the meeting were to review the status of ecosystem modeling, simulation, and analysis and identify and promote dialogue on key issues in macrosystem modeling.

Shugart (1984) used simulation models in studies of forest succession, and related successional theory to ecological systems modeling. He considered these models "extensions of the ability of ecologists to further understand the dynamic aspects of successional processes." Several books (Grant et al. 1997; Ford 1999; Odum and Odum 2000; Hannon and Ruth 2001; Grant and Swannack 2008) provide a conceptual background for much of what we discuss in this chapter.

The utility of simulation modeling comes as much from the process (problem specification, model development, and model evaluation) as from the product (final model and simulations). Ecologists and natural resource managers deal mostly with systems characterized by "organized complexity" in which there are relatively few data and little hope of ever accumulating a "complete" data set for analysis (see Grant et al. 1997: their Fig. 1). In wildlife science and ecology we deal with exactly the

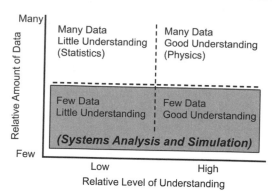

The Systems Approach to Problem Solving

Fig. 9.1.   The systems approach to problem-solving relative to the amount of data and the relative level of understanding of the problem *Source: Modified from Grant et al. (1997: Fig. 1.2).*

types of systems for which systems analysis and simulation were developed. A systems approach does not replace other methods of problem-solving, but instead provides a framework that allows integration of knowledge gained from description and classification, as well as mathematical and statistical analysis of our observations of the world (Grant et al. 1997; Fig. 9.1). Recently, systems modeling analyses have seen an increase in use in wildlife science (e.g., Walters 2001; VerCauteren et al. 2006; DeMaso et al. 2011; Rader et al. 2011; DeMaso et al. 2012) although applications of such analytical approaches are still far less frequent than more traditional modeling techniques.

Some Definitions and Concepts

To understand and facilitate a working knowledge of system analysis and simulation modeling one must be familiar with some important concepts that are implied by specific terms.

- *System.* A system can be defined as any collection of interacting objects and processes. The main attribute of a system is it can be understood only by viewing it as a whole (Grant et al. 1997). The *system-of-interest* applies to a specific problem and care must be taken in defining boundaries specific to that problem.

- *Systems Analysis.* Systems analysis is the application of the scientific method to solving problems involving complex systems. Systems analysis is not achieved by using advanced mathematical and statistical procedures or computers; rather, it is a broad-based problem-solving strategy.
- *Model.* A model is a formal description of essential elements of a problem. These essential elements are what define our system-of-interest. In other words, a model is an abstraction of reality. There are many types of models. Some of the more relevant are dichotomies and include: hysical versus abstract; dynamic versus static; empirical versus mechanistic; deterministic versus stochastic; and simulation versus analytical (Grant et al. 1997). They can all be defined in greater detail as follows:
  - *Physical*, models are replicas of study objects at a reduced scale. An example might be modeling an algae bloom in a lake, but using tanks in a laboratory to represent a smaller scale lake.
  - *Abstract*, models use symbols instead of physical devices to represent a system being studied. Abstract models can be created by a written language, verbal description, or a thought process. An example is weight loss of an individual, which is 1.5 pounds/day when temperature is 5°C and increases by 0.5 pounds/day with each 10°C increase in temperature. This can be translated to $Y = 1.5 + 0.5(X)$, where Y represents weight loss (pounds/day) and X represents temperature (°C).
  - *Dynamic* models describe a time-varying relationship. Examples are simulation models and regression models that include time as an independent variable (X).
  - *Static* models describe a relationship or set of relationships that do not change through time. An example is a regression analysis that does not include time as an independent variable (X).
  - *Empirical (correlative)* models describe a set of relationships without representation of the processes or mechanisms that operate in the real system. The goal is prediction, not explanation. For example, predicting a person's weight based only on his or her height.
  - *Mechanistic (explanatory)* models describe the internal dynamics of a system-of-interest adequately. The goal is explanation through representation of the causal mechanisms underlying system behavior. An example is a model representing the weight of a person as a function of height, age, diet, activity level, and occupation.
  - *Deterministic* models contain no random variables. That is, every time you run the model with the same set of data you get the same answer; nothing varies.
  - *Stochastic* models that contain one or more random variables. Variables are selected at random in each time step, so you get a different answer each time the model runs.
  - *Simulation* models mimic, step by step, the behavior of a system. They are composed of a series of mathematical and logical operations that represent the structure (i.e., state) and behavior (i.e., change of state) of the system-of-interest. If appropriate variables describe the system and appropriately represent the rules governing change, we should be able to trace the state of the system through time, thus simulating the behavior of the system.
  - *Analytical* models that can be solved in closed form mathematically. Most of the time, in wildlife conservation and natural resource management modeling the system-of-interest is too complex for analytical treatment.

The Phases of Systems Analysis

The systems approach emphasizes a holistic approach to problem solving as opposed to the more traditional reductionist approach (i.e., a philosophical position regarding the connections between phenomena, or theories, reducing one to another, usually considered simpler or more basic). Using mathematical models to identify and simulate important characteristics of complex systems is an important part of the systems analysis approach. As we mentioned above, a system is any set of objects that interact. A mathematical model is a set of equations that describe interrelationships among these objects. By solving equations in the mathematical model, we can simulate the dynamic behavior of the system (Grant and Swannack 2008). The four steps in the systems approach to problem solving are formulating a conceptual model, quantifying the model, evaluating the model, and using the model.

1. The first step in the process is to develop model objectives (Starfield 1997) and a conceptual model of the system-of-interest. Based on model objectives, the conceptual model should include components of the real-world system and how they relate to one another. These components and their

relationships are drawn with various symbols representing the nature of those relationships (Fig. 9.2). During this step it is helpful to keep track of patterns of behavior that we expect the model to exhibit (Grant and Swannack 2008).

Conceptualizing the model is critical to later quantification and model evaluation. This provides a visual representation of the system-of-interest constructed by the modeler, and reveals potential relationships among parameters and/or variables. Initial draft models may end up being more complex than is necessary or contain elements for which few or no data exist (which may be valuable in itself). This portion of the exercise is critical, because it forces the modeler to carefully consider the system-of-interest and incorporate only the detail necessary to represent the main drivers in the system. The iterative nature of modeling (Fig. 9.3) allows the modeler to continually revise and reconsider conceptual elements of the model until it captures the system-of-

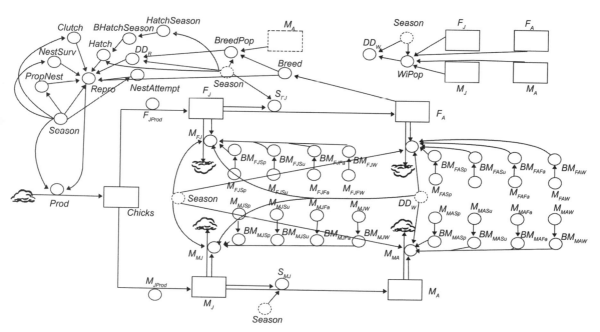

Fig. 9.2. Conceptual diagram of a northern bobwhite (*Colinus virginianus*) population model for the Rio Grande Plains, Texas based on DeMaso et al. (2011). Boxes indicate state variables (stocks); circles (converters) indicate driving variables, constants, or auxiliary variables; and arrows going from a state variable to another state variable with a circle touching the arrow are material transfers (flows).

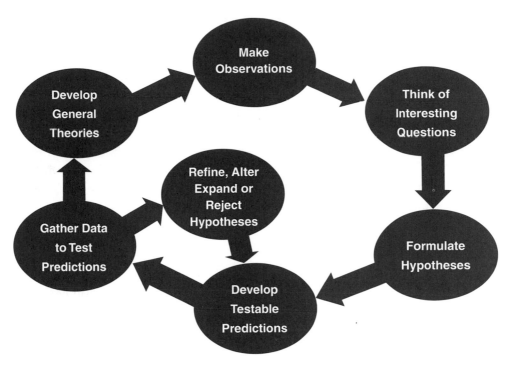

Fig. 9.3.   Conceptual diagram of the iterative nature of the scientific method.

interest. This iterative nature is very similar to the iterative nature of the scientific method (Fig. 9.3), where hypotheses are refined, altered, expanded, or outright rejected in favor of more plausible hypotheses. Similarly, model structures are refined until they mimic the system-of-interest.

2. Step 2 is to develop a quantitative model of the system-of-interest. In this step, the modeler translates the conceptual model into a series of mathematical equations that form the quantitative model. This quantification is based on consideration of various types of information about the real-world system we have at our disposal. We then use a computer to solve all model equations, at each time step over the entire period of simulated time in which we have an interest (Grant and Swannack 2008).

3. Step 3 evaluates the usefulness of the model in meeting project objectives. That is, model evaluation considers a broad array of different aspects of model structure and behavior that make it useful, rather than simply validating model predictions with real system observations. This step focuses specifically on the parts of the model (relationships of model components versus model predictions) in relation to objectives of the model. Sometimes the modeler is interested in determining how sensitive model predictions are to uncertainties that are represented in the model (Grant and Swannack 2008). Often in simulation modeling the term model "evaluation" differs from model "validation" (Grant et al. 1997). Model evaluation involves critically viewing the model structure and asking such questions as, Does the model address the project objectives? and Does the model structure mimic the structure of the real-world system? Model validation involves comparing model predictions with real-world observations and assuming the model is good if the model predictions and real-world observations are similar.

4. Step 4 uses the model to answer questions identified at the beginning of the project.

This involves designing and simulating the experiments that we would conduct in the real world to answer the questions. We then analyze, interpret, and communicate simulation results using the same general procedures used for real-world results (Grant and Swannack 2008).

An Iteration of Phases: The Modeling Process in Practice

The four steps of systems analysis are highly interconnected. Although one might think of the process as occurring sequentially in steps 1 through 4, in reality, the process is iterative, and may circle through several steps more than once (Fig. 9.4). During any phase, the modeler might have overlooked or misrepresented an important system component or process and will need to go back to an earlier phase, often to the conceptual model formulation or quantitative model specification (Grant et al. 1997).

During model evaluation, we examine the model to detect any flaws in logic that may require us to cycle back to earlier steps. For example, one or more of the following situations could occur: (1) the model structure and functional relationships may not correspond with the system-of-interest, (2) the model projections do not behave as expected, and/or (3) the model projections do not agree with independent data from the system-of-interest. Discovery of such flaws during model development usually provides additional insight into the dynamics of the system-of-interest and is an important benefit of modeling.

Reporting the Development and Use of Simulation Models

Typically, it is awkward to describe development and use of a simulation model in the standard "methods"

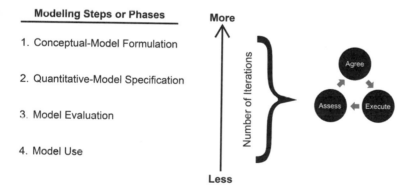

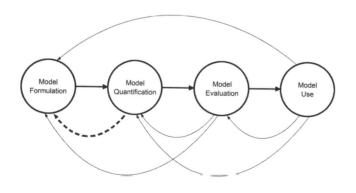

Fig. 9.4. Relative frequency of iterations needed in various phases of simulation modeling (a) and schematic of the modeling process (b). Thicker dashed lines (b) indicate that the first two phases are generally the most iterative.

and "results" format of technical reports and scientific journals. The model can be viewed both as a method and a result. Reviewers and editors not familiar with systems analysis and simulation modeling often see the justification for using certain parameter estimates in the model not as descriptive, but rather as reporting a result. "Overview of the Model," "Model Description," "Model Evaluation," and "Simulation [of Project Objective]" are more useful section heading in reports and manuscripts and can help to avoid such misunderstandings. These section headings can be used to clearly describe the model and its purpose; quantitatively (mathematically) describe the model; compare simulation to observed system dynamics; and simulate results of the project objective. Grant et al. (1997) provide a good example in their chapter 10 of how to report a systems analysis and simulation project. Another source for examples of reported results is published literature, such as *Ecological Modelling*, an *International Journal on Ecological Modelling, and Systems Ecology*.

Simulation Software

There are various software programs available for systems dynamics and simulation modeling (Table 9.1). Each modeling program has strengths and weaknesses. Modelers need to investigate which programs have the functions they desire for their modeling project(s). A starting point for beginning modelers is to evaluate different programs and decide which software program meets specific needs (Table 9.1). For instance, a program without a Geographic Information System interface may not be very useful in developing a model of waterfowl migration.

Some Examples of Systems Analysis and Simulation Modeling
Components of Systems Models

We provide a basic overview of systems model components. There are multiple systems modeling software platforms available on the market, and some

details may vary between them. Our personal experience is primarily with the STELLA platform, which we have used for the examples here (also see Richmond 2004).

- *Stocks*. Stocks are the currency of a systems model. A stock holds the materials of the system (e.g., money, individuals, energy, etc.). Stocks change only via flows(See below). In systems modeling software, stocks are often represented with a square (Figs. 9.2 and 9.5).
- *Flows*. Flows control transfer of materials (stocks) in the model. This transfer could be either into (inflow) or out of (outflow) a stock. In other words, flows are the agents of change in a systems model. Documenting the dynamics of flows in a model can be as valuable and informative in the modeling process as documenting dynamics of stocks. In systems modeling software flows are represented via thick arrows, which indicate direction of flow (Figs. 9.2 and 9.5).
- *Converters*. Converters parameterize and regulate flows. Essentially a converter is a variable that will alter the rate of a flow. Further, a converter can be affected by a stock, but cannot directly affect a stock. Converters can be fixed (a specific value) or random parameters (drawn from a distribution). In systems modeling software, stocks are often represented with a circle (Figs. 9.2 and 9.5).

A Basic Population Model

Think of the elements and characteristics of a population. A population is essentially a collection of individuals that interact in some way. Gotelli (2001:2) defined a population as "a group of plants, animals, or other organisms, all of the same species, that live together and reproduce." Four processes can cause a change in abundance of a population: births, deaths, immigration, and emigration (Gotelli 2001). In many population models, immigration and emigration are not considered explicitly (e.g., are assumed to be equal). An argument could be made that they should

Table 9.1. Available systems dynamics and simulation software programs.

Program name	Licensing	Program language	Last update	Website
AMESim	Proprietary, commercial	Unknown	2017	www.plm.automation.siemens.com/en_us/products/lms/imagine-lab/amesim/
Analytica	Proprietary, commercial, free limited version	C++	2016	www.lumina.com
AnyLogic	Proprietary, commercial, free personal learning edition (PLE) for education, formal or not	Java	2017	anylogic.com
ASCEND	Free, general public license (GPL)	C	2012	ascend4.org
Berkeley Madonna	Proprietary, shareware	C++, Java	2016	berkeleymadonna.com
Consideo	Proprietary, commercial	Unknown	2012	www.consideo-modeler.de
DYNAMO	Proprietary, no longer distributed commercially	AED, Pascal	1986	N/A
Dynaplan Smia	Proprietary, commercial	C++	2013	www.dynaplan.com
Forio Simulations	Proprietary, commercial	Unknown	2016	www.forio.com
GoldSim	Proprietary, commercial	C++	2016	www.goldsim.com
Insight Maker	Free, Insight Maker Public LicenseL	JavaScript	2017	www.insightmaker.com
JDynSim	Free, GPL	Java	2010	code.google.com/p/dynsim
MapleSim	Proprietary, commercial	Java (GUI), C, Maple (engine)	2017	www.maplesoft.com/products/maplesim
Mapsim	Free, lesser general public license (LGPL)	.NET Framework	2013	mapsim.sourceforge.net
Minsky	Free, LGPL	C++	2017	sourceforge.net/projects/minsky
NetLogo	Free, GPLv2	Java, Scala	2015	ccl.northwestern.edu/netlogo
OptiSim	Free, for education. Copyright 2009–2010 by Tomasz Zawadzki, all rights reserved.	Java	2010	www.optisim.org/QLENG
Powersim Studio	Proprietary, commercial, free trial 30 days	C++	2015	www.powersim.com
Pyndamics	Free, MIT License	Python	2013	github.com/bblais/pyndamics
PySD	Free, MIT	Python	2016	github.com/JamesPHoughton/pysd
RecurDyn	Proprietary, commercial, free trial 30 days	C++	2013	www.recurdyn.de
Simantics System Dynamics	Free, Eclipse Public License (EPL)	Java, Modelica	2015	sysdyn.simantics.org
Simile	Proprietary, commercial	C++, Prolog, Tcl	2013	www.simulistics.com
Simulink	Proprietary, commercial	Unknown	2015	www.mathworks.com/products/simulink
Simplorer	Proprietary, commercial	Unknown	2017	www.ansys.com/products/systems/ansys-simplorer
Sphinx SD Tools	Free, Apache License (ASL), version 2.0	Java	2013	sourceforge.net/projects/sphinxes
Stella, iThink	Proprietary, commercial	Unknown	2012	iseesystems.com
Sysdea	Proprietary, commercial	Unknown	2013	sysdea.com
SystemDynamics	Free, GPL	Java	2009	sourceforge.net/projects/system-dynamics

Table 9.1. (continued)

Program name	Licensing	Program language	Last update	Website
Temporal Reasoning Universal Elaboration (TRUE)	Proprietary, freeware. Copyright 2002–2014 True-World, all rights reserved.	WLanguage Windev	2014	www.true-world.com
Vensim	Proprietary, commercial, free PLE for education and personal use	C	2016	vensim.com
Ventity	Proprietary, commercial, free PLE for education and personal use	C	2017	ventity.biz
VisSim	Proprietary, commercial	C	2011	www.vissim.com
Wolfram SystemModeler	Proprietary, commercial	Unknown	2014	www.wolfram.com/system-modeler

https://en.wikipedia.org/wiki/Comparison_of_system_dynamics_software.

not always be ignored; however, for this basic example we will consider them equal.

A basic systems model of a population must contain a stock and an inflow for reproduction (Fig. 9.5a). We have labeled these two elements "Population" and "Reproduction" accordingly. However, basic biology suggests that at an appropriate scale, no population can grow indefinitely, so adding a population regulator (e.g., a carrying capacity) is appropriate (Fig. 9.5b). In this case, "Carrying Capacity" is a constant (e.g., 1,000 individuals) and affects production via the standard Verhulst equation (Box 9.1). So far, the model structure does not include an outflow (e.g., mortality or emigration). In wildlife science applications, mortality is a factor that deserves careful consideration. Adding an outflow "Mortality" to the model (Fig. 9.5c) makes the model more realistic. The model was run from 0 to 25 years with one calculation per time step ($\Delta t = 1$).

Model output (Fig. 9.6) shows differences in population dynamics as various parameters are added to the model. Essentially, population growth is exponential (despite use of the logistic model) in the absence of carrying capacity (i.e., no resource is limiting). As expected, when a form of density dependence reproduction is added, this parameter begins to regulate productivity, slowing population growth, and as mortality is added to the model, this parameter

(a)

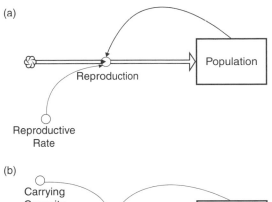

(b)

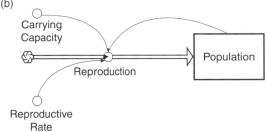

(c)

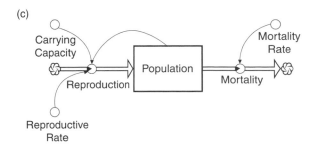

Fig. 9.5. Structures for a basic population model. A basic population requires (a) an inflow and a stock. More realism is added with a carrying capacity parameter (b), and completed with a mortality outflow (c).

STOCKS:

    Population(t) = Population(t-dt) + (Reproduction - Mortality * dtINIT Population = 50

INFLOWS:

    Reproduction= (Population*Reproductive__Rate)* (1-(Population/Carrying__Capacity))

OUTFLOWS:

    Mortality = Population*Mortality__Rate

CONVERTERS:

    Carrying__Capacity = NORMAL(1000,200)
    Mortality__Rate = 0.25
    Reproductive__Rate = 0.5

Box 9.1. Documentation of a basic population model constructed in STELLA run from 0 to 25 years with one calculation per time step ($\Delta t = 1$).

limits population growth to below carrying capacity. In effect you can observe both dampening (productivity is being reduced) and limitation (the habitat can support only X individuals, i.e., carrying capacity) of the population simultaneously.

Many generic model structures exist for a variety of population modeling applications (Grant et al. 1997; Deaton and Winebrake 2000; Richmond 2004); however, what is not generic are the data that underlie model parameters. In the example above (Fig. 9.5C), the outflow represents mortality, but if

other factors that affect population change are considered, the outflow could potentially represent emigration out of the population. Data are critical to any modeling effort, and modelers should strive to use the all available data when parameterizing models. However, one value of systems modeling is that it can frequently illustrate information gaps during the quantification stage. In the absence of empirical data for a parameter, the modeling process can still be quite useful (Starfield 1997). The modeler can test the parameter with a data gap to evaluate its effect on system dynamics. If the effect is large, then collecting data on this parameter may be quite valuable, and the model building process has been a useful activity (Starfield 1997).

The model above (Fig. 9.5) is deterministic: its output does not vary no matter how many times it is run, and while it provides an example of how a theoretical population may behave, it is generally unrealistic in wildlife management. Frequently in wildlife management there is a degree of variation in a parameter that we are most interested in, as well as potential reasons for that variation. Thus an approach to capturing dynamic behaviors of systems is to add stochasticity to model parameters. In general, a common way to do so is to have a converter variable with a given mean vary by some

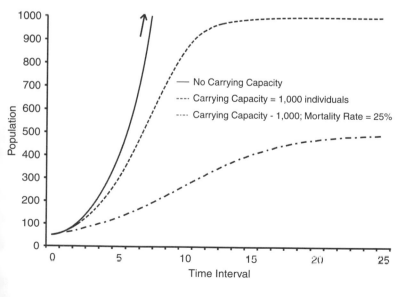

Fig. 9.6. Output from a basic model of population growth illustrating population performance for three different model structures (see Fig 9.4a–b).

probability distribution. Examples of this could be a normal, Poisson, or Weibull distribution. Most modeling platforms will have built-in distributions where means and variances can be selected, or distributions can be programmed specifically. De-Maso et al. (2011) used data-based Weibull and normal distributions in the development of a stochastic model of a northern bobwhite population, whereas Rader et al. (2011) varied stochastic parameters randomly, based on ranges of data, and used built-in functions. Each of these strategies produced models that simulated observed system dynamics in a realistic way, and were useful from a model application standpoint.

Often, probability distribution functions are not available from empirical data for modeling purposes and are assumed to be normally distributed (Guthery et al. 2000), which is a tenuous assumption when dealing with biological and ecological data (Young and Young 1998). When making a model stochastic, the modeler should always strive to use the underlying distribution of the data.

Using the basic logistic model presented above, and making carrying capacity stochastic by using a normal distribution with a mean of 1,000 and a standard deviation of 200 (using STELLA's built-in normal distribution), the population projections of 25 simulations are seen in Fig. 9.7a. The variability

(a)

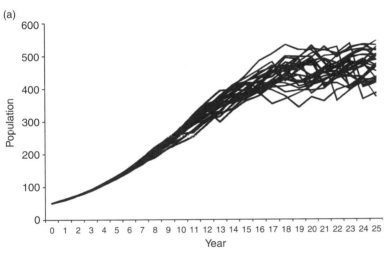

(b)

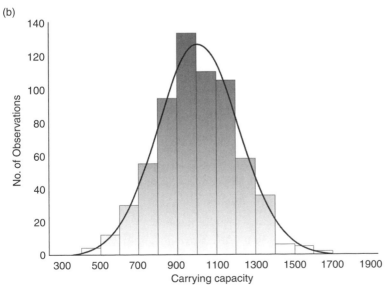

Fig. 9.7. Output from a stochastic model of population growth. Population growth (a) varies based on the density-dependent carrying parameter (b).

in population growth (Fig. 9.7a) is produced by the stochastic selection of values for carrying capacity (Fig. 9.7b), which in this case represent a normal distribution (as expected).

In a real-world application of a model, the number of simulations to run (e.g., sample size of the stochastic model) should be based on modeling objectives and the number of simulations to run (n) using the formula provided by Sokal and Rohlf (1969:247) and Grant et al. (1997:62–63) using the variance estimates from 50 preliminary stochastic baseline simulations to obtain variance estimates and calculate required sample sizes. Essentially, the modeler needs to determine the number of simulations necessary to show a statistically significant difference in magnitude of values for model parameters of interest (Grant et al. 1997). For example, the modeler may want to detect a statistical difference of 250 individuals in the fall population. The difference could be caused by different management strategies, mortality rates, or other environmental variables. Grant et al. (1997:61–64) provide a process for conducting this preliminary analysis, based on model variability at a given significance level (α) with a specific probability of detecting the difference, given that it exists.

Some Applied Examples of Simulation Modeling in Wildlife Conservation

Peterson et al. (1998) developed and evaluated a computerized model that explicitly represents prairie chicken (*Tympanuchus cupido*) clutch size, egg hatchability, nesting success, brood survival, survivorship of chicks within successful broods, and juvenile and adult survival. Sensitivity analyses of these variables suggested that the proportion of hens losing their entire brood has the greatest influence on the number of prairie chickens in the subsequent spring breeding population. They then used the model to compare the relative importance of three reproductive parameters of endangered Attwater's prairie chicken (*T. c. attwateri*) populations that are known to be significantly less productive than those

of the greater prairie chicken (*T. c. pinnatus*). When long-term nesting success, brood survival, and number of chicks per brood prior to brood breakup were individually increased, values for each parameter had to be substantially greater than typically seen in greater prairie chicken populations before the decline in Attwater's prairie chicken numbers was reversed. When these three variables were increased simultaneously, about 90% of the difference between Attwater's and greater prairie chicken values had to be closed before the decline in numbers was reversed.

Martinez et al. (2005) synthesized decades of field data on white-winged doves (*Zenaida asiatica asiatica*) in the Tamaulipan Biotic Province. They evaluated the model by comparing simulated annual productivity and long-term population trends to field data. Based on simulation results, they identified apparent inconsistencies in the database. They could not generate the observed annual production index with the model parameterized on field nest success and survivorship data. Also, they could not generate a stable long-term population trend with the model parameterized based on suggested sustainable harvest rates and empirically based estimates of migratory return rates. Their simulation results suggested that nest success might be closer to 22% (rather than 35%). A similar trend resulted when simulated hunting pressure was increased by 25% (to 31%), or return rates of migrating juveniles and adults were decreased by 5.5 and 5.0%, to 69 and 77%, respectively, with all other values at the baseline level. They concluded that until better estimates of nest success and migratory return rates are available, their model predictions must be viewed with caution.

Rader et al. (2011) examined the relative impact of altering nest-predation rate, nesting habitat, and weather (i.e., temperature and precipitation) on northern bobwhite population dynamics in a hypothetical 15,000-ha subtropical-rangeland ecosystem in South Texas using a simulation model. The systems model consisted of a three-stage (i.e., eggs, juvenile, and adult) bobwhite population with dynamics influenced by variables affecting production, recruitment, nest predation, and mortality. The baseline simulated bobwhite population dynamics corresponded closely to empirical data, with no difference between medians of simulated ($n=30$ year) and observed bobwhite age ratios over a 28-year period. They then created simulated population scenarios representing (1) baseline historical conditions, (2) predator control, (3) low precipitation, (4) low precipitation with predator control, (5) high temperature, (6) high temperature with predator control, (7) reduced nest-clump availability, and (8) reduced nest-clump availability with predator control; each of these scenarios resulted in considerably different median bobwhite densities over ten years. For example, under simulated predator control, populations increased by about 55% from the baseline scenario, whereas under simulated reduced nest-clump availability, populations decreased by about 75% from the baseline scenario. Comparisons of time-series for each scenario showed that reduced nest-clump availability, low precipitation, and high temperature reduced bobwhite densities to a greater degree than a natural nest predation rate. Reduced nest-clump availability resulted in the most substantial decline of simulated bobwhite densities. The results from these simulations suggested that management efforts should focus on maintaining adequate nest-clump availability and then possibly consider nest predator control as a secondary priority.

DeMaso et al. (2011) used radio-telemetry data from northern bobwhites in southern Texas from 2000 to 2005 to develop a stochastic simulation model for bobwhite populations. Their model is based on difference equations, with stochastic variables drawn from normal and Weibull distributions. They simulated bobwhite populations to 100 years and evaluated their model by comparing results with independent estimates of four population parameters (spring and fall density, finite rate of increase in the fall population [λ], and winter juvenile:adult age ratios). With a quasi-extinction criterion of ≤ 40 birds

(density $= \leq 0.05$ birds/ha), the probability of persistence to 100 years was 88.3% (106 of 120 simulations) for the spring population and 96.7% (116 of 120 simulations) for the fall population. With a less restrictive quasi-extinction criterion (≤ 14 birds), the probability of persistence was 93.3% (112 of 120 simulations) for the spring population and 98.3% (118 of 120 simulations) for the fall population. Simulated population parameters were similar to independent estimates for four of four population parameters. Winter age ratios differed between the model ($\bar{x} = 4:98$ juv:ad, $n = 120$, Standard Error (SE) $= 0.32$) and empirical age ratios from harvested bobwhites on the study area ($\bar{x} = 2:85$ juv:ad, $n = 25$, SE $= 0.24$). However, when they corrected harvest age ratios for bias in juvenile harvest ($\bar{x} = 3:85$ juv:ad, $n = 25$, SE $= 0.32$), simulated and empirical estimates were similar. Their model appears to be a reliable predictor of bobwhite populations in southern Texas. The simulation results indicate that bobwhite hunters and managers can expect excellent bobwhite hunting (fall populations ≥ 2.2 birds per ha) during about one of every ten years.

Sands (2010) constructed a simulation model of bobwhite population dynamics to conduct a sensitivity analysis designed to evaluate the impacts of harvest on northern bobwhite populations. He used stochastic simulations to determine optimal harvest rates, maximize the probability of long-term population persistence within the context of sustained-yield harvest (SYH), and determine minimum viable spring densities for northern bobwhite populations in the South Texas Plains. He hypothesized that northern bobwhite annual harvest of 25–30% of the pre-hunt population would represent a sustainable yield (number of northern bobwhites harvested per year) for a northern bobwhite population in the South Texas Plains, and that annual harvest rates >45% would reduce the probability of population persistence below 95% over 100 years. Contrary to his original hypothesis, only annual harvest rates ≤ 20% did not impact probability of population persistence. An annual harvest rate of 20% produced the greatest average (±SE) yields (231±10 bobwhites harvested/year) over the 100-year simulation. A 30% annual harvest rate resulted in population extinction in 75% of simulations within an average (±SE) of 47.8±2.3 years. A 40% annual harvest rate resulted in population extinction in 100% of simulations within an average of 15.5±2.6 years. Discontinuing harvest at winter densities of 0.25 bobwhites/ha reduced the probability of population extinction at 20–30% harvest rates, and maximized yield at 20% and 25% harvest rates. Spring densities of 0.60–0.80 bobwhites/ha may represent minimum viable spring densities for northern bobwhite populations in the South Texas Plains, but weak density-dependent reproduction makes harvesting at high rates to achieve these densities a less optimal strategy than harvesting at more conservative rates. Sand's modeling results indicated that harvesting northern bobwhite populations in the South Texas Plains at rates of 20–25% of the pre-hunt population should maximize long-term harvest while minimizing the probability of population extinction. Harvest rates >30% are likely excessive with respect to long-term population persistence for bobwhite populations in the South Texas Plains.

DeMaso et al. (2012) used a systems analysis approach with a compartment model based on difference equations ($\Delta t = 3$ months) for bobwhites in South Texas to simulate population behavior using 16 different functional forms of density-dependent production and overwinter mortality. During the reproductive season, a weak linear density-dependent relationship resulted in the longest population persistence (up to 100.0 years), whereas a reverse-sigmoid density-dependent relationship had the worst population persistence (2.5–3.5 years). Regarding overwinter mortality, a sigmoid or weak linear density-dependent relationship and a weak linear or no density-dependent reproduction relationship had the longest population persistence (87.5–100.0 years). Weak linear density-dependent reproduction with either sigmoid or weak linear overwinter mortality produced stable fall population

trends. Their results indicated that density dependence may have a greater influence on overwinter survival of bobwhites than previously thought. Inclusion of density-dependent functional relationships that represent both density dependent reproduction and overwinter mortality, were critical for their simulation model to function properly. Therefore, integrating density-dependent relationships for both reproductive and overwinter periods of the annual cycle of bobwhite life history events is essential for conducting realistic bobwhite population simulation analyses that can be used to test different management scenarios in an integrated and interdisciplinary manner.

Moon (2014) used systems-based modeling approach for regional mottled duck (*Anas fulvigula*) populations to elucidate the importance of individual vital rates and develop predictions regarding mottled duck persistence, while simultaneously identifying key uncertainties and priority research needs. Through the use of STELLA 10.0.3, she constructed, parameterized, and evaluated a stochastic, seasonal demographic model based on data currently available on the Western Gulf Coast (WGC) population of mottled ducks. Her model is based on difference equations, with stochastic variables drawn from normal distributions. She simulated mottled duck populations for 100 years and evaluated her model by comparing results with independent estimates of population parameters reported in technical literature. The model simulated the flow of individual mottled ducks through annual cycle events within the WGC region (i.e., the system). The flow of individuals was driven by production and began with nest survival in Season 1. The model was partitioned into three different seasons based on mottled duck life history: breeding/brooding (Season 1: February 1–July 15), post-breeding (Season 2: July 16–October 31), and winter (Season 3: November 1–January 31). Ducklings were separated into male and female classes in Season 1 at a 50:50 ratio. Mortalities were removed seasonally from each population segment (i.e., ducklings, juveniles, and adults).

Because mottled ducks are nonmigratory and her model was based on the entire population of WGC mottled ducks, she assumed no immigration and emigration. Following model evaluation a sensitivity analysis was conducted. The model was sensitive to variation in all breeding parameters, which can be influenced by quality habitat management practices. As presented, the model assumes constant habitat conditions across time and does not incorporate future degradation of habitats. This quantitative model can be used to elucidate functional relationships among demographic rates and population growth to provide input for conservation actions and long-term management of the WGC mottled duck population.

Systems Modeling and Decision-Making in Natural Resource Management

Natural resource management is a field where complex decisions are often made in a state of partial uncertainty (e.g., Nichols et al. 1995). Making good decisions in this environment generally necessitates the use of models (Kendall 2001). A systems approach to modeling lends itself as a practical tool to use in these situations (recall Fig. 9.1) through integration of relevant data, as well as expert opinion or other sources of information and can help identify sources of system uncertainty and data gaps. Costanza and Ruth (1998) presented case studies ranging from industrial systems (mining, smelting, and refining of iron and steel in the United States) to ecosystems (Louisiana coastal wetlands and Fynbos ecosystems in South Africa) to linked ecological-economic systems (Maryland's Patuxent River basin in the United States). These studies illustrate uses of dynamic modeling to include stakeholders in all stages of consensus building, ranging from initial problem scoping to model development.

The systems modeling approach described in this chapter is intended to be an introduction for wildlife scientists who are unfamiliar with this approach. This type of modeling is less intimidating to people who are beginning modelers, or agency staff who may use models less frequently than professionals in research-oriented positions. Most professionals can prepare and interpret box and arrow diagrams, which are the foundation of systems modeling. Being able to conceptualize the system-of-interest and predict how it appears to function are critical for successful development of a systems model. Regardless of your ecological and/or mathematical knowledge or computer programming ability, applying the modeling process described in this chapter will augment your understanding of your system-of-interest and will likely result in making better management decisions. Successful implementation of the systems modeling process requires clear communication among stakeholders; therefore, clearly articulating objectives, defining systems of interest, and identifying and quantifying model parameters are essential for understanding outcomes of the model and the resulting basis for management decisions.

Acknowledgments

The authors thank the editors for the invitation to contribute this chapter to this important book. The findings and conclusions in the chapter are those of the authors and do not necessarily represent the views of the U.S. Fish and Wildlife Service.

LITERATURE CITED

Costanza, R., and M. Ruth. 1998. Using dynamic modeling to scope environmental problems and build consensus. Environmental Management 22:183–195.

Deaton, M. L., and J. I. Winebrake. 2000. Dynamic modeling of environmental systems. Springer, New York.

DeMaso, S. J., W. E. Grant, F. Hernández, L. A. Brennan, N. J. Silvy, X. Ben Wu, and F. C. Bryant. 2011. A population model to simulate northern bobwhite population dynamics in southern Texas. Journal of Wildlife Management 75:319–332.

DeMaso, S. J., J. P. Sands, L. A. Brennan, F. Hernández, and R. W. DeYoung. 2012. Simulating density-dependent relationships in South Texas northern bobwhite populations. Journal of Wildlife Management 77:24–32.

Ford, A. 1999. Modeling the environment: An introduction to system dynamics modeling of environmental systems. Island Press, Washington, DC.

Gotelli, N. J. 2001. A primer of ecology, 3rd ed. Sinauer and Associates, Sunderland, MA.

Grant, W. E., E. K Pedersen, and S. L. Marín. 1997. Ecology and natural resource management: Systems analysis and simulation. John Wiley & Sons, New York.

Grant, W. E., and T. M. Swannack. 2008. Ecological modeling: A common-sense approach to theory and practice. Blackwell Publishing, Malden, MA.

Guthery, F. S., M. J. Peterson, and R. R. George. 2000. Viability of northern bobwhite populations. Journal of Wildlife Management 64:646–662.

Hannon, B., and M. Ruth. 2001. Dynamic modeling, 2nd ed. Springer-Verlag, New York.

Kendall, W. L. 2001. Using models to facilitate complex decisions. In Modeling in natural resource management: Development, interpretation, and application, edited by T. M. Shenk and A. B. Franklin, pp. 147–170. Island Press, Washington, DC.

Leopold, A. 1949. A Sand County almanac and sketches here and there. Oxford University Press, New York.

Lindeman, R. L. 1942. The trophic-dynamic aspect of ecology. Ecology 23:399–417.

Martinez, C. A., W. E. Grant, S. J. Hejl, M. J. Peterson, A. Martinez, and G. L. Waggerman. 2005. Simulation of annual productivity and long-term population trends of white-winged doves in the Tamaulipan biotic province. Ecological Modelling 181:149–159.

Meadows, D. H. 2008. Thinking in systems: A primer. Chelsea Green Publishing, White River Junction, VT.

Moon, J. A. 2014. Mottled duck (Anas fulvigula) ecology in the Texas Chenier Plain region. PhD Dissertation. Stephen F. Austin University, Nacogdoches, TX.

Nichols, J. D., F. A. Johnson, and B. K. Williams. 1995. Managing North American waterfowl in the face of uncertainty. Annual Review of Ecology and Systematics 26:177–199.

Odum, E. P. 1969. The strategy of ecosystem development. Science 164:262–270.

Odum, H. T., and E. C. Odum. 2000. Modeling for all scales: An introduction to system simulation. Academic Press, San Diego, CA.

Patten, B. C. 1971. Systems analysis and simulation in ecology: Volume I. Academic Press, New York.

Patten, B. C. 1972. Systems analysis and simulation in ecology: Volume II. Academic Press, New York.

Patten, B. C. 1975. Systems analysis and simulation in ecology: Volume III. Academic Press, New York.

Patten, B. C. 1976. Systems analysis and simulation in ecology: Volume IV. Academic Press, New York.

Peterson, M. J., W. E. Grant, and N. J. Silvy. 1998. Simulation of reproductive stages limiting productivity of the endangered Attwater's prairie chicken. Ecological Modelling 111:283–295.

Rader, M. J., L. A. Brennan, K. Brazil, F. Hernández, and N. J. Silvy. 2011. Simulating northern bobwhite population responses, nest predation, nesting habitat, and weather in South Texas. Journal of Wildlife Management 75:61–70.

Richmond, B. 2004. An introduction to systems thinking, STELLA. Isee Systems, Lebanon, NH.

Sands, J. P. 2010. Testing sustained-yield harvest theory to regulate northern bobwhite hunting. PhD Dissertation, Texas A&M University–Kingsville.

Shugart, H. H. 1984. A theory of forest dynamics. Springer-Verlag, New York.

Sokal, R. R., and F. J. Rohlf. 1969. Biometry. W. H. Freeman and Company, San Francisco, CA.

Starfield, A. M. 1997. A pragmatic approach to modeling for wildlife management. Journal of Wildlife Management 61:261–270.

Van Dyne, G. M. 1969. The ecosystem concept in natural resource management. Academic Press, New York.

VerCauteren, K. C., M. J. Lavelle, and S. E. Hyngstrom. 2006. A simulation model for determining cost-effectiveness of fences for reducing deer damage. Wildlife Society Bulletin 34:16–22.

Walters, S. 2001. Landscape pattern and productivity effects on source-sink dynamics of deer populations. Ecological Modelling 134:17–32.

Young, L. J., and J. H. Young. 1998. Statistical ecology: A population perspective. Kluwer Academic Publishers, Boston, MA.

10 — Applications of Individual-Based Models

Julie A. Heinrichs
and Bruce G. Marcot

> Individual-based models have been used for many applied or "pragmatic" issues, such as informing the protection and management of particular populations in specific locations, but their use in addressing theoretical questions has also grown rapidly, recently helping us to understand how the sets of traits of individual organisms influence the assembly of communities and food webs.
> —DeAngelis, and Grimm (2014:39)

Introduction

Modern wildlife ecology is recognizing the need for robust predictive tools to support evidence-based decision-making (Wood et al. 2015). Although ecologists have historically sought to understand wildlife systems using statistical and phenomenological models, individual-based models (IBMs) are increasingly used to predict and evaluate multi-scale patterns and processes of complex ecological systems (Grimm and Railsback 2012). IBMs found favor in the 1990s (DeAngelis and Gross 1992; Judson 1994) and quickly grew in popularity due to the variety of questions they can help answer. Many important ecological insights have resulted from the ability of IBMs to relax traditional population model assumptions, explicitly link spatial and temporal conditions with individual states and population outcomes, and simultaneously integrate many influences and conditions. Yet IBMs also present unique challenges. In this chapter, we review the types, uses, advantages, and challenges of using IBMs in wildlife science.

By definition, individuals are the focal unit in individual-based models. Although IBMs can take a wide range of forms, IBMs constructed for wildlife are generally designed to simulate movement, dispersal, habitat selection and resource use, demography, or other multi-scale interactions. Wildlife IBMs are often structured by population units or maps representing landscape conditions. Decision rules and interactions with local and regional conditions influence individual outcomes within each population or location. Rather than solving equations, most IBMs rely on stochastic processes (demographic variation, variable movement direction, stratified random selections of individuals, etc.) and produce a unique result each time the model is run. Many simulation iterations yield a distribution of model outcomes rather than a single result.

In this chapter, we limit the concept of IBMs to those using an individual, agent, or similar object-oriented programming approach to simulate actions and interactions of individuals or groups of individuals (Baggio et al. 2011). We focus on single-species (or small species assemblage) models designed to develop ecological theory and support wildlife management. We distinguish between IBMs that simulate individuals from analyses and data collection relating to individuals. Measurements or statistical models developed for individuals or groups of individuals, such as statistical descriptions of resource selection functions, are not IBMs. However, they can be important in parameterizing IBMs (Heinrichs et al. 2017). Many kinds of wildlife population models simulate population dynamics, density-dependent effects, disease transmission, competition, predation, and inbreeding depression. However, IBMs differ from population-based, phenomenological models such as those constructed in RAMAS, STELLA, VenSim, SPOMSIM, Zonation, and GAPPS. Most notably, IBMs track unique individuals with identifiers or geo-referenced locations rather than tracking populations. Further, mechanistic bottom-up processes influence individual outcomes rather than top-down states.

How Do Individual-Based Models Work?

Wildlife IBMs are constructed to fit the ecological question and available data. As such, IBMs vary widely in their design, construction, implementation, and outputs. Yet IBMs for wildlife species often contain similar elements including unique individuals, decision-based behavior, demography, habitat, and interactions among these elements. Individuals can vary in their traits (age, sex, status, etc.), experiences and memories (exposure to disease, stress), genetics, and site preferences. These individual traits or conditions can alter decision-based behavior or outcomes including movement, habitat selection, breeding, and interactions with other individuals (joining or leaving groups). For example, in a spatially explicit IBM for Ord's kangaroo rats (*Diopodomys ordii*), individual traits (age and stage) and conditions (local density and quality of habitat) influenced individual choices such as age- and density-dependent movement. The quality of habitat encountered while dispersing influenced movement paths, dispersal distances, and range establishment (Heinrichs et al. 2010).

Wildlife IBMs often include some sort of representation of space or habitat and require decisions on how mobile species move through space and select a range or population. Models variably link fitness to population and habitat conditions, and include stochastic processes.

- *Spatial Representation.* IBMs vary widely in the degree to which mapped environmental conditions are represented in the model. A-spatial models allow the structuring of individuals in populations, but do not rely on habitat maps to do so. For example, individuals can be identified by age, sex, or genetic alleles, but their locations and relationships to habitat are often generalized through population membership and population-specific carrying capacities (e.g., VORTEX; Lacy and Pollak 2014). Since wildlife model outcomes often depend on location-specific movement cues and heterogeneous environmental conditions, IBMs are increasingly including some form of mapped environment. Spatially representative, but not explicit models provide some degree of environmental realism, such as mapped patch sizes, locations, and distance-dependent movement. By contrast, spatially explicit IBMs are capable of fully representing heterogeneous landscapes and individual interactions with landscapes.

- *Movement.* In spatially explicit IBMs, movements are often based on pixel-by-pixel decision rules that use mathematical expressions (distributions, random number generators, etc.) to determine an individual's movement distance, path direction, degree of path autocorrelation, ability to cross habitat gaps, avoidance, and attraction. Many wildlife models use correlated random walks to provide forward momentum (spatial autocorrelation) in path direction. Similarly, biased random walks contain a consistent bias in direction or choice (i.e., follow habitat quality along a gradient; Mclane et al. 2011). Lévy walks, bias-correlated random walks, composite-correlated random walks, area-restricted search, exploration behaviors, and other movement routines have also been used in IBMs (Reynolds 2012; Auger-Méthé et al. 2016). Some movement routines allow individuals to interact with semi- or fully

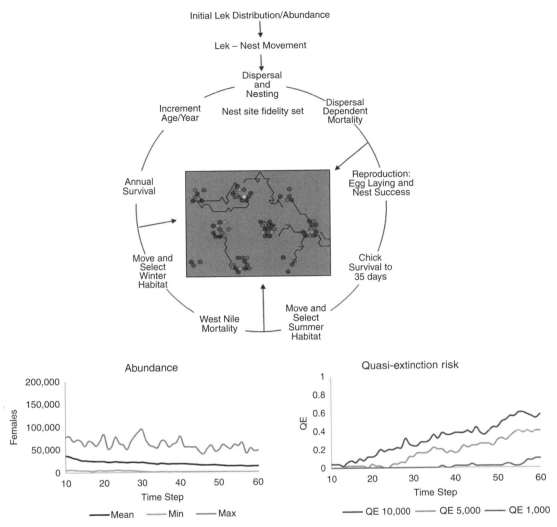

Fig. 10.1.   Example of spatially explicit individual-based model for greater sage-grouse (*Centrocercus urophasianus*) in Wyoming (Heinrichs et al. 2017), describing model events or routines, movements, and resulting abundance and quasi-extinction risk of the population falling below 10,000, 5,000, and 1,000 individuals. Simulated grouse preferred to select higher quality, but could cross nonhabitat (gray) in search of a seasonal range. Arrows indicate the direction of each individual's movement.

permeable barriers or change their movement behavior based on experiential events like encountering a habitat edge (as in HexSim; Schumaker and Brookes 2018).

- *Habitat or Population Selection.* The ability to relate landscape features with habitat selection and individual consequences allows the emergence of new individual and population states. Decisions based on an individual's perceived environment

and past experiences (location-based resources, site fidelity) can shape emergent conditions through time (Mclane et al. 2011; Fig. 10.1 and Fig 10.2). For example, limited area searches are used by simulated individuals to seek the best quality territory or range within heterogeneous habitat (Watkins and Rose 2017), with ensuing demographic consequences of the selection. Habitat selection rules can reflect realistic adaptive or

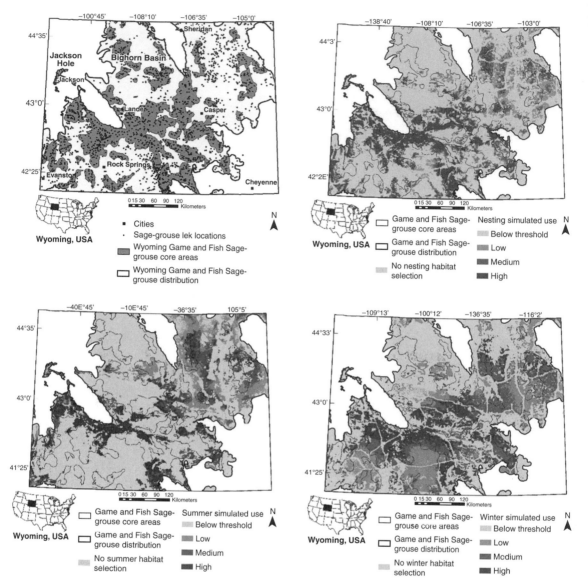

Fig. 10.2.   Emergent long-term seasonal distribution of simulated greater sage-grouse (*Centrocercus urophasianus*) in Wyoming (see Fig. 10.1) during nesting (top right), summer (bottom left), and winter (bottom right) seasons (Heinrichs et al. 2017). Individuals moved and selected habitat using resource selection function maps, and their locations were recorded for each season, through time.

maladaptive choices (Mclane et al. 2011), where data are available to link decisions with fitness. In a-spatial models, decisions on which local population to join within a regional network can be based on proximity, density, size, or other conditions.

• *Fitness.* Survival, reproduction, vulnerability to stressors or catastrophes, and the consequences of

biotic interactions are often implemented using rules, mathematical functions, or matrices of demographic variables, depending on the underlying data and IBM approach. For species in heterogeneous landscapes, characterizing differential demographic conditions among local populations can be data intensive but important in understanding metapopulation and source-sink dynamics. Biotic interactions among individuals (competition for resources, density) and species (predation, parasitism) can be directly modeled as agents or as factors that influence demographic distributions.

• *Stochastic Elements.* Individual actions and states are driven by conditional rules or probability distributions derived from empirical data. Random draws from distributions and random number seeds are often designed to change specific outcomes with each event or time step, leading to one simulation repetition being different from the next. A range of population outcomes emerges from running repetitions of the same model. In this way, simulation modeling partially reconstructs random variation in system processes and observed variance in measured conditions, such as variation in movement path lengths and survival rates.

Constructing Individual-Based Models

Individual-based models are generally constructed by defining (1) population and habitat structure, gradations of habitat quality, and movement filters or barriers; (2) starting population conditions, such as starting traits, states, and locations of individuals; (3) relationships among individuals, environments, and occupancy or fitness outcomes—e.g., age-, location- or condition-specific survivorship and fecundity; (4) important behaviors and movement characteristics; (5) alternative scenarios for comparison.

Individual-based models can be built using a variety of modeling platforms that vary in degrees of flexibility and realism. Programs for developing individual-based frameworks (Table 10.1) include HexSim, NetLogo, and Vortex (Wilensky 1999; Lacy and Pollak 2014; Schumaker et al. 2017) and vary in their flexibility and data requirements. Cellular automata models (CAMs) are a specific form of IBMs, which represent landscapes as tiled cells and movement of individuals with simple rules that guide their movement direction and response to cell contents including the presence of other individuals. IBMs

also can be developed using a range of other programming platforms and languages. In general, user-defined IBMs are more constrained by data availability than by rigid model structures or a priori assumptions. Models that do not simulate individuals are not IBMs, including population transition–based, patch occupancy, and habitat connectivity models. Such phenomenological models usually provide a deterministic analytic solution instead of a distribution of results based on dynamic simulations with emergent phenomena.

Individual-based models produce a range of outputs, tailored to meet model objectives. Common outputs include both individual-based records and population-level summaries. Extensive data logs of individual traits and experiences through time (in movements, locations, reproductive status, survival rates) can be used for model verification by tracing individuals through the simulation (Grimm et al. 1999). IBMs can also summarize individual resource or energy levels, alleles, breeding status, encounter or exposure histories, group membership, and memories of habitat through time. However, individual-based outputs are often presented as population-level summaries such as population size, density, and distribution (Heinrichs et al. 2017). Population abundance and persistence are common metrics for population viability models (Grimm 1999), wherein extinction risk is calculated by summing extinction events relative to the number of stochastic repetitions over a specified time-frame (Fig 10.1). These metrics are used to compare the possible population outcomes among alternative scenarios (Fig. 10.3). For example, Heinrichs et al. (2010) evaluated changes in extinction risk to assess the impact of removing versus restoring habitat for endangered Ord's kangaroo rats. Spatial IBMs can also summarize individual outcomes by habitat patch or other spatial unit. For example, Heinrichs et al (2015, 2018) evaluated source-sink dynamics by evaluating movement and demography through both space and time. Such spatial results can also be summarized at multiple spatial scales by resampling location-based

Table 10.1. Comparison of commonly considered software platforms for developing wildlife individual-based models (IBM)s.

IBM Platform	Description	Spatially explicit	User manual	Flexibility	Limitations	Citations
NetLogo	Programming environment to support spatially explicit IBM simulation modeling	Y	Y	Medium-high	• Limited support for read/write binary files • Limited inheritance and declaration of global variables • Requires logo scripting, can be slow	Wilensky (1999)[1]
VORTEX	Captures dynamics of age-specific cohorts, structured by population Certain aspects of the model (e.g., genetics) are individual-based	N	Y	Medium	• Designed for sexually reproducing, diploid animals • Movement is a-spatial Habitat is not spatially explicit, but represented by carrying capacities	Lacy and Pollak (2014)[2]
VORTEX Spatial	Models spatial movement	Y	N	Lower	• Limited use	Pollak (2013)[3]
HexSim	Spatially explicit IBMs are flexibly defined by the user through graphical user interface (GUI) Individuals can incorporate behavioral, movement, trait-based, genetic, and demographic processes	Y	Y	High	• Lacks 3-D functionality	Schumaker and Brooks (2018)[4]
CDMetaPop (CDPOP)	Couples genetics with spatial demography Uses both individual-based and class-based demography controls	Y	N	Medium-high	• Must supply selection coefficients for genotypes under selection • Limited detail representing density dependence	Landguth et al (2017)[5]
RANGE SHIFTER	Includes population dynamics, dispersal behavior Models plastic and evolutionary processes in a constrained way	Y	Y	Medium	• Limited linkage among demography, dispersal, climate, and location No interspecific interactions	Bocedi et al. (2014)
ALADYN	Individuals' locations are resolved in continuous space. Incorporates genetics of populations and demographic parameters at the population level	Y		Lower	• Limited to diploid organisms Set sequence of life history events	Schiffers & Travis (2014)[6]

Non-IBM	Modeling emphasis			Non-IBM example		
Population-based	Operates on discrete populations No genetics or individual traits No explicit movement			RAMAS; Akcakaya (2005)[7]		
Patch occupancy	Requires discrete populations No genetics, individual traits, or explicit movement Predicts occupancy rather than abundance			SPOMSIM; Moilanen (2004)[8]		
Habitat corridors	No explicit modeling of population processes or movement			Corridor mapping software[9]		

[1]https://ccl.northwestern.edu/netlogo/

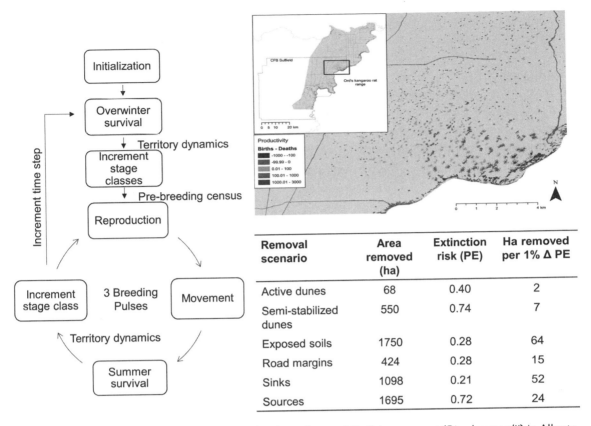

Removal scenario	Area removed (ha)	Extinction risk (PE)	Ha removed per 1% Δ PE
Active dunes	68	0.40	2
Semi-stabilized dunes	550	0.74	7
Exposed soils	1750	0.28	64
Road margins	424	0.28	15
Sinks	1098	0.21	52
Sources	1695	0.72	24

Fig. 10.3. Spatially explicit individual-based model for the endangered Ord's kangaroo rat (*Dipodomys ordii*) in Alberta, Canada, describing the annual sequence of model events that represent three reproduction pulses, followed by dispersal, territory searching, and survival (Heinrichs et al. 2010). Births and deaths were tracked through space and time to identify sources and sinks. Alternative habitat scenarios tested the removal of types of habitats as well as sources and sinks to quantify the contribution of each kind of habitat to persistence.

individual data (Schumaker et al. 2014). Spatially explicit movement and connectivity analyses can yield predicted movement paths and intensities of use, and can help to identify constraints to movement that are conditional on individual traits and dynamic population states. Others have evaluated site occupancy relative to stressors (Tuma et al. 2016), as well as many other pattern-based outcomes.

What Are Wildlife Individual-Based Models Used For?

Individual-based models have been used in developing both theory and applied predictions for focal species and communities. Wildlife IBMs have been used for a wide range of purposes including evaluating the key population and habitat factors influencing wildlife abundance, distribution, and persistence. For species at risk of decline or extinction, IBMs have been used to assess the influences of individual behavior, population structure, and genetic diversity on regional population outcomes. In addition, spatially explicit IBMs have been used to explore the influences of habitat change on individual and population outcomes (habitat loss, restoration), to test the sufficiency of habitat for population recovery, to identify dispersal barriers and movement corridors, and to prioritize restoration sites. IBMs have also been used to assess the effects of natural disturbance (i.e., vegetation succession, fire, disease), anthropogenic stressors (e.g., land-use and climate change), changes in intra- and interspecific competition, reserve design (Marcot et al. 2013), and management practices (translocations, food supplementation, etc.) on collections of individuals.

IBMs can be flexible tools for exploring different possibilities. Alternative scenarios compare the influences of different possible actions, conditions, data inputs, and model structures on IBM results. These comparisons increase the understanding of ecological systems by quantifying and visualizing the direct, interactive, and cumulative effects of alternative influences on population outcomes. Alternative models can be as simple as a baseline versus an altered condition. In habitat-based scenarios, landscape maps can be manipulated in IBMs to explore the implications of different amounts, qualities, and configurations of habitat for species recovery or stability. For example, researchers have used IBMs to explore the potential for recovery of at-risk wildlife species, such as for black bears (*Ursus americanus*) in Louisiana (Morzillo et al. 2011), the reintroduction of sea otters (*Enhydra lutris*) in British Columbia, Canada (Espinosa-Romero et al. 2011), and the need for habitat and population connectivity for reintroduced Eurasian lynx (*Lynx lynx*) in Germany (Kramer-Schadt et al. 2004). Although our focus is on non-fish vertebrates, IBMs have also been used successfully with a wide variety of other taxa including fish, invertebrates, and plants (e.g., Brown et al. 2017; Halsey et al. 2017; Zhang et al. 2017).

Strengths and Challenges of Wildlife Individual-Based Models

Individual-based models (IBMs) present several distinct advantages and unique challenges over top-down phenomenological models such as traditional state-based mathematical modeling or statistical analyses.

1. *Individual-based models are intuitively constructed.* Model construction emphasizes an animal's life cycle and behavior, making IBMs easy to conceptualize. Because individuals are the basic unit, IBM construction also focuses on the behavior of individuals, which can be easier to understand than the collective behavior of a population. IBMs use explicitly defined event sequences describing how individuals behave, and such event sequences clearly and intuitively display the operations conducted in the simulations. IBMs also provide insight into the state of individuals, adding another dimension by which the ecological systems can be assessed and understood (Cartwright et al. 2016). As such, they provide a means of investigating questions that are difficult or impossible to address with classical state-variables approaches (Grimm and Railsback 2005).

2. *Ecological realism is possible.* The flexibly to represent complex ecological systems has driven the recent rise of wildlife IBMs. IBMs can embrace ecological complexity to the degree to which there is supporting data and reason. They often include greater realism in representing life cycles than top-down models (Grimm and Railsback 2005) and include individual differences beyond age or stage class. Further, spatially explicit IBMs can represent any kind of spatial detail that is important (Le Page et al. 2013). Mapped environments can be used to represent habitat heterogeneity and variability, animal locations, and differential outcomes (location-based fitness), making IBMs more like real systems (Cartwright et al. 2016).

3. *Individual-based models relax simplistic population assumptions.* IBMs are based on tangible mechanisms, processes, and realistic assumptions rather than on the more abstract and less realistic assumptions of classical population models (Cartwright et al. 2016). IBMs allow the relaxation of overly simplistic and empirically implausible assumptions (Le Page et al. 2013) by representing unique individuals (Grimm et al. 1999) and their responses to ecological complexity. Theoretical wildlife analyses and classical models generally assume population stability or equilibrium (Lamberson 2012), limiting inferences for declining or variable populations. Further,

unique traits and states of individuals are often generalized. For example, traditional population models using mathematical formulas or statistics often assume that individuals of a given age are the same, equally mixed, or uniformly spatially distributed (Grimm et al. 1999; Le Page et al. 2013). Simplistic models based on overly general assumptions can reduce the usefulness of model outputs for informing conservation and management.

4. *Individual-based models can relax spatial assumptions.* Despite widespread recognition of the importance of spatial structure in habitat and population dynamics, many wildlife analyses generalize landscape complexity (Lamberson 2012). Many, but not all, metapopulation and statistical models assume equal accessibility of habitat patches, homogeneous patch quality (Lamberson 2012), and static environmental conditions. However, these factors are not equal across space or time, and heterogeneity can dramatically shape local and regional population outcomes. For example, differential equations and IBMs of the same system can produce different results when ecological complexity is explicitly considered (Rahmandad and Sterman 2008). Hence, such assumptions can limit the usefulness of models when these conditions are not met in heterogeneous environments. Spatially explicit IBMs link individual conditions or traits with population and environmental conditions to trigger spatially dependent processes, such as movement, in heterogeneous landscapes. In doing so, spatially explicit IBMs can include relevant ecological processes that are important for understanding the system. Ultimately, however, a balance must be struck between representing ecological complexity and model tractability. Both overly simplistic and overly complicated models are difficult to use as decision support tools and can limit ecological insights (Cartwright et al. 2016).

5. *Individual-based models produce emergent insights.* Empirical data are used to develop starting conditions and relationships among individuals and the environment, but model outcomes are an emergent property of all factors and conditions. The dynamic interaction of individuals and their environment alter emergent patterns through time (Lamberson 2012). Nonequilibrium system conditions (nonstable age or spatial distributions, directional habitat change) influence individual decisions and states such as breeding age and density-dependent emigration. Cascading through the system, individual-level patterns affect multi-scale population patterns including regional population outcomes such as abundance and distribution. Emergent outcomes are also influenced by the degree to which an animal grows, learns, perceives, and seeks fitness and is exposed to and can adjust to changing conditions (Grimm and Railsback 2005). The nonlinear interactions of these factors yield spatially and temporally varying ecological contexts. Predicting the outcomes of emergent phenomena is difficult without both ecological realism and a bottom-up approach.

6. *Individual-based models yield insightful predictions.* Mechanistic models often out-perform phenomenological models in making predictions beyond the conditions in which data were collected (Stillman et al. 2015). Environmental change can affect populations in ways that are not predictable from past measurements. However, many of the decision rules and processes that form the basis of IBMs are not expected to change (Wood et al. 2015), which supports predictions about population behavior in novel conditions. For example, many IBMs are designed to allow individuals to seek fitness within constraints of their perceptual ranges. In novel conditions, fitness-seeking behavior is more likely to retain its predictive strength than static relationships that are contingent on a measured range of conditions (Lamberson 2012). Similarly, when species consistently display maladaptive behaviors and seek ecological traps, IBMs can produce robust predictions outside of the range of environmental conditions in which the model was parameterized (Grimm and Railsback 2005, Wood et al. 2015). In this way, IBM projections can support evaluations of the implications of management decisions in future environments (DeAngelis and Mooij M. 2005; Mclane

et al. 2011). Nonetheless, the reliability of predicting future conditions can still depend on the constancy of existing behavior in novel conditions, such as the same fitness-seeking behaviors, as well as the accuracy of projected future conditions (Norris 2004). In any case, future scenarios are best viewed as relative comparisons or what-if scenarios rather than absolute predictions.

7. *Individual-based models uniquely integrate information.* All models assemble data and ideas; however, IBMs are ambitious in their scale of data integration and unique in their flexibility. For well-studied species, IBMs often include extensive information on the physical environment, life history, behaviors, demography, and interacting factors. Realistic IBMs often assemble and integrate previously disconnected pieces of information, conceptually advancing the understanding of ecological systems (Schumaker and Brookes 2018). For example, northern spotted owl (*Strix occidentalis caurina*) habitat and demographic data collected at the scale of regional demographics units were unified in a single IBM, allowing the prediction of species-level movement and source-sink dynamics (Schumaker et al. 2014). This kind of system-level knowledge surpasses the understanding of any one expert or data set and can be clearly communicated (Parrott et al. 2012). However, it can be time-consuming to locate, assess, and integrate large amounts or types of information for realistic IBMs. It can also be challenging to translate empirical data collected for other purposes into individual-based conditions and relationships.

For less-studied species, the exercise of evaluating and integrating existing data provides an opportunity to identify and prioritize key information gaps in responding behaviors or relationships among key variables. Simple and informative IBMs can be constructed for species with minimal data and projects with appropriate objectives. Yet for some questions, generalizations of environmental or individual conditions and responses can lead to overly simplistic models with little ecological insight. Individual-based

modelers strive to develop parsimonious IBMs with sufficient but minimal required detail.

8. *Individual-based models require clear communication.* As with other kinds of modeling, the clear communication of model objectives, constraints, and appropriate interpretation is an important but often under-resourced task. Although the construction of wildlife IBMs is often biologically intuitive, the computer code or instructions required to execute the conceptual model can be complex. Difficulties in communicating overly complex or poorly documented models (Grimm et al. 1999) have prompted simulation modelers to more extensively document and communicate their IBMs using standardized protocols. Such standardized communication protocols are increasingly used for developing, testing, and documenting IBMs and include POM (pattern-oriented modeling; Grimm and Railsback 2012), the ODD protocol (Overview, Design, concepts and Details; Grimm et al. 2010), and TRACE (transparent and comprehensive ecological modeling) documentation (Grimm et al. 2014). Individual-based modelers are also emphasizing the importance of effectively communicating simulation models to stakeholders, as many quantitative ecologists and practitioners are unfamiliar with simulation models have little experience evaluating IBM inputs, outcomes and results.

Evaluating Individual-Based Models

Like most models, IBMs can be developed and used for a variety of objectives, including prediction, forecasting, projection, scenario planning, diagnosis, and mitigation (Marcot 2017). Wildlife IBMs developed to represent real-world situations use strong links to empirical data and can incorporate expert knowledge as needed. Realistic, but not exact, representations of the phenomena can be achieved by evaluating model inputs and outputs with multiple data sets and criteria during model development, calibration, and validation (Parrott et al. 2012; MacPherson and Gras 2016). A variety of data sources and statistical methods can be used to develop models, choose and calibrate parameter values (e.g., goodness-of-fit tests of movement; Frost et al. 2009), and verify that the model is performing as intended (e.g., past occurrence locations; Guillaumet et al. 2017). One approach to calibrating model parameters is to compare hindcasted model predictions to empirical time series and assess the degree of fit. Model input parameters are then adjusted to achieve the best fit (Lagarrigues et al. 2015).

Validation can use multiple approaches to demonstrate that the model is a reasonable representation of the system and addresses the model objectives. Validation is generally conducted by assessing model assumptions, input parameter values, and outputs against expert opinion and empirical data and by evaluating alignment with ecological theory or expectations. Sensitivity analysis is a common approach to evaluating the influence of model assumptions, structure, and input parameter values on model outcomes. Alternative model structures can be compared or input parameter values can be incrementally increased or decreased, individually or in some logical tandem (Marcot et al. 2015). Values are varied to test how much change in the value of a parameter is required before a meaningful change in the outcome. This exploration helps inform how precise (or imprecise) and accurate (or inaccurate) parameter values need to be to get reasonable and useful results. For wildlife IBMs, changes in demographic or landscape conditions, movement, perceptual ranges, and search algorithms (Cooper et al. 2002; Pe'er and Kramer-Schadt 2008) are commonly evaluated to determine their influence on population outcomes such as abundance or persistence. Marcot et al. (2015) outlined a general methodology and conducted a broad sensitivity analysis of an IBM of northern spotted owl population responses to habitat quality, movement and dispersal, and model architecture. In a two-species IBM, Schumaker et al (2014) found that source-sink dynamics were sensitive to habitat connectivity and awareness of habitat availability and quality. In model validation it is important to remember that no one model represents true reality in all its detail, and that congruence with all empirical data is not generally a realistic expectation. In some cases, a multi-model approach may be useful (Brook et al. 1999).

Novel Insights and Future Trends

Important ecological insights have been gained through individual-based modeling including the interdependency of ecological conditions and influences of landscape composition and structure on population outcomes. Decades ago, spatial IBMs were developed to assess the influences of habitat amount and fragmentation on habitat connectivity and population persistence (Fahrig 1997; With and King 1999). Spatial IBMs continue to highlight the importance of resource amount, quality, and structure in evaluating population outcomes (Wiegand et al. 2005; Fordham et al. 2014; Heinrichs et al. 2016). IBMs have also drawn attention to the ecological importance of floaters, or individuals without territories or fixed ranges. These animals can be highly mobile and are often undetected or excluded from many traditional population analyses, yet their influences on density-dependent population regulation can be important to understand. For example, Marcot et al. (2013) modeled the hypothetical responses of northern spotted owl populations to alternative landscape structures and found that floaters played a key role in habitat recolonization but were disproportionately reduced as abundance declined, accelerating regional population decline. Empirical studies of other owl species have confirmed the importance of floaters on population dynamics (Rohner 1997; Aebischer et al. 2010). Further, the importance of mate-finding in small populations, landscape heterogeneity, and movement constraints (Stronen et al. 2012), as well as the importance of site selection for reintroduction and release (Huber et al. 2014), have all been demonstrated with spatial IBMs. Although IBMs can require more effort, time, and different skills than traditional population modeling, they provide the advantage of producing models that are more reusable than models based purely on empirical parameters and static equations. Once constructed and evaluated, IBMs can be used to efficiently test a wide range of specific and general questions (Stillman et al. 2015).

Wildlife IBMs are increasingly including spatial information in the form of GIS-based mapping products (e.g., habitat or distribution maps; Heckbert et al. 2010). Realistic environments are represented with Cartesian coordinate systems with spatial projections and data maintained within the model (Mclane et al. 2011). Remotely sensed data and derived mapping products or other spatial data are imported into the IBM environment. With these maps, resources can vary spatially, and the IBM can maintain the desired level of landscape complexity. Advances in spatial ecology with improved remote sensing and habitat maps, better tracking of individuals, and better knowledge of movement ecology are all paving the way for larger-scale IBM experiments (Lamberson 2012). Concurrently, advances in computing power and resources and IBM software are facilitating the construction of wildlife IBMs to track larger numbers of animals across larger geographic extents. Freely available software with graphical user interfaces (GUI) are also making IBM modeling accessible to those without prior programming skills. New approaches to acquiring large amounts of high resolution data (i.e., spatial, individual behavior) could further advance the development of data-rich IBMs (e.g., citizen science observations; Wood et al. 2015) in the future.

Many IBMs are able to represent dynamic environmental or interspecific conditions (Grimm and Railsback 2005) and are coupling temporally varying landscapes with dynamic population processes. Although this dynamic coupling increases model complexity, it can be fundamental to adequately representing natural systems. Pattern-based models that provide future projections in the context of dynamic landscape-population coupling may be uniquely positioned to provide insight into how conservation and management strategies may fare in nonstationary environments (Synes et al. 2016).

As IBMs are increasingly representing human influences on wildlife, diverse teams are being constructed to access and evaluate data in support of species management (Le Page et al. 2013). Multidisciplinary teams of habitat, behavior, population, and management experts are being assembled to assess the ecological implications of management (in)actions. There is also a growing need to integrate social, political, and economic factors and expertise into ecological assessments to support viable management decisions (Synes et al. 2016). Like many other decision support approaches, there is increasing recognition of the benefits of co-producing IBMs for ensuring that the IBMs will directly address the problem and increase the chance that IBMs will be used in decision-making (Wood et al. 2015). The process of creating an IBM can create a reason or path for engaging stakeholders by involving them in acquiring data, evaluating assumptions and outputs, and developing scenarios (Parrott et al. 2012).

Individual-based models are increasing in popularity and have become a well-established and accepted method for evaluating wildlife systems and questions (Mclane et al. 2011). They have become more sophisticated in the last decade and have become an important tool for studying population dynamics in heterogeneous landscapes (Dunning et al. 1995). Reviews of IBMs highlight improving methods, the refinement of tools and applications, communicable models, and an experienced research community (Heckbert et al. 2010). Many analytical approaches are needed to make sense of ecological systems, and IBMs can be particularly insightful and necessary tools (DeAngelis and Grimm 2014) for understanding wildlife in complex environments and supporting important conservation and management decisions.

Acknowledgments

The authors appreciate discussions and collaborations on IBM-related modeling projects with numerous colleagues, including C. Aldridge, J. Lawler, M. O'Donnell, M. Raphael, N. Schumaker, and P. Singleton. The authors thank the Natural Resource Ecology Laboratory at Colorado State University, and The Computational Ecology Group and acknowledge support from the Pacific Northwest Research Station of U.S. Forest Service. Mention of commercial products does not imply endorsement by the U.S. Government.

LITERATURE CITED

Aebischer, A., P. Nyffeler, and R. Arlettaz. 2010. Wide-range dispersal in juvenile Eagle Owls (Bubo bubo) across the European Alps calls for transnational conservation programmes. Journal of Ornithology 151:1–9.

Akcakaya, R., and W. Root. 2005. RAMAS GIS: Viability analysis for stage-structured metapopulations (Version 5). Applied Biomathematics, Setauket NY, USA.

Auger-Méthé, M., A. E. Derocher, C. A. DeMars, M. J. Plank, E. A. Codling, and M. A. Lewis. 2016. Evaluating random search strategies in three mammals from distinct feeding guilds. Journal of Animal Ecology 85:1411–1421.

Baggio, J. A., K. Salau, M. A. Janssen, M. L. Schoon, and Ö. Bodin. 2011. Landscape connectivity and predator-prey population dynamics. Landscape Ecology 26:33–45.

Brook, B. W., J. R. Cannon, R. C. Lacy, C. Mirande, and R. Frankham. 1999. Comparison of the population viability analysis packages GAPPS, INMAT, RAMAS and VORTEX for the whooping crane (Grus americana). Animal Conservation 2:23–31.

Bocedi, G., S. C. F. Palmer, G. Pe'er, R. K. Heikkinen, Y. G. Matsinos, K. Watts, and J. M. J. Travis. 2014. Range-Shifter: a platform for modelling spatial eco-evolutionary dynamics and species' responses to environmental changes. Methods in Ecology and Evolution 5(4): 388–396.

Brown, L. M., R. K. Fuda, N. Schtickzelle, H. Coffman, A. Jost, A. Kazberouk, E. Kemper, E. Sass, and E. E. Crone. 2017. Using animal movement behavior to categorize land cover and predict consequences for connectivity and patch residence times. Landscape Ecology 32:1657–1670.

Cartwright, S. J., K. M. Bowgen, C. Collop, K. Hyder, J. Nabe-Nielsen, R. Stafford, R. A. Stillman, R. B. Thorpe, and R. M. Sibly. 2016. Communicating complex ecological models to non-scientist end users. Ecological Modelling 338:51–59.

Cooper, C. B., J. R. Walters, and J. Priddy. 2002. Landscape patterns and dispersal success: Simulated population dynamics in the brown tree creeper. Ecological Applications 12:1576–1587.

Deangelis, D. L., and V. Grimm. 2014. Individual-based models in ecology after four decades. F1000 Prime Reports 6:39.

DeAngelis, D. L., and J. L. Gross. 1992. Individual-based models and approaches in ecology: populations, communities and ecosystems. Routledge, Chapman and Hall, New York.

DeAngelis, D. L., and W. Mooij M. 2005. Individual-based modeling of ecological and evolutionary processes. Annual Review of Ecology, Evolution, and Systematics 36:147–168.

Dunning, J. B., J., D. J. Stewart, B. J. Danielson, B. R. Noon, T. L. Root, R. H. Lamberson, and E. E. Stevens. 1995. Spatially explicit population models: Current forms and future uses. Ecological Applications 5:3–11.

Espinosa-Romero, M. J., E. J. Gregr, C. Walters, V. Christensen, and K. M. A. Chan. 2011. Representing mediating effects and species reintroductions in Ecopath with Ecosim. Ecological Modelling 222:1569–1579.

Fahrig, L. 1997. Relative effects of habitat loss and fragmentation on population extinction. Journal of Wildlife Management 61:603–610.

Fordham, D. A., K. T. Shoemaker, N. H. Schumaker, H. R. Akcakaya, N. Clisby, and B. W. Brook. 2014. How interactions between animal movement and landscape processes modify local range dynamics and extinction risk. Biology Letters 10(5):20140198.

Frost, C. J., S. E. Hygnstrom, A. J. Tyre, K. M. Eskridge, D. M. Baasch, J. R. Boner, G. M. Clements, J. M. Gilsdorf, T. C. Kinsell, and K. C. Vercauteren. 2009. Probabilistic movement model with emigration simulates movements of deer in Nebraska, 1990–2006. Ecological Modelling 220:2481–2490.

Grimm, V. 1999. Ten years of individual-based modelling in ecology: What have we learned and what could we learn in the future? Ecological Modelling 115:129–148.

Grimm, V., J. Augusiak, A. Focks, B. M. Frank, F. Gabsi, A. S. A. Johnston, C. Liu, B. T. Martin, M. Meli, V. Radchuk, P. Thorbek, and S. F. Railsback. 2014. Towards better modelling and decision support: Documenting model development, testing, and analysis using TRACE. Ecological Modelling 280:129–139.

Grimm, V., U. Berger, D. L. DeAngelis, J. G. Polhill, J. Giske, and S. F. Railsback. 2010. The ODD protocol: A review and first update. Ecological Modelling 221:2760–2768.

Grimm, V., and S. F. Railsback. 2005. Individual-based modeling and ecology. Princeton University Press, Princeton, NJ.

Grimm, V., and S. F. Railsback. 2012. Pattern-oriented modelling: A "multi-scope" for predictive systems ecology. Philosophical Transactions of the Royal Society B: Biological Sciences 367:298–310.

Grimm, V., T. Wyszomirski, D. Aikman, and J. Uchmanski. 1999. Individual-based modelling and ecological theory: Synthesis of a workshop. Ecological modelling 115:275–282.

Guillaumet, A., W. A. Kuntz, M. D. Samuel, and E. H. Paxton. 2017. Altitudinal migration and the future of an iconic Hawaiian honeycreeper in response to climate change and management. Ecological Monographs 87:410–428.

Halsey, S. J., S. Cinel, J. Wilson, T. J. Bell, and M. Bowles. 2017. Predicting population viability of a monocarpic perennial dune thistle using individual-based models. Ecological Modelling 359:363–371.

Heckbert, S., T. Baynes, and A. Reeson. 2010. Agent-based modeling in ecological economics. Annals of the New York Academy of Sciences 1185:39–53.

Heinrichs, J. A., C. L. Aldridge, D. L. Gummer, A. P. Monroe, and N. H. Schumaker. 2018. Prioritizing actions for the recovery of endangered species: Emergent insights from Greater Sage-grouse simulation modeling. Biological Conservation 218:134–143.

Heinrichs, J. A., C. L. Aldridge, M. S. O'Donnell, and N. H. Schumaker. 2017. Using dynamic population simulations to extend resource selection analyses and prioritize habitats for conservation. Ecological Modelling 359:449–459.

Heinrichs, J. A., D. J. Bender, D. L. Gummer, and N. H. Schumaker. 2010. Assessing critical habitat: Evaluating the relative contribution of habitats to population persistence. Biological Conservation 143:2229–2237.

Heinrichs, J. A., D. J. Bender, and N. H. Schumaker. 2016. Habitat degradation and loss as key drivers of regional population extinction. Ecological Modelling 335:64–73.

Heinrichs, J. A., J. J. Lawler, N. H. Schumaker, C. B. Wilsey, and D. J. Bender. 2015. Divergence in sink contributions to population persistence. Conservation Biology 29:1674–1683.

Huber, P. R., S. E. Greco, N. H. Schumaker, and J. Hobbs. 2014. A priori assessment of reintroduction strategies for a native ungulate: Using HexSim to guide release site selection. Landscape Ecology 29:689–701.

Judson, O. P. 1994. The rise of the individual-based model in ecology. Trends in Ecology and Evolution 9:9–14.

Kramer-Schadt, S., E. Revilla, T. Wiegand, and U. Breitenmoser. 2004. Fragmented landscapes, road mortality and patch connectivity: Modelling influences on the dispersal of Eurasian lynx. Journal of Applied Ecology 41:711–723.

Lacy, R. C., and J. P. Pollak. 2014. Vortex: A stochastic simulation of the extinction process. Chicago Zoological Society, Brookfield, IL.

Lagarrigues, G., F. Jabot, V. Lafond, and B. Courbaud. 2015. Approximate Bayesian computation to recalibrate individual-based models with population data: Illustra-

tion with a forest simulation model. Ecological Modelling 306:278–286.

Lamberson, R. H. 2012. A brief and biased look at spatial structure in ecological models: A route to individual-based models. Natural Resource Modeling 25:145–167.

Landguth, Erin L., A. Bearlin, C. C. Day, and J. Dunham. CDMetaPOP: an individual-based, eco-evoluationary model for spatially explicit simulation of landscape demogenetics. Methods in Ecology and Evolution 8(1):4–11.

Le Page, C., D. Bazile, N. Becu, P. Bommel, F. Bousquet, M. Etienne, R. Mathevet, V. Souchère, G. Trébuil, and J. Weber. 2013. Agent-based modelling and simulation applied to environmental management. Simulating social complexity. Springer-Verlag, Berlin Heidelberg. 499–540.

MacPherson, B., and R. Gras. 2016. Individual-based ecological models: Adjunctive tools or experimental systems? Ecological Modelling 323:106–114.

Marcot, B. G. 2017. Common quandaries and their practical solutions in Bayesian network modeling. Ecological Modelling 358:1–9.

Marcot, B. G., M. G. Raphael, N. H. Schumaker, and B. Galleher. 2013. How big and how close? Habitat patch size and spacing to conserve a threatened species. Natural Resource Modeling 26:194–214.

Marcot, B. G., P. H. Singleton and N. H. Schumaker. 2015. Analysis of sensitivity and uncertainty in an individual-based model of a threatened wildlife species. Natural Resource Modeling 28:37–58.

Mclane, A. J., C. Semeniuk, G. J. Mcdermid, and D. J. Marceau. 2011. The role of agent-based models in wildlife ecology and management. Ecological Modelling 222:1544–1556.

Moilanen, A. 2004. SPOMSIM: software for stochastic patch occupancy models of metapopulation dynamics. Ecological Modelling 179:533–550.

Morzillo, A. T., J. R. Ferrari, and J. Liu. 2011. An integration of habitat evaluation, individual based modeling, and graph theory for a potential black bear population recovery in southeastern Texas, USA. Landscape Ecology 26:69–81.

Norris, K. E. N. 2004. Managing threatened species: The ecological toolbox, evolutionary theory and declining-population paradigm. Journal of Applied Ecology 41:413–426.

Parrott, L., C. Chion, R. Gonzalès, and G. Latombe. 2012. Agents, individuals, and networks: Modeling methods to inform natural resource management in regional landscapes. Ecology and Society 17:32.

Pe'er, G., and S. Kramer-Schadt. 2008. Incorporating the perceptual range of animals into connectivity models. Ecological Modelling 213:73 85.

Pollak, J. P. 2013. Spatial model of animal movement on landscapes. JP Pollak, New York, NY.

Rahmandad, H., and J. Sterman. 2008. Heterogeneity and network structure in the dynamics of diffusion: Comparing agent-based and differential equation models. Management Science 54:998–1014.

Reynolds, A. M. 2012. Distinguishing between Lévy walks and strong alternative models. Ecology 93:1228–1233.

Rohner, C. 1997. Non-territorial "floaters" in great horned owls: Space use during a cyclic peak of snowshoe hares. Animal Behaviour 53:901–912.

Schumaker, N. H., and A. Brookes. 2018. HexSim: A modeling environment for ecology and conservation. Landscape Ecology 33(2):197–211.

Schumaker, N. H., A. Brookes, J. R. Dunk, B. Woodbridge, J. A. Heinrichs, J. J. Lawler, C. Carroll, and D. LaPlante. 2014. Mapping sources, sinks, and connectivity using a simulation model of northern spotted owls. Landscape Ecology 29:579–592.

Schumaker, N. H., A. Brooks, K. Djang, and M. Armour. 2017. HexSim, Corvallis, OR. Available from http://www.hexsim.net.

Stillman, R. A., S. F. Railsback, J. Giske, U. T. A. Berger, and V. Grimm. 2015. Making predictions in a changing world: The benefits of individual-based ecology. BioScience 65:140–150.

Stronen, A. V, N. H. Schumaker, G. J. Forbes, P. C. Paquet, and R. K. Brook. 2012. Landscape resistance to dispersal: Simulating long-term effects of human disturbance on a small and isolated wolf population in southwestern Manitoba, Canada. Ecological Monitoring and Assessment 184:6923–6934.

Synes, N. W., C. Brown, K. Watts, S. M. White, M. A. Gilbert, and J. M. J. Travis. 2016. Emerging opportunities for landscape ecological modelling. Current Landscape Ecology Reports 1:146–167.

Thierry, H., D. Sheeren, N. Marilleau, N. Corson, M. Amalric, and C. Monteil. 2015. From the Lotka-Volterra model to a spatialised population-driven individual-based model. Ecological Modelling 306:287–293.

Tuma, M. W., C. Millington, N. Schumaker, and P. Burnett. 2016. Modeling Agassiz's desert tortoise population response to anthropogenic stressors. Journal of Wildlife Management 80:414–429.

Watkins, K. S., and K. A. Rose. 2017. Simulating individual-based movement in dynamic environments. Ecological Modelling 356:59–72.

Wiegand, T., E. Revilla, and K. A. Moloney. 2005. Effects of habitat loss and fragmentation on population dynamics. Conservation Biology 19:108–121.

With, K. A., and A. W. King. 1999. Dispersal success on fractal landscapes: A consequence of lacunarity thresholds. Landscape Ecology 14:73–82.

Wood, K. A., R. A. Stillman, and J. D. Goss-Custard. 2015. Co-creation of individual-based models by practitioners and modellers to inform environmental decision-making. Journal of Applied Ecology 52:810–815.

Zhang, B., D. L. DeAngelis, M. B. Rayamajhi, and D. Botkin. 2017. Modeling the long-term effects of introduced herbivores on the spread of an invasive tree. Landscape Ecology 32:1147–1161.

11 — Detecting and Analyzing Density Dependence

Zachary S. Ladin and
Christopher K. Williams

> The phenomenon of density dependence lurks quietly but crucially in the chaos
> of quail demography.
> —Guthery (2002:54).

Introduction

For millennia, long before ecologists' fascination with the dynamics of density dependence, humanity has observed, acknowledged, and investigated phenomena related to the density-dependent behavior of populations. After the advent of plant (agricultural) and animal domestication roughly 10,000–20,000 years ago, and perhaps much earlier than that, our motivation to pay attention to density dependence was an important aspect of our survival. For example, our awareness of and yearning to understand the causes of diseases within our own human communities (local outbreaks) and populations (epidemics) inevitably led to health-improving behaviors. Additionally, these self-preserving behaviors also fostered our interest in population dynamics related to the natural resources we depended on for our survival (e.g., food, fuel, shelter, tools, medicine, water). Unsurprisingly, early biblical writings contain numerous accounts of pestilence and disease outbreaks. The writings of ancient Greek philosophers including Aristotle, Hippocrates, and Varro depicted their understanding of density dependence as a natural factor that could affect population dynamics (and cause disease), and in one example suggested optimal herd sizes for shepherds in light of connections between herd density and disease susceptibility and spread (Sallares 1991).

In more recent times during the eighteenth century European colonial period, effects of increasing population growth and overcrowding of humans in Europe became apparent. This led to speculation about how human (Franklin and Jackson 1760; Malthus 1798; Edmonds 1832) and natural populations (Darwin 1861) are inextricably linked with population density and limited resources. During the early twentieth century, the mathematical formalization of the theoretical underpinnings of density dependence and population behavior (e.g., cycling) had begun to take shape. Elton (1924) made notable observations of periodic population cycles in the snowshoe hare (*Lepus americanus*), while Grinnell (1917) published his seminal work on niche theory and competitive exclusion, as well as mathematical descriptions of population oscillations (Lotka 1925; Volterra 1928). The resulting work from studying predator-prey dynamics, and its empirical demonstration by Gause (1935) laid the foundation for understanding how density-dependent effects were related to population regulation. Around the same time, researchers at the School of Hygiene and Public Health at Johns Hopkins University also demonstrated the phenomenon (Pearl and Parker 1922). Through the serendipitous experimentation and model fitting of data in their efforts to raise fruit flies (*Drosophila melanogaster*) more efficiently for research, Pearl and Parker (1922) derived the form for the equation:

$$\log y = a - bx - c \log x, \qquad (11.1)$$

where y is the birth rate (in this case, fruit fly offspring produced per female per day) and x is the population density. They recognized how this model was the inverse of Farr's Law, which was originally an equation developed through Edmonds' (1832) epidemiological research:

$$\log D = \log a - k \log d, \qquad (11.2)$$

where D is the mortality rate, d is the population density, and both a and k are constants. These equations describe how density-dependent effects can be positive (in relation to mortality rates) and negative with respect to birth rates, and imply a third idealized state of equilibrium (Fig. 11.1). When these processes occur simultaneously, they can give rise to highly predictable (after accounting for components of variance resulting from environmental stochasticity and sources of measurement error), law-like emergence of density dependence within population dynamics (Turchin 2001). Within the field of population ecology, we have gained much information from studies that at first uncovered novel cyclical oscillations through long-term observational studies of wild populations (Elton and Nicholson 1942; Errington 1945; Keith 1963; Royama 1977; Boyce and Miller 1985). Subsequently, researchers sought to understand the underlying processes and density-dependent factors driving demographic patterns (Bulmer 1974; Krebs et al. 1995, 2001; Webster 2004).

What Is Density Dependence?

Density dependence is the generalized way that population demographic parameters (such as birth and death rates) are a function of population density. All populations are regulated, in part, by a requisite component of density-dependent mechanisms. Population vital rates are also influenced by density-independent factors (those that are not a function of population size). Hence, the combination and interplay of these two component factor types are what ultimately drive observable population growth rates and longer-term population dynamics (see Fig. 11.1 for conceptual model). Murdoch and Walde (1989)

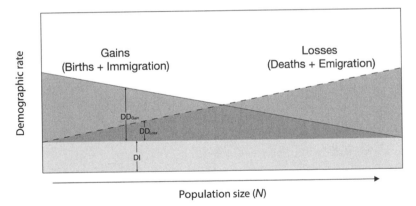

Fig. 11.1. Classical diagram showing the functional relationship among population demographic rates, depicted as either population gains (births + immigration; solid line) or losses (deaths + emigration; dashed line), and population size (density). Note that for any given combination of demographic rate and population size, the summation of both density dependent (DD) and density independent (DI; light gray shaded region below y-intercepts) contributing factors gives rise to resultant demographic rates.

define density dependence more specifically as the dependence of the per capita growth rate on past and/or present population densities. For those interested in nuances associated with how a practitioner could potentially interpret density-dependent (or independent) factors, Turchin (1999) and Berryman et al. (2002) discuss concepts from dynamic systems theory and apply a more generalized idea of negative feedback (either biotic or abiotic) to aid in the broader definition of population regulation. Berryman et al. (2002: 603) state: "1. Negative feedback is the necessary condition for the regulation of a dynamic variable, 2. Negative feedback occurs whenever the rate of change of a dynamic variable is inversely proportional to its current or past states, and 3. Biological systems are subject to the same rules as any other dynamic system."

For continuity in this chapter we will continue to use the term "density dependence," acknowledging our support of the less ambiguous manner of defining population regulation. By way of its functional relationship, density-dependent effects can influence any one of the fundamental components of population demographic rates, including births, immigration, deaths, and emigration. Moreover, it is this phenomenon that sets the lower and upper bounds within which the tendency of populations to oscillate around and return to some equilibrium point (i.e., regulation) can be observed. When density dependence no longer persists as demonstrated by models of species coexistence (e.g., competitive exclusion, predator-prey), component populations are predicted to become extinct (Gause 1935; Chesson 1982; Lampert and Hastings 2016).

It is not our goal within this chapter to recount the details about the controversy surrounding whether general laws exist or not within ecology (Berryman et al. 2002). Rather, we are interested in drawing on ideas and concepts that have an important role in the applications for describing, modeling, and understanding mechanisms that can give rise to observed dynamics of populations. For example, we find much utility within Turchin's (2001, 2003) tenable advocacy of how fundamental laws do reasonably exist in the field of population ecology, where he proposes that it makes sense for ecologists to build upon first principles in a logical and stepwise manner. The first foundational principle is a law of exponential growth (Malthus 1798), whereby all populations will increase (or decrease) exponentially in all situations where the environmental conditions (including all abiotic and biotic factors) are constant for each individual within the population. Turchin (2003) uses two fundamental postulates to derive the exponential law: *postulate 1*, which states that population size is a function of births, immigration, deaths, or emigration, and *postulate 2*, which describes methodological individualism, whereby all population processes (e.g., births, deaths) that can influence changes in the population size must act on individuals. A simple closed population (lacking immigration or emigration) can then be described with the ordinary differential equations:

$$\frac{dN}{dt} = B - D = bN - dN = (b-d)N = r_0 N, \quad (11.3)$$

where the per capita rate of change within a population (r_0) follows from the change in population size (N) over time (t). This is equivalent to the number of births (B) minus the number of deaths (D), which is also equivalent to relationships of birth rate (b) and death rate (d) to N (Turchin 2003).

Another foundational principle Turchin (2003) terms *postulate 3* is the concept of population self-limitation by which populations are inherently bounded by upper and lower thresholds. It can be expressed via the partial derivative of the population growth rate:

$$\frac{\partial r(t)}{\partial N} < 0 \text{ for } N > \bar{N}, \quad (11.4)$$

where $r(t)$ is the growth rate, N is the population density, and $\bar{N}$ is the long-term mean density. According to this statement, when varying only the population density (N) with respect to the growth rate, $r(t)$, and keeping all other factors that might influence $r(t)$

constant (hence the use of the partial derivative ½), as population density increases, population growth rates will decrease. The simplest example of density-dependent population self-limitation can be modeled using the logistic equation (Verhulst 1845):

$$\frac{dN}{dt} = \frac{r_0 N(K - N)}{K}, \tag{11.5}$$

where $r_0 N$ is the familiar resulting term from the law of exponential growth (Eq. 11.3), N is the population density, and K is the term defined as the equilibrium abundance (here akin to the carrying capacity), which indicates the approximated equilibrium point a given population density may reach.

While the logistic equation provides an important starting point for formulating more realistic (and generalizable) models of self-limitation within populations, it must be expanded upon to overcome inherent shortcomings (e.g., no time-lags) that make it unsuitable for modeling anything beyond simplified single-species patterns of growth. Alternatively, discrete-time forms of the logistic equation can add more biological realism by prohibiting spurious negative population densities and have been successfully applied in real-world situations. For example, Ricker (1954) used the following equation to estimate fisheries stocks:

$$N_{t+1} = N_t \exp\left[r_0 \left(1 - \frac{N_t}{K}\right)\right], \tag{11.6}$$

where the population density at the discrete time-step is (N_{t+1}), N_t is the population density at the previous time (t), r_0 is the per capita growth rate, and K is carrying capacity. The ability of simple nonlinear difference equations and their discrete-time counterpart models of various forms to give rise to complex population dynamics has long excited the interest of theoretical ecologists, who have also applied them in modeling and understanding populations (Ricker 1954, Beverton and Holt 1957, May 1974). Despite the logistic equation's ability to model such complex processes and the foundational role it plays in population ecology, it is in many cases plagued by oversimplification and an inability to adequately model more realistic cases that we observe in nature (May 1976). Hence, population ecologists have sought to expand upon the logistic growth equation by including terms describing characteristics of naturally occurring effects of time-lags (May 1973), age-structure (King and Anderson 1971), overlapping generations (Leslie 1966; Hassell et al. 1976; Benhabib and Day 1982), demographic or environmental stochasticity (Lewontin and Cohen 1969; Lande 1993), and interspecific interactions (Gause 1935) on density-dependent population dynamics.

Modeling single-species population dynamics is useful in understanding how complex density-dependent behavior can emerge from even the most simplified cases. However, these are largely abstractions from real-world populations, as no species exists in true isolation. It follows, then, that we must incorporate both intra- and interspecific interactions within models of population dynamics to improve their realism and subsequent utility in fitting data from natural populations. Several pair-wise interactions between hypothetical individuals are widely acknowledged as being either positive (+), negative (−), or neutral (0), and include: mutualism (+, +), commensalism (+, 0), amensalism (−, 0), interference competition (−, −), and consumer-resource (+, −), which may include host-parasitism interactions (Turchin 2003). However, it is the lattermost type of trophic interactions that has captivated much of ecologists' attention with respect to its role in population regulation (Lotka 1925; Volterra 1928; MacArthur 1970; DeAngelis et al. 1975; Oksanen et al. 1992; Murdoch et al. 2003; Holland and DeAngelis 2010). Nevertheless, it is important to note that none of the aforementioned types of interactions is mutually exclusive. However, given the importance of consumer-resource interactions and their intuitive linkages with laws of thermodynamics (Odum 1988; Hairston and Hairston 1993), population ecology has spent a considerable amount of time and effort studying this particular class of interactions.

Although it may be difficult to know for certain whether trophic interactions play a primary role in the majority of regulating populations, this pattern finds support in the abundance of evidence accrued from empirical studies (Chitty 1960; Andersson and Erlinge 1977; Gliwicz 1980; Krebs et al. 1995; Turchin et al. 2000). Based on these, Turchin (2003) proposes *postulate 4*, which is analogous to the law of mass action, as the proportional relationship between the number of captured individuals (resource) by a single consumer and the density of the resource in situations where the density of a resource is low. This can be expressed by:

$$\text{capture rate} = aN \text{ as } N \to 0, \qquad (11.7)$$

where N is the density of the resource population, and a is a proportionality constant. Turchin (2003) goes on to define *postulate 5*, which states that the rates of conversion of biomass to energy used by a consumer for growth, maintenance, and reproduction, is a function of the captured biomass amount (Maynard-Smith 1978, Getz 2001). Finally, *postulate 6* proposed by Turchin (2003) states that the consumption of resources by a consumer is limited (via physiological constraints) and will be bounded by a maximum rate independent of the density of a resource, thereby ultimately leading to a maximum threshold for reproductive rates (Chapman 1931; Birch 1948; Yasuda 1990; Ginzburg 1998). Through the combination of postulates 4, 5, and 6, the hyperbolic functional response can be derived (Type II; Maynard-Smith 1978). Combining the resulting relationships leading to the functional response curves with the previously described assumptions (postulates 1, 2, and 3) of single-species population growth, Turchin (2003) derives the seminal Lotka-Volterra model (Lotka 1925; Volterra 1926). This is represented by equations:

$$\frac{dN}{Ndt} = r_0 - aP,$$
$$\frac{dP}{Pdt} = -\delta + \chi a, \qquad (11.8)$$

where the population densities of the consumer and resource are designated as N and P, respectively. The per capita growth rate of the resource population independent of predation is r_0, the per capita decline rate of the consumers independent of resources is δ, and the proportionality constants for consumers and resources that relate the rate of consumption of resources (capture rate) with resource population density, and the fecundity rate for the consumer population are represented by a and χ, respectively.

We thought it was important to first focus on the emerging consensus of the existence of general ecological laws and their utility (Lawton 1999; Turchin 2001, 2003) to build a strong theoretical foundation for the role of density dependence in population ecology. Interest in the application and extension of these foundational principles in studying density dependence continues to have important implications for the study of the ecology, population dynamics, and conservation of wild populations (Errington 1945; Lande 1993; Webster 2004; Williams et al. 2004; Coulon et al. 2008). We will now shift our focus to more applied examples to give practitioners insight into how they might approach their own research questions.

Detecting Density Dependence

Detecting the influence of density dependence within populations can be complicated by many factors that contribute to variance in data from measurement error, environmental and demographic stochasticity, complex multi-species interactions, and life-history characteristics of species. In spite of these known challenges, the study of density dependence in populations is important for several reasons. For one, understanding the regulatory mechanisms of populations remains a central focus of ecology. Our interest in estimating population sizes and understanding drivers of population dynamics requires the study of the balance and interplay among density-dependent and independent factors contributing to population change over time. Another pragmatic reason to study density dependence is to aid in the conservation of species. In an effort to manage for species of conservation concern and to prevent common species from becoming scarce, we need to understand how abiotic and biotic drivers influence density-dependent mechanisms. We must then use this information to guide management actions to preserve these mechanisms, thereby enhancing the conditions for the long-term persistence of populations.

Several approaches for detecting density dependence exist, but in general, they all include some aspect of the analysis of relationships among the influence of demographic rates (births, immigration, deaths, and emigration) as a function of the per capita rate of population change. Consideration of life-history characteristics associated with r/K selection theory (e.g., evolutionary trade-offs associated with reproduction and resultant population growth rates) of species can play an important role when attempting to detect density-dependent effects within populations (MacArthur and Wilson 1967; Pianka 1972; Taylor et al. 1990). MacArthur (1972) describes fundamental theoretical differences between r- and K-selected species in terms of the predicted ways that density dependence contributes to population dynamics (Fig. 11.2). Moreover, both theoretical and empirical studies of density dependence within wild populations of r- and K-selected species show how nonlinear density-dependent relationships (see Fig. 11.3) can occur (Murdoch 1966; Fowler 1981; Sæther and Bakke 2000; Johst et al. 2008; Williams 2013).

More recently, incorporation of the theta-logistic model has expanded beyond the constraints of the logistic equation to incorporate more biological realism associated with nonlinear density dependence (Gilpin and Ayala 1973; Fowler 1981; Turchin 2003; Sibly et al. 2005; Brook and Bradshaw 2006; Williams 2013). The theta-logistic model can be expressed as:

$$\frac{dN}{dt} = r_0 N \left[1 - \left(\frac{N}{k} \right)^\theta \right], \quad (11.9)$$

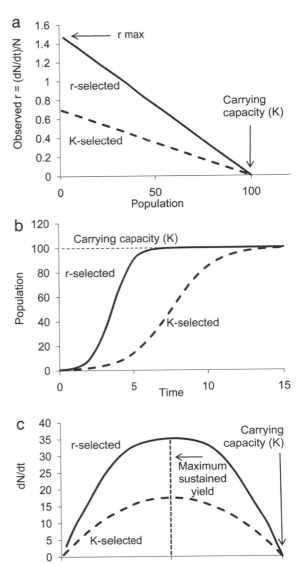

Fig. 11.2. Comparison of differential theoretical responses between r- and K-selected species from MacArthur (1972) population models depicting (a) the second derivative population growth rate as a function of population density (dN/dt)N, where r_{max} (the maximum growth potential or exponential growth) is represented by the y-intercept, (b) logistic-growth where r-selected species grow at faster rates toward carrying capacity (K), and (c) changes in population size (dN/dt), which will have a symmetric parabolic shape that peaks at K/2 (also maximum sustained yield) and approach zero as the populations reach carrying capacity. *Source: Adapted from Williams (2012), and used with the permission of John Wiley & Sons.*

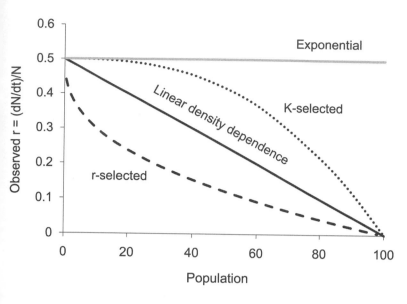

Fig. 11.3. Theoretical influence of nonlinear density dependence on the observed rate of population change, (dN/dt)/N, having a constant maximum observed intrinsic rate of increase (r_{max}) and carrying capacity, in this case $K = 100$. This depiction shows how r- and K-selected species exhibit inverse patterns of how density dependence can affect population growth rates. *Source: Depiction used with permission from Williams (2012).*

where the familiar terms of the logistic model, r_0 (the intrinsic rate of population growth), N (population density), and k (carrying capacity) are modified by the exponent theta (θ), which governs and enables nonlinearity of the shape of the density dependence relationship (Turchin 2003).

The advantage in using the theta (θ)-logistic model lie in its ability to incorporate variation in the density-dependent responses of species with different life-history characteristics. For example, r-selected species will typically have values of $\theta < 1$, whereas K-selected species will have values of $\theta > 1$ (Figs. 11.3 and 11.4). Additionally, we can calculate the stationary distribution and corresponding variance given our well-founded understanding of the θ-logistic model's statistical properties (Gilpin et al. 1976; Sæther et al. 1996; Diserud and Engen 2000; Sæther et al. 2000). However, issues with the θ-logistic model raised by Clark et al. (2010) can arise under circumstances where both θ and r_{max} are simultaneously estimated from time-series data. In these cases, there is the potential for spurious parameter estimation that can produce a mathematical model that may "fit" data yet lead to confounding results when using predictive models relying on regression, autoregression, or time-series analytical techniques (Williams 2012). Existing workarounds for this issue include transforming data to fit assumptions of linearity of density-dependent responses. However, a more satisfactory alternative is for researchers to estimate r_{max} and θ independently. For instance, while θ can be estimated from time-series data, r_{max} may be estimated from allometric relationships or demographic models (Sæther et al. 2002; Clark et al. 2010).

Williams (2012) demonstrated the importance of incorporating interactions between life-history strategy characteristics and environmental stochasticity on the nature of density dependent responses of populations. This holds important implications for building more realistic models that can more accurately predict local extinction risks or aid in harvest management applications. The analytically predicted threshold at which nonlinear θ-logistic models can maintain stable equilibrium occurs when $0 < r\theta < 2$ (Thomas et al. 1980). However, this case doesn't explicitly account for stochasticity. The inclusion of environmental stochasticity in modeling nonlinear patterns of density dependence has improved model predictions of the likelihood of local extinction and parameter estimation (Sæther et al. 2002). Building upon Case's (1999) simple logistic equation that includes stochastic variation, Williams (2012) included the θ exponent to allow for the modeling of the rela-

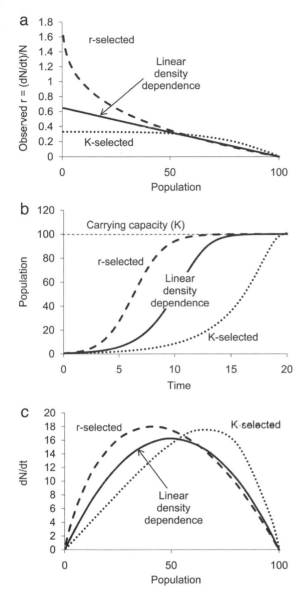

Fig. 11.4. Differential patterns of nonlinear density dependence for *r*- and *K*-selected species that account for (a) variable r_{max} (maximum biological growth potential) associated with different life-history strategies, which can lead to (b) different rates of growth toward carrying capacity (*K*), and (c) asymmetries in the parabolic patterns associated with per capita population change. *Source: Adapted from Williams (2012), and used with the permission of John Wiley & Sons.*

tionship between density independent environmental stochasticity and life-history patterns in the formula:

$$N_{t+1} = N_t + r_{max}N^t\left[1 - \left(\frac{N}{K}\right)^{\theta}\right] + \left[N_t(VRnd_t)\right],$$

(11.10)

where $VRnd_t$ is a random number drawn at each time period, *t*. Using this equation to run simulations (see Williams 2012; Fig. 11.5), we can draw *Rnd* from a uniform distribution between −0.5 and 0.5 (to account for all possible interacting stochasticities when prior knowledge of their distributions is unknown). Then, *Rnd* is multiplied by a constant *V* to compute a per capita density independent change in the population, where *V* represents the magnifier of the amount of environmental noise. Temporal noise is absent when $V=0$; however, as *V* increases, the density-independent term becomes increasingly large compared to the density-dependent term. Although the relative impact of *V* will depend on a species' growth rate, Case (2000) illustrated how values that are approximately 0.01–0.30 will generally mimic low noise, values of 0.31–0.60 will mimic moderate noise, and values > 0.60 will mimic large amounts of noise. As values for *V* become larger, the likelihood of extinction increases, except for the most slowly growing populations, since low reproduction can dampen the magnitude of population fluctuations (Case 2000).

Indeed, when detecting and analyzing density dependence, we should also consider the potential influence of spatial and temporal scales at which populations are being measured (data collection) and modeled (e.g., local population, metapopulation). Theoretical examples of models provided thus far have ignored immigration and emigration and represent idealized closed populations. If we consider a large enough spatial scale, we can assume immigration and emigration are simply local movements within a closed population and ignore immigration and emigration terms within models. Many studies

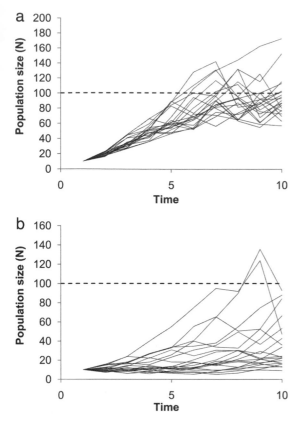

Fig. 11.5. Comparative results from 20 iterations (each) simulating the Williams (2012) theta-logistic equation that accounts for interactions between life-history traits and environmental stochasticity for (a) *r*-selected species with large environmental stochasticity (growth rate (r) = 2.2, carrying capacity (K) = 100, stochasticity constant (V) = 0.80, and the shape parameter (θ) = 0.25), and (b) K-selected species (r = 0.33, K = 100, V = 0.80, and θ = 4.0). *Source: Adapted from Williams (2012), and used with the permission of John Wiley & Sons.*

of wild populations are rarely performed at such a spatial scale and must either modify models to include immigration and emigration terms (e.g., Birth, Immigration, Death, Emigration models) or recognize that the scale at which inference may be made regarding the contribution of density dependence to demographic rates will be somewhat limited.

In their review of density dependence in populations of sea-floor dwelling (demersal) coral reef fishes, Hixon and Webster (2002) point out the importance of defining local populations and metapop-

ulations and that different analytical approaches (e.g., per capita rates versus total rates) should be chosen with care given constraints due to the spatial scale at which studies are conducted. Additionally, consideration must be given to the duration of studies and how this can affect our ability to detect density dependence. Studies aimed at testing for population regulatory patterns from density dependence must include, at a minimum, multiple generations. However, it has been found from observational studies that the ability to detect density dependence increases as the length of study duration increases (Cappuccino and Price 1995; Turchin 2003).

Several formal statistical tests have been developed for detecting density dependence within populations (Vickery and Nudds 1984; Pollard et al. 1987; Dennis and Taper 1994; Nisbet and Gurney 2003). While these tests rely on simplified population models of limited application to actual populations (e.g., not accounting for delayed density dependence, overlapping generations, stochasticity, nonlinearity), they can be useful for the generalized statistical assessment of the presence of density dependence. For example, Cappuchino and Price (1995) introduce the following model:

$$r_t \equiv \ln N_t - \ln N_{t-1} = f(N_{t-1}) + \varepsilon_t, \quad (11.11)$$

where the intrinsic growth rate (r_t) is equal by definition to either a linear function of N_{t-1} (Dennis and Taper 1994) or $\ln N_{t-1}$ (Pollard et al. 1987), and where ε_t are random density independent factors. Here, we provide an example analysis using the randomization approach favored by Pollard et al. (1987), where we simulated data sets for both density-dependent and density-independent populations and tested the hypothesis that the population was density independent. For both sets of population data, we used an initial population size of 10, and simulated data over 30 time steps. Additionally, we used values of 0.4 for *r*, a drift parameter, and 0.8 for β (slope parameter in the density-dependent model), and for the stochasticity term ε_t, we used a sampled distribution with a mean of 0 and variance of 0.01. For each of these "observed" populations, which we knew to be either density independent or density-dependent, we then simulated 200 new populations to compare to the observed populations. For each we then calculated the test statistic rdx, which is equivalent to the correlation coefficient between a set of population sizes (x_i), and corresponding randomized changes in population size ($x_{i+1} - x_i$) for a given population (see Pollard et al. 1987 for details). In the final step of evaluation (considering an $\alpha = 0.05$), we tallied the number of times the null hypothesis was rejected, and if the tally was below 5% of N simulations (in this case, 200), we rejected the null hypothesis and concluded that the observed population was density-dependent (Fig. 11.6). We ran the analysis in program R (R Core Team 2016), and developed a script that readers can use to test whether populations exhibit density dependence, which can be accessed freely at the following GitHub repository: https://github.com/zachladin/Density_Dependence_Test.

While tests exist as to whether or not density dependence is detectable within populations, it is more informative and pertinent to go beyond mere detection and seek to understand the mechanisms by which density dependence influences population regulation (Cappuccino and Price 1995). Additionally, limiting assumptions such as linearity and absence of delayed density-dependent effects are not captured in the tests described above, which diminish their utility for use with actual population data. However, these limitations can be overcome using models that account for delayed density dependence and nonlinearity and have the general form:

$$r_t = f(N_{t-1}, N_{t-2}, \ldots, N_{t-n,}) + \varepsilon_t, \quad (11.12)$$

where the rate of population change is a function of both the population density at the nth previous time step (N_{t-n}) plus ε_t, representing other density independent factors (including contributing variance) modeled as some random normally distributed inde-

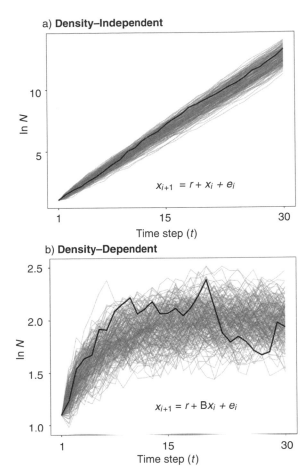

Fig. 11.6. Results from randomization test for density dependence developed by Pollard et al. (1987) that compares observed population data (black line) to simulated population data from either (a) density-independent population model or (b) density-dependent population model. Tests were run in program R; and code is provided by the authors and can be downloaded at https://github.com/zachladin/Density_Dependence_Test.

pendent variable. These types of models that can be fit as nonlinear functions have been successfully used to analyze ecological time-series data (Turchin and Taylor 1992; Ellner and Turchin 1995; Berryman and Turchin 2001; Brook and Bradshaw 2006).

Density Dependence in Wild Populations

Currently, there is a general consensus among ecologists that population regulation involving density dependence is likely commonplace in nature. Yet, its mechanisms may not always be observable due a combination of potentially confounding factors related to complex cross-scale interactions between local and metapopulation dynamics. Due to these challenges, there is a lack of studies with long enough time-series data, which has led to contrary evidence based on the previous use of unreliable methods of detection. Here, we focus on the application of currently available tools for analyzing ecological time-series data to better understand the underlying drivers of population regulation. Additionally, we discuss the importance of taking advantage of natural experiments through comparative studies, as well as the use of density-manipulation experiments, which can enhance our understanding of specific drivers of population regulation and have implications for conservation and management of wildlife populations.

Ecologists agree that gaining a full understanding of population regulation requires both theoretical and empirical approaches (Royama 1977; Harrison and Cappuccino 1995; Krebs et al. 1995; Lawton 1999; Hixon and Webster 2002; Turchin 2003). Both taking advantage of natural "experiments," in the MacArthurian tradition of comparative ecological studies and designing manipulations to test specific hypotheses can provide novel insight into population regulation. For example, in Krebs et al.'s (1995) study on snowshoe hares, the simultaneous manipulation of food availability and predation helped demonstrate a tri-trophic (food-hare-predator) interaction, which generated the classic decadal cycles in hare populations. In another example of using field experiments, Forrester and Steele (2004) found biological drivers of density dependence over multiple spatial scales by manipulating the density of small coral reef–dwelling gobies and the amount of available refuges to test if goby population density (by mortality rates across treatments) was influenced by the interaction between predator density and refuge availability. In general, studies that manipulate the density of organisms in natural populations to measure subsequent changes in population demo

graphic rates are valuable because they control for potentially confounding factors when seeking to determine mechanisms and causality (Murdoch 1970; Harrison and Cappuccino 1995).

We recommend Turchin's (2003: part 3) systematic approach for analyzing population data, which highlights several case studies for analyzing population time-series data and determining the role of density dependence within observed population dynamics. These logical steps include: (1) attempting to understand the patterns in time-series data by fitting potentially (mathematically) appropriate models, using a phenomenological approach, (2) considering proposed hypotheses (and mechanistic model counterparts) to understand potential factors that would give rise to such population patterns, and (3) thinking about experimentation (comparative studies or designed manipulations) that can help provide support for competing hypotheses.

Alternatively, ecologists can actively take advantage of using comparative approaches in designing studies to try to tease out the corresponding influences of density-dependent and density-independent effects on population dynamics. Williams et al. (2003) took advantage of a 36-year Rural Mail Carrier Survey dataset conducted in Kansas (and other Midwestern states) to monitor populations of game birds and small game mammals. The analysis of these long-term survey data collected four times per year (i.e., in January, April, July, and October) included northern bobwhite (*Colinus virginianus*), ring-necked pheasants (*Phasianus colchicus*), and eastern cottontails (*Sylvilagus floridanus*). These data offered a unique opportunity to study the relative contributions of density-independent and dependent effects on population growth rates, while also explicitly accounting for intra-annual patterns. While Williams et al. (2003) did use a comparative approach by analyzing multiple species from the same study area and dataset, we highlight how the study's comparisons of density dependence between peripheral versus core areas of each of the species' distribution ranges (spatial), as well as among within-season (temporal) periods, demonstrate the utility in taking advantage of "natural" experiments.

A commonly used and important method for analyzing time-series data are models that include an autocorrelation function and include both a density-dependent and independent component (Turchin 2003). To analyze the Rural Mail Carrier Survey data, Williams et al. (2003) used the following model within a state-space framework:

$$x_t^* = a_s + x_{t-1}^* + b_s x_{t-1}^* + c_{st} + d_{st2} + \varepsilon_\sigma$$
$$x_t = x_t^* + \gamma_{\sigma t}, \tag{11.13}$$

where x_t is the log of the observed population density for a given region at time (t), including the measurement error term ($\gamma_{\sigma t}$), x_t^* is the true log (population density) without any measurement error, and the process error (ε_σ) is a random variable with mean of 0 and variance of σ_s^2 that represents random variability (not explained by the population density at t–1) in the per capita growth rate. The autoregression coefficients (a_s, b_s, c_{st}, and, d_{st2}) with subscript s indicating the season in which surveys were conducted. The strength of density dependence can be estimated by the slope of the autoregressive coefficient (b_s). Using the above autoregressive population model to account for delayed density dependence and measurement error, Williams et al. (2003) found that while peripheral populations exhibited greater variability in per capita growth rates due to density independent effects (i.e., environmental stochasticity), when compared to populations located within core areas, they also showed surprisingly greater density dependent effects in winter/spring months (January–April), which were potentially driven by reduced food availability.

The exploration of consumer-resource interactions that can give rise to population cycles has also aided in teasing apart the relative effects that direct and delayed density dependence can have in sustaining those cycles, as well as leading to their collapse (Keith and Rusch 1989; Hanski et al. 1991; Hansen et al. 1999). The observation and study of broad-scale geographic patterns in cyclical population dynamics across latitudinal gradients has also provided insight into underlying mechanisms (Bjornstad et al. 1995; Turchin and Hanski 1997; Klemola et al. 2002; Williams et al. 2004). Interestingly, when moving from northern to southern latitudes in Europe, suggested drivers involving a shift from specialist to generalist predators can lead to cycle shortening and the eventual collapse of cycles (Hansson and Henttonen 1985; Ims et al. 2008; Cornulier et al. 2013). However, Williams et al. (2004) demonstrated that along a north–south latitudinal gradient in North America, cycles may collapse through cycle period lengthening. In their study using 27 long-term data sets from 1939 to 2001 on ruffed grouse (*Bonasa umbellus*), sharp-tailed grouse (*Tympanuchus phasianellus*), and greater prairie chicken (*Tympanuchus cupido*), Williams et al. (2004) employed an autoregressive model with a stochastic moving average process to estimate effects from both direct and delayed density dependence (Reinsel 2003):

$$X_1(t) = \beta_0 + \beta_1 X_1(t-1) + \beta_2 X_1(t-2) + \varepsilon(t)$$
$$\varepsilon(t) = \alpha(t) + a^*\alpha(t-1), \tag{11.14}$$

where β_1 and β_2 coefficients indicate the associated strengths of direct and delayed density dependence, respectively. The stochastic term in the model, $\varepsilon(t)$, is then defined as a moving average process, where component $\alpha(t)$ is an uncorrelated random variable, and the coefficient a is dependent on a matrix of interaction coefficients between consumer and resource species, as well as potential covariance structures between $\varepsilon_1(t)$ and $\varepsilon_2(t)$ error terms. Given the mathematical structure of this model, which includes simultaneous additive effects between direct and delayed density-dependent factors, it can be further shown how the relative strength of both direct and delayed density-dependent terms can lead to the collapse of population cycles by way of either period shortening or period lengthening (Fig. 11.7). Contrary to the patterns of period shortening leading to

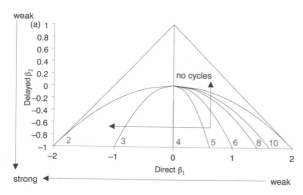

Fig. 11.7. Conceptual diagram of the theoretical interactive relationship between direct (β_1; x-axis) and delayed (β_2; y-axis) density-dependent effects and corresponding period lengths (numbers shown in gray). Arrows show two modes of potential cycle collapse as direct density dependence becomes stronger and period lengths shorten, or as delayed density dependence weakens and period lengths increase. *Source: Adapted from Williams et al. (2004), and used with the permission of John Wiley & Sons.*

the collapse of cycles in Europe, Williams et al. (2004) concluded that in North American grouse populations, cycle collapses at more southern latitudes were due to period lengthening. These results highlight important complexities that the effects of time-lags and period lengthening can play in population cycles, particularly in longer-lived species.

Another vital and complementary (to experimentation, pure or otherwise) component in studying drivers of population dynamics involves modeling long-term time-series data from biological systems where unique population patterns (e.g., periodic cycles) naturally arise (Krebs 1991). Several notable ecological studies (see Turchin 2003: part 3) have collected long-term population data via regular monitoring efforts in concert with key environmental (abiotic) and biological covariates include the larch bud moth (*Zeiraphera diniana*) in the Swiss Alps (Fischlin and Baltensweiler 1979; Anderson and May 1980; Turchin et al. 2002), the spruce bud worm (*Choristoneura fumifer*) in Canada (Royama 1984; Royama et al. 2005; Swetnam and Lynch 1993), the southern pine beetle (*Dendroctonus frontalis*) extending from southern United States into parts of Central

America (Coulson 1975; Lorio 1986; Turchin et al. 1991), the northern bobwhite quail (Errington 1945; Williams et al. 2003; Thogmartin et al. 2002), the red grouse (*Lagopus lagopus scotica*) in Great Britain and Ireland (Potts et al. 1984; Watson et al. 2000; Moss and Watson 2001), arctic vole and lemming populations (Elton 1924; Chitty 1960; Turchin et al. 2000), and the snowshoe hare (Gilpin 1973; Krebs et al. 1995, 2001).

Through the thoughtful consideration of the mechanisms by which differentially (and not mutually exclusive) acting exogenous (density independent), first-order (e.g., population self-limitation), and more complex second-order (e.g., trophic coupling) drivers influence population dynamics, we can test competing hypotheses. It is also important to note it may not be pragmatic or appropriate for population ecologists to search for one "best" hypothesis to explain variance within population data, and that we should instead be interested in building useful models that can reasonably capture the dynamics of complex ecological relationships given the best available information.

Conclusion

Although we have come a considerable way in our study of population ecology and in our ability to model and predict population patterns, there remains much to learn as we continue moving forward into a data-rich future. The past century of studying population ecology has included an initially thought-provoking controversy over whether population regulation by density dependence is a truly mechanistic biological phenomenon (Nicholson 1933) or is driven by purely environmental limitations of space and time (Andrewartha and Birch 1954; Murray 1999), and we continue to have a healthy and stimulating dialogue about the foundational properties that govern population dynamics, which remain central to both theoretical and applied ecology. As ecologists continue to engage in these issues and recognize and account for potential ambiguities in formal definitions and differing perspectives (Berryman et al. 2002), more novel insights into population regulatory mechanisms will be produced.

Our treatment and discussion of density dependence and its role in population regulation here is by no means exhaustive, and we encourage readers to delve more deeply into the seminal papers and books to which we have referred. Furthermore, there is a deep body of literature that we have been unable to include, and we similarly encourage readers to seek out and explore these vital contributions to our collective understanding of density dependence as well. It was our goal to provide an introductory overview of the way ecologists have thought about and approached the study of density dependence and its role in population dynamics. Given the importance of this topic to theoretical ecology, applied ecology, and conservation, we hope to have provided a brief background as to its theoretical development and underpinnings, as well as some insight into and key examples of how practical models can be applied to analyzing population data. Ultimately, we hope to have helped new students and practitioners in their understanding of how and why populations persist and behave over time in a complex and dynamic environment.

LITERATURE CITED

Anderson, R. M., and R. M. May. 1980. Infectious diseases and population cycles of forest insects. Science 210:658–661.

Andersson, M., and S. Erlinge. 1977. Influence of predation on rodent populations. Oikos 29:591–597.

Andrewartha, H. G., and L. C. Birch. 1954. The distribution and abundance of animals. University of Chicago Press, Chicago.

Benhabib, J., and R. H. Day. 1982. A characterization of erratic dynamics in the overlapping generations model. Journal of Economic Dynamics and Control 4:37–55.

Berryman, A. A., M. Lima Arce, and B. A. Hawkins. 2002. Population regulation, emergent properties, and a requiem for density dependence. Oikos 99:600–606.

Berryman, A., and P. Turchin. 2001. Identifying the density-dependent structure underlying ecological time series. Oikos 92:265–270.

Beverton, R. J. H., and S. J. Holt. 1957. On the dynamics of exploited fish populations. Fisheries Investigation Series 2, vol. 19. UK Ministry of Agriculture, London.

Birch, L. 1948. The intrinsic rate of natural increase of an insect population. Journal of Animal Ecology 17:15–26.

Bjornstad, O. N., W. Falck, and N. C. Stenseth. 1995. A geographic gradient in small rodent density fluctuations: A statistical modelling approach. Proceedings of the Royal Society of London B: Biological Sciences 262:127–133.

Boyce, M. S., and R. S. Miller. 1985. Ten-year periodicity in whooping crane census. The Auk 102:658–660.

Brook, B. W., and C. J. Bradshaw. 2006. Strength of evidence for density dependence in abundance time series of 1198 species. Ecology 87:1445–1451.

Bulmer, M. G. 1974. A statistical analysis of the 10-year cycle in Canada. Journal of Animal Ecology 43:701–718.

Cappuccino, N., and P. W. Price. 1995. Population dynamics: New approaches and synthesis. Academic Press, Cambridge, MA.

Case, T. J. 1999. Illustrated guide to theoretical ecology. Ecology vol. 80 pg. 2848.

Chapman, R. N. 1931. Animal ecology with especial reference to insects. McGraw-Hill, New York.

Chesson, P. L. 1982. The stabilizing effect of a random environment. Journal of Mathematical Biology 15:1–36.

Chitty, D. 1960. Population processes in the vole and their relevance to general theory. Canadian Journal of Zoology 38:99–113.

Clark, F., B. W. Brook, S. Delean, H. Reşit Akçakaya, and C. J. Bradshaw. 2010. The theta-logistic is unreliable for modelling most census data. Methods in Ecology and Evolution 1:253–262.

Cornulier, T., N. G. Yoccoz, V. Bretagnolle, J. E. Brommer, A. Butet, F. Ecke, D. A. Elston, E. Framstad, H. Henttonen, B. Hörnfeldt, O Huitu, C. Imholt, R. A. Ims, J. Jacob, B. Jedrzejewska, A. Milon, S. J. Petty, H. Pietiäinen, E. Tkadlec, K. Zub, and X. Lambin. 2013. Europe-wide dampening of population cycles in keystone herbivores. Science 340:63–66.

Coulon, A., J. W. Fitzpatrick, R. Bowman, B. M. Stith, C. A. Makarewich, L. M. Stenzler, and I. J. Lovette. 2008. Congruent population structure inferred from dispersal behaviour and intensive genetic surveys of the threatened Florida scrub-jay (*Aphelocoma coerulescens*). Molecular Ecology 17:1685–1701.

Coulson, R. N. 1975. Techniques for sampling the dynamics of southern pine beetle populations. Texas Agricultural Experimental Station Miscellaneous Publication 1185. College Station, TX.

Darwin, C. 1861. On the origin of species by means of natural selection, or the preservation of favoured races in the struggle for life. Murray, London.

DeAngelis, D. L., R. A. Goldstein, and R. V. O'Neill. 1975. A model for tropic interaction. Ecology 56:881–892.

Dennis, B., and M. L. Taper. 1994. Density dependence in time series observations of natural populations: Estimation and testing. Ecological Monographs 64:205–224.

Diserud, O. H., and S. Engen. 2000. A general and dynamic species abundance model, embracing the lognormal and the gamma models. American Naturalist 155:497–511.

Edmonds, T. R. 1832. An enquiry into the principles of population, exhibiting a system of regulations for the poor. J. Duncan, London.

Ellner, S., and P. Turchin. 1995. Chaos in a noisy world: New methods and evidence from time-series analysis. American Naturalist 145:343–375.

Elton, C., and M. Nicholson. 1942. The ten-year cycle in numbers of the lynx in Canada. Journal of Animal Ecology 11:215–244.

Elton, C. S. 1924. Periodic fluctuations in the numbers of animals: Their causes and effects. Journal of Experimental Biology 2:119–163.

Errington, P. L. 1945. Some contributions of a fifteen-year local study of the northern bobwhite to a knowledge of population phenomena. Ecological Monographs 15:1–34.

Fischlin, A., and W. Baltensweiler. 1979. Systems analysis of the larch bud moth system: 1. The larc-larch bud moth relationship. *Mitteilungen der Schweizerischen Entomologischen Gesellschaft.= Bulletin de la Societe entomologique suisse.* Department of Entomology, Swiss Federal Institute of Technology, ETH-Zentrum, CH-8092 Zurich 52:273.

Forrester, G. E., and M. A. Steele. 2004. Predators, prey refuges, and the spatial scaling of density-dependent prey mortality. Ecology 85:1332–1342.

Fowler, C. W. 1981. Density dependence as related to life history strategy. Ecology 62:602–610.

Franklin, B., and R. Jackson. 1760. The Interest of Great Britain Considered with Regard to Her Colonies and the Acquisitions of Canada and Guadaloupe, To which are Added, Observations Concerning the Increase of Mankind, Peopling of Countries, Etc., pp. 50–58, T. Beckett, London.

Gause, G. F. 1935. Experimental demonstration of Volterra's periodic oscillations in the numbers of animals. Journal of Experimental Biology 12:44–48.

Getz, W. M. 2001. Population and evolutionary dynamics of consumer-resource systems. In Advanced Ecological Theory, edited by J. McGlade, pp. 194–231. Blackwell Scientific Press, Oxford, UK.

Gilpin, M. E. 1973. Do hares eat lynx? American Naturalist 107:727–730.

Gilpin, M. E., and F. J. Ayala. 1973. Global models of growth and competition. Proceedings of the National Academy of Sciences 70:3590–3593.

Gilpin, M. E., T. J. Case, and F. J. Ayala. 1976. Beta-selection. Mathematical Biosciences 32:131–139.

Ginzburg, L. R. 1998. Assuming reproduction to be a function of consumption raises doubts about some popular predator-prey models. Journal of Animal Ecology 67:325–327.

Gliwicz, J. 1980. Island populations of rodents: Their organization and functioning. Biological Reviews 55:109–138.

Grinnell, J. 1917. The niche-relationships of the California thrasher. Auk 34:427–433.

Guthery, F. S. 2002. The technology of bobwhite management: The theory behind the practice. Iowa State University Press, Ames, IA.

Hairston, N. G., Jr, and N. G. Hairston Sr. 1993. Cause-effect relationships in energy flow, trophic structure, and interspecific interactions. American Naturalist 142:379–411.

Hansen, T. F., N. C. Stenseth, H. Henttonen, and J. Tast. 1999. Interspecific and intraspecific competition as causes of direct and delayed density dependence in a fluctuating vole population. Proceedings of the National Academy of Sciences 96:986–991.

Hanski, I., L. Hansson, and H. Henttonen. 1991. Specialist predators, generalist predators, and the microtine rodent cycle. Journal of Animal Ecology 60:353–367.

Hansson, L., and H. Henttonen. 1985. Gradients in density variations of small rodents: The importance of latitude and snow cover. Oecologia 67:394–402.

Harrison, S., and N. Cappuccino. 1995. Using density-manipulation experiments to study population regulation. In Population dynamics: New approaches and synthesis, edited by N. Cappuccino and P. W. Price, pp. 131–147. Academic Press, Cambridge, MA.

Hassell, M. P., J. H. Lawton, and R. M. May. 1976. Patterns of dynamical behaviour in single-species populations. Journal of Animal Ecology 45:471–486.

Hixon, M. A., and M. S. Webster. 2002. Density dependence in reef fish populations. In Coral reef fishes: Dynamics and diversity in a complex ecosystem, pp. 303–325. Academic Press, San Diego, CA.

Holland, J. N., and D. L. DeAngelis. 2010. A consumer-resource approach to the density-dependent population dynamics of mutualism. Ecology 91:1286–1295.

Ims, R. A., J. A. Henden, and S. T. Killengreen. 2008. Collapsing population cycles. Trends in Ecology & Evolution 23:79–86.

Johst, K., A. Berryman, and M. Lima. 2008. From individual interactions to population dynamics: Individual resource partitioning simulation exposes the causes of nonlinear intra-specific competition. Population Ecology 50:79–90.

Keith, L. B. 1963. Wildlife's ten-year cycle, vol. 9. University of Wisconsin Press, Madison, WI

Keith, L. B., and D. H. Rusch. 1989. Predation's role in the cyclic fluctuations of ruffed grouse. International Ornithological Congress 19:699–732.

King, C. E., and W. W. Anderson. 1971. Age-specific selection II. The interaction between r and K during population growth. American Naturalist 105:137–156.

Klemola, T., M. Tanhuanpää, E. Korpimäki, and K. Ruohomäki. 2002. Specialist and generalist natural enemies as an explanation for geographical gradients in population cycles of northern herbivores. Oikos 99:83–94.

Krebs, C. J. 1991. The experimental paradigm and long-term population studies. Ibis 133:3–8.

Krebs, C. J., R. Boonstra, S. Boutin, and A. R. Sinclair. 2001. What drives the 10-year cycle of snowshoe hares? Bioscience 51:25–35.

Krebs, C. J., S. Boutin, R. Boonstra, A. R. E. Sinclair, J. N. M. Smith, M. R. T. Dale, K. Martin, and R. Turkington. 1995. Impact of food and predation on the snowshoe hare cycle. Science 269:1112.

Lampert, A., and A. Hastings. 2016. Stability and distribution of predator-prey systems: Local and regional mechanisms and patterns. Ecology Letters 19:279–288.

Lande, R. 1993. Risks of population extinction from demographic and environmental stochasticity and random catastrophes. American Naturalist 142:911–927.

Lawton, J. H. 1999. Are there general laws in ecology? Oikos 84:177–192.

Leslie, P. H. 1966. The intrinsic rate of increase and the overlap of successive generations in a population of Guillemots (Uria aalge Pont.). Journal of Animal Ecology 35:291–301.

Lewontin, R. C., and D. Cohen. 1969. On population growth in a randomly varying environment. Proceedings of the National Academy of Sciences 62: 1056–1060.

Lorio, P. L. 1986. Growth-differentiation balance: A basis for understanding southern pine beetle–tree interactions. Forest Ecology and Management 14:259–273.

Lotka, A. J. 1925. Elements of physical biology. Baltimore, Williams, and Wilkins Company, Philadelphia, PA.

MacArthur, R. 1970. Species packing and competitive equilibrium for many species. Theoretical Population Biology 1:1–11.

MacArthur, R. H. 1972. Geographical ecology: Patterns in the distribution of species. Princeton University Press, Princeton, NJ.

MacArthur, R. H., and E. O. Wilson. 1967. Theory of island biogeography, Princeton University Press, Princeton, NJ.

Malthus, T. 1798. An essay on the principle of population, as it affects the future improvement of society with remarks on the speculations of M. R. Godwin, M. Condorcet, and other writers. J. Johnson, London.

May, R. M. 1973. Time-delay versus stability in population models with two and three trophic levels. Ecology 54:315–325.

May, R. M. 1974. Biological populations with nonoverlapping generations: Stable points, stable cycles, and chaos. Science 186:645–647.

May, R. M. 1976. Simple mathematical models with very complicated dynamics. Nature 261:459–467.

Maynard-Smith, J. 1978. Models in ecology. Cambridge University Press, Cambridge, UK.

Moss, R., and A. Watson. 2001. Population cycles in birds of the grouse family (Tetraonidae). Advances in Ecological Research 32:53–111.

Murdoch, W. W. 1966. Population stability and life history phenomena. American Naturalist 100:5–11.

Murdoch, W. W. 1970. Population regulation and population inertia. Ecology 51:497–502.

Murdoch, W. W., C. J. Briggs, and R. M. Nisbet. 2003. Consumer-resource dynamics, vol. 36. Princeton University Press, Princeton, NJ.

Murdoch, W. W., and S. J. Walde. 1989. Analysis of insect population dynamics. In Towards a more exact ecology, edited by P.J. Grubb and J.B. Whittaker pp. 113–140. Blackwell, Oxford, UK.

Murray, B. G. Jr. 1999. Can the population regulation controversy be buried and forgotten? Oikos 84: 148–152.

Nicholson, A. J. 1933. Supplement: The balance of animal populations. Journal of Animal Ecology 131–178.

Nisbet, R. M., and W. Gurney. 2003. Modelling fluctuating populations, repr. Blackburn Press, Caldwell, NJ.

Odum, H. T. 1988. Self-organization, transformity, and information. Science 242:1132–1139.

Oksanen, T., L. Oksanen, and S. D. Fretwell. 1992. Habitat selection and predator-prey dynamics. Trends in Ecology & Evolution 7: 313.

Pearl, R., and S. L. Parker. 1922. On the influence of density of population upon the rate of reproduction in Drosophila. Proceedings of the National Academy of Sciences 8: 212–219.

Pianka, E. R. 1972. r and K selection or b and d selection? American Naturalist 106:581–588.

Pollard, E., K. H. Lakhani, and P. Rothery. 1987. The detection of density-dependence from a series of annual censuses. Ecology 68:2046–2055.

Potts, G. R., S. C. Tapper, and P. J. Hudson. 1984. Population fluctuations in red grouse: Analysis of bag records and a simulation model. Journal of Animal Ecology 53:21–36.

R Core Team. 2016. R: A language and environment for statistical computing (Version 3.0. 2). R Foundation for Statistical Computing, Vienna, Austria.

Reinsel, G. C. 2003. Elements of multivariate time series analysis. Springer Science & Business Media, Berlin, Germany.

Ricker, W. E. 1954. Stock and recruitment. Journal of the Fisheries Board of Canada 11:559–623.

Royama, T. 1977. Population persistence and density dependence. Ecological Monographs 47:1–35.

Royama, T. 1984. Population dynamics of the spruce budworm *Choristoneura fumiferana*. Ecological Monographs 54:429–462.

Royama, T., W. E. MacKinnon, E. G. Kettela, N. E. Carter, and L. K. Hartling. 2005. Analysis of spruce budworm outbreak cycles in New Brunswick, Canada, since 1952. Ecology 86:1212–1224.

Sæther, B.-E., and Ø. Bakke. 2000. Avian life history variation and contribution of demographic traits to the population growth rate. Ecology 81:642–653.

Sæther, B.-E., S. Engen, R. Lande, P. Arcese, and J. N. Smith. 2000. Estimating the time to extinction in an island population of song sparrows. Proceedings of the Royal Society B: Biological Sciences 267:621–626.

Sæther, B.-E., S. Engen, R. Lande, C. Both, and M. E. Visser. 2002. Density dependence and stochastic variation in a newly established population of a small songbird. Oikos 99:331–337.

Sæther, B.-E., T. H. Ringsby, and E. Røskaft. 1996. Life history variation, population processes and priorities in species conservation: Towards a reunion of research paradigms. Oikos 77:217–226.

Sallares, R. 1991. The ecology of the ancient Greek world. Cornell University Press, Ithaca, NY.

Sibly, R. M., D. Barker, M. C. Denham, J. Hone, and M. Pagel. 2005. On the regulation of populations of mammals, birds, fish, and insects. Science 309:607 610.

Swetnam, T. W., and A. M. Lynch. 1993. Multicentury, regional-scale patterns of western spruce budworm outbreaks. Ecological Monographs 63:399–424.

Taylor, D. R., L. W. Aarssen, and C. Loehle. 1990. On the relationship between r/K selection and environmental carrying capacity: A new habitat templet for plant life history strategies. Oikos 58:239–250.

Thogmartin, W. E., J. L. Roseberry, and A. Woolf. 2002. Cyclicity in northern bobwhites: A time-analytic review of the evidence. Proceedings of the National Quail Symposium 5: 192–200.

Thomas, W. R., M. J. Pomerantz, and M. E. Gilpin. 1980. Chaos, asymmetric growth and group selection for dynamical stability. Ecology 61:1312–1320.

Turchin, P. 1999. Population regulation: A synthetic view. Oikos 84:153–159.

Turchin, P. 2001. Does population ecology have general laws? Oikos 94:17–26.

Turchin, P. 2003. Complex population dynamics: A theoretical/empirical synthesis, vol. 35. Princeton University Press, Princeton, NJ.

Turchin, P., C. J. Briggs, S. P. Ellner, A. Fischlin, B. E. Kendall, and E. McCauley. 2002. Population cycles of the larch budmoth in Switzerland. In Population cycles: The case for trophic interactions edited by A. Berryman, pp. 130–141. Oxford, UK.

Turchin, P., and I. Hanski. 1997. An empirically based model for latitudinal gradient in vole population dynamics. American Naturalist 149:842–874.

Turchin, P., P. L. Lorio, A. D. Taylor, and R. F. Billings. 1991. Why do populations of southern pine beetles (Coleoptera: Scolytidae) fluctuate? Environmental Entomology 20:401–409.

Turchin, P., L. Oksanen, P. Ekerholm, T. Oksanen, and H. Henttonen. 2000. Are lemmings prey or predators? Nature 405:562–565.

Turchin, P., and A. D. Taylor. 1992. Complex dynamics in ecological time series. Ecology 73:289–305.

Verhulst, P. F. 1845. Mathematical research on the law of population growth. New memoirs of the Royal Academy of Sciences and belles-lettres de Bruxelles. 18:14–54.

Vickery, W. L., and T. D. Nudds. 1984. Detection of density-dependent effects in annual duck censuses. Ecology 65:96–104.

Volterra, V. 1926. Fluctuations in the abundance of a species considered mathematically. Nature 118:558–560.

Volterra, V. 1928. Variations and fluctuations of the number of individuals in animal species living together. International Council for the Exploration of the Sea (ICES) Journal of Marine Science 3:3–51.

Watson, A., R. Moss, and P. Rothery. 2000. Weather and synchrony in 10-year population cycles of rock ptarmigan and red grouse in Scotland. Ecology 81:2126–2136.

Webster, M. S. 2004. Density dependence via intercohort competition in a coral-reef fish. Ecology 85:986–994.

Williams, C. K. 2012. Accounting for wildlife life-history strategies when modeling stochastic density-dependent populations: A review. Journal of Wildlife Management 77:4–11.

Williams, C. K., A. R. Ives, and R. D. Applegate. 2003. Population dynamics across geographical ranges: Time-series analyses of three small game species. Ecology 84:2654–2667.

Williams, C. K., A. R. Ives, R. D. Applegate, and J. Ripa. 2004. The collapse of cycles in the dynamics of North American grouse populations. Ecology Letters 7:1135–1142.

Yasuda, H. 1990. Effect of population density on reproduction of two sympatric dung beetle species, *Aphodius haroldianus* and *A. Elegans* (Coleoptera: Scarabaeidae). Researches on Population Ecology 32:99–111.

PART IV ANALYSIS OF SPATIALLY BASED DATA ON ANIMALS AND RESOURCES

12 — Resource Selection Analysis

Julianna M. A. Jenkins,
Damon B. Lesmeister,
and Raymond J. Davis

Different taxa use resources differently across a single day, season, or life cycle and these dependencies must be quantified in order to understand management.
—Mathewson and Morrison (2015:4)

Introduction

The process of an animal selecting resources involves a series of behavioral choices. Understanding these behaviors is a foundational and cross-cutting theme in wildlife research and management. Resource selection analyses (RSAs) represent a broad class of statistical models for identifying underlying environmental correlates of animal resource selection and space use patterns. Many other chapters of this book describe analyses that can be encompassed by the general RSA definition. RSAs encompass several categories of habitat analyses: methods focused on testing for disproportionate use of habitat components, often referred to as "use-availability" or "presence-absence" models (Manly et al. 2002; Johnson et al. 2006), and "presence-only" models (McDonald 2013; Warton and Aarts 2013). RSAs have been used to improve our basic understanding of wildlife ecology (e.g., hypothesis testing, ecological niche models or species distribution models), to predict animal space use, to inform habitat management and conservation (e.g., critical habitat delineation), and to assess impacts of environmental change on wildlife (e.g., wildlife spatial response to wildfire).

Volumes have been written on the subject of RSAs and despite—or because of—this, choosing the most appropriate method of analysis for any given study question can be an overwhelming task. The goal of this chapter is to dispel the confusion and to present a basic introduction to resource selection concepts and an overview of common RSA methods used by wildlife professionals. This is not meant to be a comprehensive review, but rather an introductory guide. Other reviews and primary sources are referenced throughout the chapter, which we strongly suggest reading for a broader understanding of RSA theory and practice. Manly et al. (2002), the March 2006 special section of the *Journal of Wildlife Management* (Strickland and McDonald 2006), and the November 2013 special feature in the *Journal of Animal Ecology* (Mcdonald et al. 2013) provide more extensive discussions and reviews of specific RSAs.

Foundational Concepts
Resources and Resource Units

Habitat is a comprehensive term describing an area that encompasses the necessary combination of environmental conditions and resources that promote occupancy, reproduction, and survival of a species

(Morrison et al. 2006). Therefore, wildlife biologists interested in understanding habitat are typically focused on the identification, availability, and relative importance of resources. The definition of a *resource* in wildlife biology is broad and includes: matter taken up by an animal (e.g., food items), objects with which animals associate (e.g., nest tree), and conditions that influence the use of places and ultimately affect fitness (e.g., vegetation cover type; Buskirk and Millspaugh 2006). *Resource units* are quantifiable items or areas that can be observed as used (or not) by an animal, sometimes also called sample units (Lele et al. 2013; Rota et al. 2013). We focus our discussion on resource units comprised of spatial areas (e.g., quadrats) since these are most common in wildlife ecology. *Available resource units* are those units that are accessible to animals during a period of interest (Johnson 1980). *Used resources* are by definition a subset of available resources that are encountered and utilized, while *unused resource units* can be defined as either available or unavailable based upon study designs and assumptions.

Resource units can be described by a single attribute (e.g., canopy cover) or multiple attributes (e.g., land unit's slope and soil type) that differ among units. The method used to describe resource units affects the appropriate analysis, model interpretation, and study costs. Improvements in remotely sensed environmental attributes and geographic information systems (GIS) technology have greatly expanded our ability to accurately describe some resource units over large spatial and temporal scales, but many studies still rely upon detailed ground surveys. Although acquiring and classifying resource data into broad categories may be quicker and more cost effective than producing detailed measurements in the field, general classifications may miss important ecological mechanisms of selection. For example, height of groundcover may be the most important variable for a generalist bird's selection of a nest site, and broad vegetation classifications such as grassland versus forest would not capture the impor-

tance of groundcover. However, if the purpose of an analysis is to determine how animals respond to a general management action or disturbance, such as a forest harvest, then categorizing resource units as either harvested or unharvested, or by their distance to a harvest, may be sufficient.

The distribution of available resources describes the variation of resource types in environmental space (Lele et al. 2013). For example, when resource units are described by a single categorical attribute, say cover type, the available resource distribution represents the proportion of available units in each cover type category within some given area. When a resource is used disproportionately to its availability we infer selection (Johnson 1980). A statistical model used to estimate the probability of selecting a resource unit as a function of resource attributes is often called a *resource selection function* (RSF) or *resource selection probability function* (RSPF; Manly et al. 2002; Lele et al. 2013).

Availability

Defining what resources are available to individuals or a population is a primary concern for anyone interested in resource selection (Lele et al. 2013; Manly et al. 2002). By definition, available resources are assumed to be accessible to focal animals during a period of interest (Johnson 1980). Available resources can be completely accounted for, or a random sample of available units can be sampled from the available extent (Manly et al. 2002). In practice accessibility is rarely empirically studied partly because the focus of most studies is on the used resources, and quantifying availability is often constrained by logistics. Used units are typically compared with units occurring (existence or abundance) in some portion of an animal's environment. Boundaries of available resource occurrence are based upon assumptions from animal movement studies (e.g., home range polygons), management area boundaries, and mapping extents (Elith et al. 2011), or they are limited by budget or personnel. Consequently, when available resource units and distributions are summarized, results likely represent a greater diversity of attributes and quantity of resource units than what was truly accessible to the focal animal(s), and the selectivity metrics calculated can be invalid or applicable only to the study area investigated (Buskirk and Millspaugh 2006). The determination and underlying assumptions of available resources are important factors in determining how well a model actually represents the population of interest, regardless of the statistical method.

Use

When food items are resource units, the definition of use is straightforward: a used unit is one that is eaten. When spatial areas are considered resource units, interpretations of "use" are more variable. The presence of an animal on a resource unit typically corresponds to "use"; however, spatial units can be selected more than once and areas may be visited at different rates depending upon the size of resource units, timing of sampling, and animal's behavior during an observation. Use can be defined as binary, such as used versus unused or used versus available, or use can be defined by a measure of use intensity. In many studies, unused areas are difficult to determine because observations are snapshots in time. The method of observation affects how use is measured. When individual animals are not identified, use could be inferred as the presence or abundance of a species or individuals in a sampled area or unit and measured at each location or in a buffered area or plot containing each location. For example, Neu et al. (1974) recorded use as the number of moose tracks in four plots with variable burn history over seven time periods. When individual animals are identified with repeated observations of use generated for each animal, the location of individual relocations, clusters of relocations, or the area encompassing all relocations can be used to define use.

Aebischer et al. (1993) defined use as the proportion of resources within an individual animal's entire home range area, while Millspaugh et al. (2006) divided each home range into a grid and defined use as the mean utilization distribution (UD) value within each grid cell. Rate of use can also vary by the time of year and by animal behavior. For example, resources selected for foraging in spring may be vastly different from resources selected for nesting or from resources selected in winter. Some analyses can be designed to accommodate these dynamics. The type of use data collected is a primary determinant of which RSA is appropriate.

Scales of Selection and Use

Resource selection was described by Johnson (1980) as a hierarchical process, whereby each order of selection is conditional on a selection made at preceding levels. *First-order selection* is defined at the level of the physical or geographic range of a species; *second-order selection* is defined as selection of a use-area conditional on the species range (e.g., home range); *third-order selection* is defined as use of a component within the second order resource (e.g., cover type); and *fourth-order selection* refers to use of a particular resource within the third order resource (e.g., nest tree; Johnson 1980). The resources important to selection may change between or among orders of selection. For example, the geographic range of a species (first-order) may be limited by climatic extremes, while individual use areas (second-order) may be highly correlated with the composition and mosaic of distinctive vegetative types available within the geographic range. When spatial areas are considered resource units, the used area at a larger order of selection (e.g., a territory) constrains the scope of availability for the lower order of selection (e.g., resting site). RSAs can be designed to investigate selection at one or several orders of selection, and resource attributes can be characterized at multiple geographic scales. The August 2016

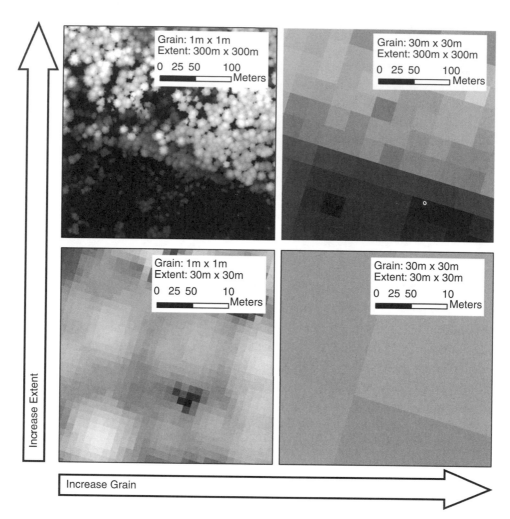

Fig. 12.1. Effect of changing geographical scale of grain and extent in heterogeneous landscapes. Expanding the study extent introduces new resources, and expanding the grain may dilute effects of rare resources.

special issue of *Landscape Ecology* is a good source for further explanations of multi-order and multi-scale resource selection analyses (McGarigal et al. 2016b).

Resources in natural systems are rarely uniformly distributed, and thus studies of resource selection are closely linked to spatial and temporal patterns of the landscape. The scale of a RSA is defined relative to the extent (or domain) and grain (or resolution) of used and available resources (Johnson 1980; Wiens 1989). When resource units represent spatial areas defined by the researcher, the grain chosen should consider the abundance and spatial distribution of use locations in the landscape, as well as the spatial

distribution of available resources (Boyce 2006). Interpretations of resource selection can vary widely depending upon the resolution of spatial resource units and what researchers define as being available to animals (Fig. 12.1). For further reading on the importance of scale in RSAs, see Johnson (1980), Wiens (1989), Boyce (2006), Gaillard et al. (2010), and McGarigal et al. (2016a).

Selecting a Resource Selection Analysis

Resource selection analyses represent a diverse collection of statistical methods ranging from simple

Table 12.1. Common resource selection analyses used by wildlife biologists.

Method	Complexity	Attribute data		Study design[a]	Citations
Chi-squared test	Simple	Univariate	categorical	I	Neu et al. (1974)
Rank based tests	Simple	Univariate	categorical	I, II	Friedman (1937); Johnson (1980)
Compositional analysis	Simple to moderate	Multivariate	categorical	II, III	Aebischer et al. (1993)
Logistic regression	Moderate to complex	Multivariate	categorical/continuous	I, II	Manly et al. (2002)
Polytomous logistic regression	Moderate to complex	Multivariate	categorical/continuous	I, II	North and Reynolds (1996)
Poisson regression	Moderate to complex	Multivariate	categorical/continuous	I, II	Manly et al. (2002)
Discrete choice	Moderate to complex	Multivariate	categorical/continuous	III, IV	Cooper and Millspaugh (1999)
Step-selection analysis	Moderate to complex	Multivariate	categorical/continuous	III, IV	Thurfjell et al. (2014)
Occupancy models	Moderate to complex	Multivariate	categorical/continuous	I, II, III	MacKenzie et al. (2002)
MaxEnt	Moderate to complex	Multivariate	categorical/continuous	I, II	Elith et al. (2011)

[a] These categorizations are based on generalities; there are exceptions to every rule, depending upon specific data collection and analysis structure.

metrics for quantifying use to complex multivariate techniques that incorporate spatial and temporal dynamics and variation among groups. Selecting the appropriate analysis can be challenging and should be an integral component of study design. Understanding assumptions inherent to use and availability data collected, methods to collect those data, the objective of the research (e.g., animal space use prediction versus hypothesis testing), and accessibility of methods affect decisions regarding which RSA to use. Here we describe some of the more common or historically influential RSAs in wildlife ecology (Table 12.1), discuss assumptions common in those analyses, and categorize resource selection study designs.

Common Analyses

The chi-square test (Neu et al. 1974) and univariate rank-based methods (Friedman 1937; Johnson 1980) are arguably the most simplistic RSAs in model structure. These analyses compare univariate categorical attribute values to test or rank attribute categories based on null hypothesis of no difference between proportions of used and available units (Alldredge and Ratti 1992; Manly et al. 2002). These methods are used less often today because models have since

been developed that rely on fewer assumptions and allow multivariate descriptions of resource units incorporating both continuous and categorical resource attributes.

Compositional analysis of resource selection is common in telemetry studies and other study designs where individual animals are repeatedly observed (Thomas and Taylor 2006). Compositional analysis is an extension of multivariate analysis of variance (MANOVA), which uses individual animals as replicates and can accommodate categorical differences between individuals (e.g., sex, group). The distribution of resources within each animal's home range boundary or other area encompassing relocations is compared to a larger available extent (Aebischer et al. 1993). Compositional analysis is appropriate only when observation sample sizes are large enough for individual use areas to stabilize (Aebischer et al. 1993). This method assumes independence between individuals sampled and normality of covariates (Manly et al. 2002). Results from standard compositional analysis cannot be used to calculate relative probability of use for spatial units (Manly et al. 2002), so weighted compositional analysis refines the method to assign resource use values based on utilization distributions of activity within use areas (Millspaugh et al. 2006). Several reviews have compared compositional analysis with other RSAs (Millspaugh and Marzluff 2001; Manly et al. 2002).

Logistic regression is a type of generalized linear model that relates a linear function of resource attributes to a binary response of used versus available. Logistic regression analyses dominated the resource selection literature in the 1990s and early 2000s. Keating and Cherry (2004) argue that it was often misused. Logistic regression models can incorporate continuous and/or categorical variables, but it is important that they be tested for noncollinearity. Odds ratios are used to interpret influence of attribute coefficients. Standard logistic regression does not account for variable use frequencies; however, polytomous (multinomial) logistic regression and Poisson (log-linear) regression, both generalized linear models, can accommodate variability in the intensity of use frequencies. Polytomous logistic regression relates a nominal (categorical) rather than binary response variable to a linear function of resource attributes (North and Reynolds 1996), and Poisson regression utilizes counts of use (Manly et al. 2002).

Improvements to statistical programs and computing power have led to an increase in hierarchical models (including random effect models) able to support complex model structures accommodating dependencies in use data (e.g., Gillies et al. 2006), multiple orders and scales of selection (McGarigal et al. 2016a), and ecological dynamics (McLoughlin et al. 2010). The incorporation of random effect terms to generalized linear models such as logistic regression models (Gillies et al. 2006) accommodates datasets where samples are not independent or cases where sampling of groups was unequal. Random effects also allow researchers interested in population-level effects to examine variation between individuals (e.g., Thomas et al. 2006).

Conditional availability models are a flexible class of RSAs that can be used whenever available resources differ within a sample of used resources. These models are particularly useful for studies that make repeated observations of marked individuals. Depending upon sampling structure, these models include conditional logistic regression models, matched-case control regressions, discrete choice models, and step-selection models. In discrete choice models, the combination of use location(s) and their matched available units is called a "choice set" (McCracken et al. 1998). These models range from the relatively simple to those that are complex; in the latter random effects in a hierarchical framework can be incorporated in either Bayesian or maximum likelihood approaches (Cooper and Millspaugh 1999; Manly et al. 2002; Thomas et al. 2006; Kneib et al. 2011). Step-selection models can be considered hierarchical extensions of basic conditional availability models in that they combine models of animal movement, which designate available resources for each use location, with integrated models of resource selection (Thurfjell et al. 2014).

Profile models apply statistical distribution measures from observed use locations to infer species use in similar environmental gradients elsewhere. Mahalanobis distance modeling is a profiling technique that uses vectors of average use to assigned values applicable to mapping animal-resource selection based on how similar other areas are to the multivariate mean (Clark et al. 1993). This method can incorporate numerous multivariate attributes and does not depend upon a sample of available habitat (Manly et. al. 2002).

Machine-learning approaches have been adapted to study resource selection as computer systems have improved. These methods are very flexible and can handle highly nonlinear relationships better than more traditional functions such as logistic regression. The maximum entropy method (e.g., MaxEnt) has increasingly been used to evaluate resource selection (Phillips et al. 2017). MaxEnt modeling software uses a machine-learning process to analyze environmental conditions and fit resource selection functions based on observed use contrasted against a large random sample of available resource units (Merow et al. 2013). Many of these more recent analytical methods were developed in conjunction with or for use in GIS. GIS programs are increasingly relevant due to their ability to handle high-volume datasets collected from automated animal and vegetation monitoring systems and for their ability to generate spatially explicit predictions (e.g., habitat maps).

Detection of any wildlife species is imperfect, and for most species detection probabilities will vary among sites. Incorporating detection probabilities in the analysis of site surveys (i.e., probabilistic models of occupancy) provides the most robust and appropriate analytical methodology for population-level studies of resource selection involving nonmarked animals (MacKenzie et al. 2006). It has been repeatedly demonstrated that failing to incorporate imperfect detection can alter forecasted population trends and estimated species distributions, especially for species with low to moderate detection probabilities (e.g., Field et al. 2005; Martin et al. 2005; Rota et al. 2011). There are many site- or time-specific factors that can impede an observer's ability to detect a focal species. For example, sight surveys may have lower detection probabilities during inclement weather if animals are less active or observer visibility is reduced. Further, even if occupancy status is equal among sites, an observer will be less likely to detect the focal species at a site with very dense vegetation compared to a site with greater visibility. MacKenzie et al. (2002) developed a flexible single-season, single-species occupancy modeling framework that has since been built upon to model multi-season (MacKenzie et al. 2003), multi-species co-occurrence (MacKenzie et al. 2004; Rota et al. 2016), abundance (Royle and Nichols 2003), and demographic vital rates (Rossman et al. 2016).

Common Assumptions

All statistical models contain inherent simplifying assumptions to describe complex ecological processes. Two assumptions common to all models of resource selection are: (1) animals display varying degrees of selection for resources at a range of spatial and temporal scales, and (2) use of a resource provides evidence for its importance for the animal's ecology. These assumptions are necessary to infer underlying drivers for selection from observations of use. The choice of additional simplifying assumptions depends upon study designs, study questions, and the method of analysis. Assumptions should be considered when planning or interpreting results in any study of resource selection. Further reading regarding assumptions in RSA using categorical resource classifications can be found in Alldredge and Griswold (2006), assumptions applicable to RSA using telemetry data can be found in Millspaugh and Marzluff (2001), and assumptions inherent in species distribution analyses can be found in Guisan and Thuiller (2005). Some common assumptions used in RSAs include:

1. *Available and used resource units are correctly classified.* This assumption may be violated if there are scale issues or biases in data collection methods. For example, used resources may be misclassified if the location error from a telemetry triangulation is larger than the grain of the categorized resource units. Inherent bias in observations or census data can also be problematic if observed use data is clumped within the available resource area due to nonenvironmental reasons. For example, spatial bias is likely to occur if surveyed areas are not representative of the area considered available to the individual or populations. This potential bias is common in studies that rely on surveys of areas most accessible by observers (e.g., road surveys) and then extend inference to inaccessible areas.

2. *Availability is constant over the period of study.* This assumption may be violated if resource availability changes between years or throughout the seasons, but inference is made more broadly.

3. *Resources or resource units are equally available to individuals within the population.* An examination of species natural history should help determine if this assumption is appropriate and guide data collection of available areas for observed use locations. For example, since dominant individuals

of a territorial species often exclude other individuals from their territory, that area is not available to the entire population.

4. *Selection criteria are constant across individuals.* This assumption is common in studies that do not identify individuals and in studies where repeated samples from individuals are lumped without random effects. This assumption may be violated when selection changes with characteristics of individuals, such as age class, sex, or breeding status.

5. *A random selection of animals was sampled, and those individuals are representative of the population.* This is a basic assumption for most studies interested in population level questions; however, it is not always the case due to the difficulty of sampling populations. This assumption could also be violated if detectability is unequal across sampling areas.

6. *Resource use or selections made by individuals are independent from other individuals.* This assumption may be violated when family members (e.g., mother and offspring) or territorial competitors are included in the same dataset without accounting for dependencies. Some analyses bypass this assumption through the use of random effects on coefficient slopes or by partitioning data.

7. *Relocations of individual animals are not spatially or temporally correlated.* This assumption may be violated when repeated observations are collected in rapid succession.

Study Designs

Resource selection studies can be classified into four general study designs (see Thomas and Taylor 1990; Erickson et al. 2001; Manly et al. 2002). Classification is based primarily on whether resource availability and use are measured at the population or individual level, and whether at least some animals in the population are identifiable. Here we summarize each study design category, list common analyses, and provide an example for each design.

DESIGN I

Individual animals are not identified in design I studies, and available areas are sampled on a population level. This design was common in early RSAs (e.g., Neu et al. 1974) and remains popular for answering questions of selection across large spatial extents. Roadside surveys, such as the North American Breeding Bird Survey (https://www.pwrc.usgs.gov/bbs/), are an example of this study design, but many assumptions are not met so inference is limited. Common analyses include chi-square, logistic regression, log-linear modeling, occupancy modeling, and MaxEnt.

Here we present an analysis that incorporates results from studies by Davis et al. (2016) and Glenn et al. (2017) that used MaxEnt to produce predictive maps (Fig. 12.2) of forests suitable for nesting and roosting by northern spotted owls (*Strix occidentalis caurina*) at two spatial scales. A primary objective of the studies was to generate models to inform regional monitoring and conservation planning, and Glenn et al. (2017) demonstrated an effective method to estimate carrying capacity of dynamic landscapes. Northern spotted owl nesting and roosting locations were collected during long-term demographic research (see Dugger et al. 2016) and land management agency surveys. A quality control process was conducted to ensure use locations were correctly identified and geographically dispersed throughout the entire modeling region. Spatial autocorrelation and sampling bias were addressed by limiting location data to only one location per territorial pair and randomly spacing them no nearer than the estimated median nearest neighbor distance (a.k.a., spatial filtering).

Two orders of selection were investigated. The third-order selection was analyzed at the scale of a forest stand, regardless of patch size or patterns. Available resource unit attributes (predictor variables) were chosen based on forest stand structure and species composition attributes (Davis et al. 2016). An RSF was fit to these data through the use of response functions (e.g., linear, product, qua-

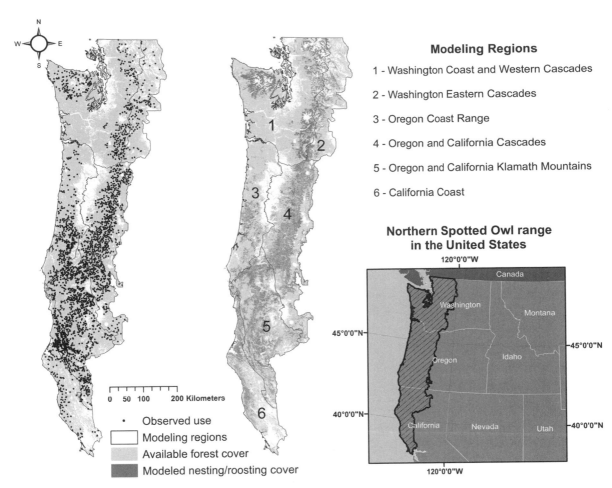

Fig. 12.2. Example of a design I resource selection analysis. Observed use locations and modeling extents used to create northern spotted owl (*Strix occidentalis caurina*) nesting and roosting habitat suitability maps in Glenn et al. (2017) and Davis et al. (2016). *Source: Adapted from Glenn et al. (2017), and used with the permission of Springer Nature.*

dratic) that were determined plausible based on species expert knowledge. Modeling encompassed the entire range of the subspecies, which varied widely in available resources (e.g., redwood forests occurred only in one region of the geographic range). Therefore, the range was subdivided into six modeling regions based on similarity of resources important for northern spotted owls. The models produced represented the relative likelihood of selection of forest types suitable for nesting and roosting use in each modeling region and were mosaicked to produce a range-wide map (Fig. 12.2). At the territory scale (second-order selection) Glenn et al. (2017) used a classified binary version of the Davis et al. (2016) maps to produce predictor variables that represented the amount and spatial arrangement of nesting/roosting forest cover, and also included topographic and climate variables. Forest cover variables were: percentage of forest cover likely to be used for nesting and roosting within various radii (Fig. 12.3a and b), the distance from contiguous large patches of nesting/roosting cover (Fig. 12.3c), and the amount of diffuse (intermixed with edge) nesting/roosting cover (Fig. 12.3d).

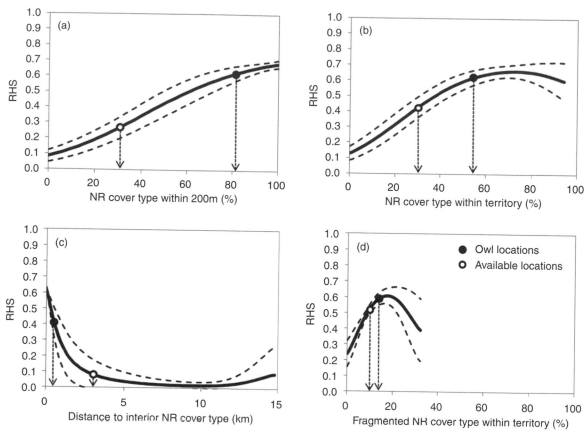

Fig. 12.3. Northern spotted owl (*Strix occidentalis caurina*) relative habitat suitability response functions (solid line) with 95% confidence intervals (dashed lines) based on nesting/roosting cover averaged across modeling regions. The dots represent the average conditions at known spotted owl locations and available locations. *Source: Adapted from Glenn et al. (2017), and used with the permission of Springer Nature.*

The study found selection for landscapes with high cover of nesting/roosting forest types at multiple scales, with a small amount of fragmentation (~20% of the territory) having a positive influence on selection (Fig. 12.3). The same observed use data were utilized for both models, yet the RSFs from each depict the relative likelihood of use at two different spatial scales. It is possible to find forest stands that have all the attributes of heavily used nesting/roosting stands, but have low likelihood of territorial use due their size and juxtaposition in the landscape. Thus, suitability for use at one scale does not always imply suitability for use at another, and in this situation, caution is warranted when representing suitable cover type as suitable habitat.

DESIGN II

In studies with design II, individual animals are identified and resource use is defined for each individual, but resource availability is defined at the population level. This design is common for radio-telemetry studies where relocations are used to describe resource use by tagged individuals, and remotely sensed data are used to describe habitat availability across a study area. Common analyses used with this study design include: Friedman's test, Johnson's method, compositional analysis, discrete choice modeling, logistic regression, log-linear modeling, and multiple regression.

In a telemetry-based study of eastern spotted skunk (*Spilogale putorius*) selection of discrete cover types, Lesmeister et al. (2009) identified resource availability based on proportional coverage of delineated cover types within the study area defined as the maximum convex polygon of all home ranges. The study was conducted in a forested landscape managed primarily for herbaceous understory and an older, open canopy forest. Home ranges were estimated for each individual from 95% fixed-kernel utilization distributions of locations. Selection was determined by a weighted compositional analysis of the proportion of utilization distribution by cover type within each individual's home range (Aebischer et al 1993; Millspaugh et al. 2006). Most available cover types were older pine forest with high herbaceous cover, but eastern spotted skunks showed strong selection for hardwood and young pine stands over other available cover types. In this case, selection for complex forest with dense overstory was likely driven by behaviors to reduce predation risk from avian predators (Lesmeister et al. 2009).

DESIGN III

Studies that identify individuals and quantify resource use and availability for each individual separately are categorized as design III. This study design is common in studies of territorial animals or other situations where monitored individuals would not have equal access to study areas. Common analyses include: compositional analysis, discrete choice modeling, logistic regression, and multiple regression.

Here we present a simplified example of the most common type of discrete choice model used in RSAs, the multinomial logit model, using a subset of northern spotted owl nesting and roosting locations collected on the Klamath, Oregon, demographic study area during long-term demographic research (see Dugger et al. 2016). The primary objective in this example was to determine if fine scale canopy structure attributes generated from remote sensing light detection and ranging (lidar) maps were useful for predicting the selection of a nest site within territories (third-order selection; Johnson 1980). The extent of the analysis was limited to owl territories that overlapped

with canopy height map coverages in both space and time. Northern spotted owls are territorial and thus the assumption that areas are equally available for all pairs at the population or regional level is not appropriate; we generated unique choice sets for each territory. The resource units for this analysis were 200-m radius circular plots centered on nest sites (use area) and two random points within the bounds of each territory (sample of available areas; Fig. 12.4).

Discrete choice analysis can accommodate many covariates with variable structure (e.g., linear, quadratic, interactions); however for simplicity, we considered the contribution of just two linear attributes to selection: percentage of mature forest (canopy >80 years) and standard deviation in canopy height. We chose to use a Bayesian framework, similar to that of Jenkins et al. (2017), but these models can also be fit with a maximum likelihood approach (Manly et al. 2002). We used an information theoretic approach, with the Watanabe-Akaike information

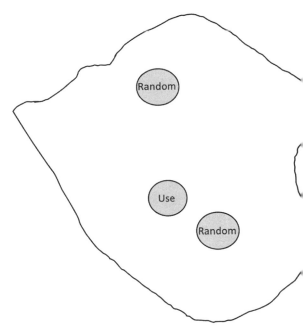

Fig. 12.4. Data collection structure for discrete choice analysis of nest site selection within northern spotted owl (*Strix occidentalis caurina*) breeding territories. Each choice set contained one used location and two randomly selected locations within owl territories.

criterion (see Chapter 4 in this volume), to evaluate single attribute models, a model containing both attributes, and a null model of random selection. The predictive fit of models was calculated using a likelihood ratio test, Estrella's R^2 (Estrella 1998), although a k-fold cross-validation method would have also been appropriate (Boyce et al. 2002). We determined that both covariates were useful in predicting nest locations of northern spotted owls (Table 12.2). Since no interaction effects were present, we can also interpret covariate selection ratios; the selection ratio (exp[coefficient]) measures the multiplicative change in probability of selection when a covariate increases by one unit while others remain the same (McDonald et al. 2006). We found support for positive effects of both percentage of mature forest within 200 m (mean selection ratio = 2.76) and standard deviation in canopy height within 200 m (mean selection ratio = 3.74) on the probability of nest site selection.

DESIGN IV

Studies that make repeated observations of identified individuals, in which resource use and availability are quantified for each observation separately, are categorized as design IV. Resource use is defined for each individual, and availability is uniquely defined for each point of use. Common analyses include discrete choice models and step-selection models.

In the same telemetry-based study of eastern spotted skunk spatial ecology highlighted for study design II, Lesmeister et al. (2008) quantified microhabitat characteristics of sites selected for denning and resting (fourth order of selection). Paired with each used site was a nearby and putatively unused (i.e., available) site suitable for resting or denning. These used and available sites were compared based on habitat characteristics measured at each site. Given the paired used/available design and because reuse of sites by the same study animal was common, Lesmeister et al. (2008) used multinomial discrete choice analysis to fit site selection models in a maximum likelihood framework. In this case discrete choice analysis was most appropriate because the researchers were able to define resource availability separately for each observation (Cooper and Millspaugh 1999). Contrasting used and available sites, they found that selection was based primarily on increased cover of dense vegetation, with additional support for higher rock densities, younger pine forest stands, older hardwood stands, steeper slope, and smaller site entrances. Their results supported the hypothesis that eastern spotted skunks select structurally complex sites for denning and resting partially for thermal regulation, but primarily to enhance protection from predators. This underlying driver in resource selection was also found at the sec-

Table 12.2. Model selection results for discrete choice models evaluating the utility of percent mature forest (mature) and the standard deviation in canopy height (SD canopy) to predict nest area selection by northern spotted owls within territories. Estrella's R^2 is a likelihood ratio test of model fit (1 = perfect prediction, 0 = random).

Model	K^a	$WAIC^b$	$\Delta WAIC$	Estrella's R^2
Mature + SD canopy height	2	85.05	0.00	0.66
SD canopy height	1	96.95	11.90	0.53
Mature	1	112.86	27.81	0.33
Null (random selection)	0	134.04	48.99	0.00

[a] Number of model parameters

[b] Watanabe-Akaike Information criterion

ond and third orders of selection (Lesmeister et al. 2009, 2013).

Evaluation and Validation of Resource Selection Analyses

Various methods are used to rank and evaluate the fit of individual models or model sets (e.g., Akaike information criterion (AIC), goodness of fit, etc.). These evaluations are based on the data used to develop (train) the model. However, once the most supported model is decided upon, it is then necessary to validate how well the model predicts use (Fig. 12.5). Using the same observation data to train

and test the model may lead to overly optimistic test results and is thus not advised for model validation (Howlin et al. 2003). Using observation data independent of the model training data is considered the best approach. For example, Glenn et al. (2017) used independently collected location data to validate their range-wide model of forests suitable for northern spotted owl nesting and roosting. When independent data are not an option, an often used alternative approach is to subset observation data into model "training" and "testing" replica subsets (e.g., k-fold cross-validation), which can be done in several ways, from geographic partitioning to random selections. A model is then built from each training subset and

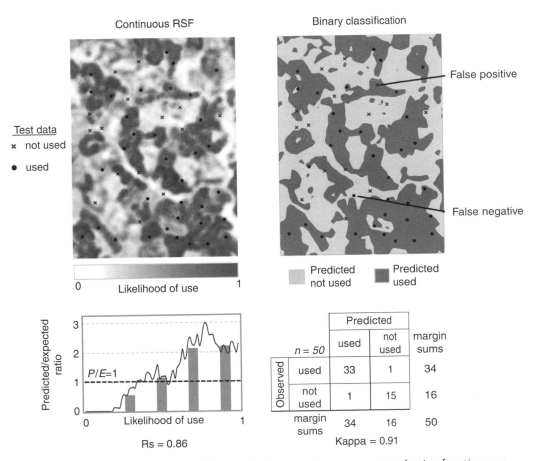

Fig. 12.5. Resource selection analysis map validations showing a continuous resource selection function map (upper-left) and a classified map (upper-right). Beneath each map are examples of proper validation methods, showing predicted-to-expected ratio curve used for continuous RSFs (lower-left) and the confusion matrix used for classified RSFs (lower-right).

its prediction tested with the held-out subset(s). The appropriate model prediction validation method is often dependent on the RSA used. Three commonly used techniques include (1) the confusion matrix, (2) area under the curve (AUC), and (3) area-adjusted frequency (Boyce et al. 2002). Another recent method uses linear regressions between predicted use and observed use, where the slope of the line provides information on how well the model predicted use (Howlin et al. 2003).

Resource selection functions are continuous representations of selection, from low to high. While this gradient of selection is informative, it is sometimes necessary to simplify it into discrete classes for mapping and analysis purposes (e.g., unsuitable or suitable). Where to divide the RSF into these classes is not always clear, and often arbitrary class breaks (e.g., 0.5) are used (Morris et al. 2016). The predicted-to-expected ratio curve method developed by Hirzel et al. (2006) produces information of a model's predictive qualities similar to the area-adjusted frequency method (Boyce et al. 2002), but also provides information on how to classify the model into ecologically meaningful classes in a nonarbitrary fashion (Fig. 12.5). Once an RSA model is divided into discrete classes of use, the confusion matrix can be used for evaluating the model's predictability (Boyce et al. 2002).

Uncertainty is inherent in all models, since they are just representations of observed, complex, and not fully understood processes. Because of this, model validations are a necessary last step in the RSA work flow. Validation methods inform us of the model's usefulness and, just as importantly, weaknesses, for applying the model's predictions of resource selection across space or time.

The science of resource selection is rapidly advancing along with improvements in technology and analytical methods, and new methods should be embraced. Hirzel and Le Lay (2008) outlined a dozen useful steps for habitat modelers. A few that we consider most useful to RSA are:

1. Cleary define the goal of the study, particularly in regard to scale and generalization.
2. Preselect resource attributes (variables) considering species ecology and the scale of use data.
3. Carefully delineate the study area to encompass only those resources accessible to the species.
4. Know your chosen analysis method's assumptions, caveats, and strengths.
5. Interpret the model results to ensure they make sense ecologically.
6. Evaluate and assess the model's predictive power and variance.

Regardless of the method, all RSAs are designed to address similar questions and provide useful insights and predictions for wildlife management and conservation. Each method has its strengths and weaknesses, but all are tools that can be used successfully if care is taken in their application.

Acknowledgments

The authors are grateful to the USDA Forest Service Pacific Northwest Research Station and Pacific Northwest Region (Region 6) for supporting their time to contribute to this chapter. They also thank the northern spotted owl demographic team for use of northern spotted owl nest data and E. Glenn for permission to use figures 12.2 & 12.3.

Summary

The study of resource selection encompasses a vast assortment of modeling techniques and study designs. We have focused our discussion on some fundamental concepts and a few of the most commonly used RSAs from the last few decades to the present.

LITERATURE CITED

Aebischer, N., P. Robertson, and R. Kenward. 1993. Compositional analysis of habitat use from animal radio-tracking data. Ecology 74:1313–1325.

Alldredge, J. R., and J. Griswold. 2006. Design and analysis of resource selection studies for categorical resource variables. Journal of Wildlife Management 70:337–346.

Alldredge, J. R., and J. T. Ratti. 1992. Further comparison of some statistical techniques for analysis of resource selection. Journal of Wildlife Management 56:157–165.

Boyce, M. S. 2006. Scale for resource selection functions. Diversity and Distributions 12:269–276.

Boyce, M. S., P. R. Vernier, S. E. Nielsen, and F. K. A. Schmiegelow. 2002. Evaluating resource selection functions. Ecological Modelling 157:281–300.

Buskirk, S. W., and J. J. Millspaugh. 2006. Metrics for studies of resource selection. Journal of Wildlife Management 70:358–366.

Clark, J. D., J. E. Dunn, and K. G. Smith. 1993. A multivariate model of female black bear habitat use for a geographic information system. Journal of Wildlife Management 57:519–526.

Cooper, A. B., and J. J. Millspaugh. 1999. The application of discrete choice models to wildlife resource selection studies. Ecology 80:566–575.

Davis, R. J., B. Hollen, J. Hobson, J. E. Gower, and D. Keenum. 2016. Northwest Forest Plan——The first 20 years (1994–2013): Status and trends of northern spotted owl habitats. PNW- GTR-929. USDA Forest Service, Pacific Northwest Research Station, Portland, OR.

Dugger, K. M., E. D. Forsman, A. B. Franklin, R. J. Davis, G. C. White, C. J. Schwarz, K. P. Burnham, J. D. Nichols, J. E. Hines, C. B. Yackulic, P. F. Doherty, L. Bailey, D. A. Clark, S. H. Ackers, L. S. Andrews, B. Augustine, B. L. Biswell, J. Blakesley, P. C. Carlson, M. J. Clement, L. V. Diller, E. M. Glenn, A. Green, S. A. Gremel, D. R. Herter, J. M. Higley, J. Hobson, R. B. Horn, K. P. Huyvaert, C. McCafferty, T. McDonald, K. McDonnell, G. S. Olson, J. A. Reid, J. Rockweit, V. Ruiz, J. Saenz, and S. G. Sovern. 2016. The effects of habitat, climate, and Barred Owls on long-term demography of Northern Spotted Owls. Condor 118:57–116.

Elith, J., S. J. Phillips, T. Hastie, M. Dudík, Y. E. Chee, and C. J. Yates. 2011. A statistical explanation of MaxEnt for ecologists. Diversity and Distributions 17:43–57.

Erickson, W. P., T. L. McDonald, K. G. Gerow, S. Howlin, and J. W. Kern. 2001. Statistical issues in resource selection studied with radio-marked animals. In Radio tracking and animal populations, edited by J. J. Millspaugh and J. M. Marzluff, pp. 211–242. Academic Press, San Diego, CA.

Estrella, A. 1998. A new measure of fit for equations with dichotomous dependent variables. Journal of Business and Economic Statistics 16:198–205.

Field, S. A., A. J. Tyre, K. H. Thorn, P. J. O'Connor, and H. P. Possingham. 2005. Improving the efficiency of wildlife monitoring by estimating detectability: A case study of foxes (*Vulpes vulpes*) on the Eyre Peninsula, South Australia. Wildlife Research 32:253–258.

Friedman, M. 1937. The use of ranks to avoid the assumption of normality implicit in the analysis of variance. Journal of the American Statistical Association 32:675–701.

Gaillard, J.-M., M. Hebblewhite, A. Loison, M. Fuller, R. Powell, M. Basille, and B. Van Moorter. 2010. Habitat-performance relationships: Finding the right metric at a given spatial scale. Philosophical Transactions of the Royal Society of London. Series B, Biological Sciences 365:2255–65.

Gillies, C. S., M. Hebblewhite, S. E. Nielsen, M. A. Krawchuk, C. L. Aldridge, J. L. Frair, D. J. Saher, C. E. Stevens, and C. L. Jerde. 2006. Application of random effects to the study of resource selection by animals. Journal of Animal Ecology 75:887–898.

Glenn, E. M., D. B. Lesmeister, R. J. Davis, B. Hollen, and A. Poopatanapong. 2017. Estimating density of a territorial species in a dynamic landscape. Landscape Ecology 32:563–579.

Guisan, A., and W. Thuiller. 2005. Predicting species distribution: Offering more than simple habitat models. Ecology Letters 8:993–1009.

Hirzel, A. H. and G. Le Lay. 2008. Habitat suitability modelling and niche theory. Journal of Applied Ecology. 45:1372–1381.

Hirzel, A.H., G. Le Lay, V. Helfer, C. Randin, and A. Guisan. 2006. Evaluating the ability of habitat suitability models to predict species presences. Ecological Modelling 199:142–152.

Howlin, S., W. Erickson, and R. Nielson. 2003. A validation technique for assessing predictive abilities of resource selection functions. In Proceedings of the First International Conference on Resource Selection, pp. 40–50. Laramie, WY.

Jenkins, J. M. A., F. R. Thompson, and J. Faaborg. 2017. Species-specific variation in nesting and postfledging resource selection for two forest breeding migrant songbirds. PLoS ONE 12:1–14.

Johnson, C. J., S. E. Nielsen, E. H. Merrill, T. L. McDonald, and M. S. Boyce. 2006. Resource selection functions based on use-availability data: Theoretical motivation and evaluation methods. Journal of Wildlife Management 70:347–357.

Johnson, D. D. H. 1980. The comparison of usage and availability measurements for evaluating resource preference. Ecology 61:65–71.

Keating, K. A., and S. Cherry. 2004. Use and interpretation of logistic regression in habitat-selection studies. Journal of Wildlife Management 68:774–789.

Kneib, T., F. Knauer, and H. Küchenhoff. 2011. A general approach to the analysis of habitat selection. Environmental and Ecological Statistics 18:1–25.

Lele, S. R., E. H. Merrill, J. Keim, and M. S. Boyce. 2013. Selection, use, choice and occupancy: Clarifying concepts in resource selection studies. Journal of Animal Ecology 82:1183–1191.

Lesmeister, D. B., R. S. Crowhurst, J. J. Millspaugh, and M. E. Gompper. 2013. Landscape ecology of eastern spotted skunks in habitats restored for red-cockaded woodpeckers. Restoration Ecology 21:267–275.

Lesmeister, D. B., M. E. Gompper, and J. J. Millspaugh. 2008. Summer resting and den site selection by eastern spotted skunks (*Spilogale putorius*) in Arkansas. Journal of Mammalogy 89:1512–1520.

Lesmeister, D. B., M. E. Gompper, and J. J. Millspaugh. 2009. Habitat selection and home range dynamics of Eastern spotted skunks in the Ouachita Mountains, Arkansas, USA. Journal of Wildlife Management 73:18–25.

MacKenzie, D. I., L. L. Bailey, and J. D. Nichols. 2004. Investigating species co-occurrence patterns when species are detected imperfectly. Journal of Animal Ecology 73:546–555.

MacKenzie, D. I., J. D. Nichols, J. E. Hines, M. G. Knutson, and A. B. Franklin. 2003. Estimating site occupancy, colonization, and local extinction when a species is detected imperfectly. Ecology 84:2200–2207.

MacKenzie, D. I., J. D. Nichols, G. B. Lachman, S. Droege, J. A. Royle, and C. A. Langtimm. 2002. Estimating site occupancy rates when detection probabilities are less than one. Ecology 83:2248–2255.

MacKenzie, D. I., J. D. Nichols, J. A. Royle, K. H. Pollock, L. L. Bailey, and J. E. Hines. 2006. Occupancy estimation and modeling: Inferring patterns and dynamics of species occurrence. Academic Press, New York.

Manly, B. F. J., L. L. McDonald, D. L. Thomas, T. L. McDonald, and W. P. Erickson. 2002. Resource selection by animals: Statistical design and analysis for field studies, 2nd ed. Chapman and Hall, London.

Martin, T. G., B. A. Wintle, J. R. Rhodes, P. M. Kuhnert, S. A. Field, S. J. Low-Choy, A. J. Tyre, and H. P. Possingham. 2005. Zero tolerance ecology: Improving ecological inference by modelling the source of zero observations. Ecology Letters 8:1235–1246.

Mathewson, H. A., and M. L. Morrison. 2015. The misunderstanding of habitat. In Wildlife habitat conservation: Concepts, challenges and solutions, edited by M. L. Morrison and H. A. Mathewson, pp. 3–8. Johns Hopkins University Press, Baltimore, MD.

McCracken, M. L., B. F. J. Manly, and M. Vander Heyden. 1998. The use of discrete-choice models for evaluating resource selection. Journal of Agricultural, Biological, and Environmental Statistics 3:268–279.

McDonald, L., B. Manly, F. Huettmann, and W. Thogmartin. 2013. Location-only and use-availability data:

Analysis methods converge. Journal of Animal Ecology 82:1120–1124.

McDonald, T. L., B. F. J. Manly, R. M. Nielson, and L. V. Diller. 2006. Discrete-choice modeling in wildlife studies exemplified by northern spotted owl nighttime habitat selection. Journal of Wildlife Management 70:375–383.

McDonald, T. L. 2013. The point process use-availability or presence-only likelihood and comments on analysis. Journal of Animal Ecology 82:1174–1182.

McGarigal, K., H. Y. Wan, K. A. Zeller, B. C. Timm, and S. A. Cushman. 2016a. Multi-scale habitat selection modeling: A review and outlook. Landscape Ecology 31:1161–1175.

McGarigal, K., K. A. Zeller, and S. A. Cushman. 2016b. Multi-scale habitat selection modeling: Introduction to the special issue. Landscape Ecology 31:1157–1160.

McLoughlin, P. D., D. W. Morris, D. Fortin, E. Vander Wal, and A. L. Contasti. 2010. Considering ecological dynamics in resource selection functions. Journal of Animal Ecology 79:4–12.

Merow, C., M. J. Smith, and J. A. Silander. 2013. A practical guide to MaxEnt for modeling species' distributions: What it does, and why inputs and settings matter. Ecography 36:1058–1069.

Millspaugh, J. J., and J. M. Marzluff, eds. 2001. Radio tracking and animal populations. Academic Press, San Diego, CA.

Millspaugh, J. J., R. M. Nielson, L. McDonald, J. M. Marzluff, R. A. Gitzen, C. D. Rittenhouse, M. W. Hubbard, and S. L. Sheriff. 2006. Analysis of resource selection using utilization distributions. Journal of Wildlife Management 70:384–395.

Morris, L. R., K. M. Proffitt, and J. K. Blackburn. 2016. Mapping resource selection functions in wildlife studies: Concerns and recommendations. Applied Geography 76:173–183.

Morrison, M. L., B. G. Marcot, and R. W. Mannan. 2006. Wildlife-habitat relationships: Concepts and applications, 3rd ed. Island Press, Washington, DC.

Neu, C. W., C. R. Byers, and J. M. Peek. 1974. A technique for analysis of utilization-availability data. Journal of Wildlife Management 38:541.

North, M. P., and J. H. Reynolds. 1996. Microhabitat analysis using radiotelemetry locations and polytomous logistic regression. Journal of Wildlife Management 60:639–653.

Phillips, S. J., R. P. Anderson, M. Dudík, R. E. Schapire, and M. E. Blair. 2017. Opening the black box: An open-source release of MaxEnt. Ecography 40:887–893.

Rossman, S., C. B. Yackulic, S. P. Saunders, J. Reid, R. Davis, and E. F. Zipkin. 2016. Dynamic N-occupancy models: Estimating demographic rates and local abundance from detection-nondetection data. Ecology 97:3300–3307.

Rota, C. T., M. A. R. Ferreira, R. W. Kays, T. D. Forrester, E. L. Kalies, W. J. McShea, A. W. Parsons, J. J. Millspaugh, and D. Warton. 2016. A multispecies occupancy model for two or more interacting species. Methods in Ecology and Evolution 7:1164–1173.

Rota, C. T., R. J. Fletcher, J. M. Evans, and R. L. Hutto. 2011. Does accounting for imperfect detection improve species distribution models? Ecography 34:659–670.

Rota, C. T., J. J. Millspaugh, D. C. Kesler, C. P. Lehman, M. A. Rumble, and C. M. B. Jachowski. 2013. A re-evaluation of a case-control model with contaminated controls for resource selection studies. Journal of Animal Ecology 82:1165–1173.

Royle, J. A., and J. D. Nichols. 2003. Estimating abundance from repeated presence-absence data or point counts. Ecology 84:777–790.

Strickland, M. D., and L. L. McDonald. 2006. Introduction to the special section on resource selection. Journal of Wildlife Management 70:321–323.

Thomas, D. L., D. Johnson, and B. Griffith. 2006. A Bayesian random effects discrete-choice model for resource selection: Population-level selection inference. Journal of Wildlife Management 70:404–412.

Thomas, D. L., and E. J. Taylor. 2006. Study designs and tests for comparing resource use and availability II. Journal of Wildlife Management 70:324–336.

Thurfjell, H., S. Ciuti, and M. S. Boyce. 2014. Applications of step-selection functions in ecology and conservation. Movement Ecology 2:4.

Warton, D., and G. Aarts. 2013. Advancing our thinking in presence-only and used available analysis. Journal of Animal Ecology 82:1125–1134.

Wiens, J. A. 1989. Spatial scaling in ecology. Functional Ecology 3:385–397.

ANDREW N. TRI

13 — Spatial Statistics in Wildlife Research

> Spatial data are everywhere. —Bivand et al. (2008:1)

Introduction

Space influences nearly every ecological process (Tilman and Kareiva 1998; Hanski 1999). The spatial arrangement of landscape features affects movement, dispersal, survival, and habitat selection of wildlife (Krebs 1999). Until the past few decades, spatial data were not explicitly included in wildlife management; spatial relationships in wildlife research were implied (i.e., Leopold's law of edge [1933], island biogeography models [Wilson and MacArthur 1967]) but not explicitly modeled. Since the 1970s, geospatial technologies—geographic information systems (GIS), global positioning systems (GPS), and remote sensing—have revolutionized the field of geostatistics and the way that we understand wildlife ecology. Spatial data allow us to identify where wildlife occurs and what factors affect the choices that wildlife make. Furthermore, spatial data can be used to augment traditional analyses to better understand the underlying ecological process. The purpose of this chapter is to illustrate the components of spatial data and conventional analysis techniques, as well as how to select an appropriate model for your spatial data and utilize the benefits of spatial data while minimizing the pitfalls. Above all else, the analysis technique depends on the question being asked.

Why Are Spatial Data Different?

Heterogeneity is ever present in all data. It occurs spatially, temporally, and from individual traits being measured. *Spatial heterogeneity*—the uneven distribution of features on the landscape—arises from the patchy or inconsistent spatial arrangement. Heterogeneous landscapes have mixes of different densities of animals, plants, nutrients, weather patterns, and soils. Spatial data are spatially heterogeneous; these data differ in their statistical properties from plot-based agricultural or ecological data. The data consist of thousands or millions of measurements taken from locations adjacent to or near neighboring observations.

Remote sensing enables researchers to quantify variables on the landscape, but subsequent data analysis can be problematic. Two major problems arise with large spatial datasets: sample size and statistical properties (Plant 2012). Traditional null hypothesis depends on sample size and significance level (α-value). The hypothesis test is much more sensitive to rejecting H_0, as sample size increases (Matloff 1991). Traditional statistical inference is rarely meaningful when spatial datasets have thousands of samples; the null hypothesis is almost always rejected in favor of the alternate (Matloff 1991; Plant 2012).

Spatial Autocorrelation

Spatial data often do not satisfy traditional statistical assumptions (Plant 2012), but rather follow Tobler's (1970:236) first law of geography: "Everything is related to everything else, but near things are more related than distant things." In statistical terms, a measurement at one location covaries with other points; the strength of the covariance is dictated by proximity. Spatial dependency among nearby locations means that these data are *spatially autocorrelated*. Features are redundantly counted, which results in the over-counting of features. For example, measuring the grass biomass in a prairie will be very similar when plots are adjacent, but different when plots are spaced far apart. The adjacent plots are not representative of the whole prairie, which results in over-counting. Such over-counting violates the assumption of independence in statistical tests. Regression techniques that do not account for spatial autocorrelation have unreliable significance tests and unstable parameter estimates (Knegt et al. 2010), but some regression techniques, such as spatial regression or geographically weighted regression, do not suffer from these weaknesses (Fotheringham et al. 2002).

Spatial autocorrelation can be loosely defined as "the coincidence of value similarity with locational similarity" (Anselin and Bera 1998:241). In simple terms, data are spatially autocorrelated if the values at locations i and j covary. Imagine a map of mean annual temperature; nearby locations have similar values of temperature and locations farther apart are similar. Temperatures at locations close together have high covariance (temperatures in Phoenix, Arizona, are more similar to those in Tucson, Arizona, than temperatures in Phoenix, and Minneapolis, Minnesota). Mean annual temperature is a random variable—a quantity sampled whose values represent a probability distribution. A set of locations at which the random variable is measured is called a *random field*.

Data with spatial heterogeneity are some realization of a random field. A random field is some generalization of a stochastic process that is defined in Euclidean space (x and y values), with a dimensionality of at least 1 (Cressie 1991:8). For example, a random field could be temperature values measured from a grid of weather stations on a mountain. Random fields are composed of three parts: (1) T, the deterministic trend in the data, (2) η, the spatially autocorrelated random process, and (3) ε, the uncorrelated random variation and measurement error (Burrough and McDonnell 1998:134). Using the (x, y) coordinates, this is formulated as:

$$Y(x, y) = T(x, y) + \eta(x, y) + \varepsilon(x, y).$$

Using this relationship, we can better understand the underlying parts of random fields. Mean annual temperature at a point $Y(x, y)$ is modeled as a deterministic trend ($T[x,y]$). Temperature trends would vary with large-scale features, such as slope, aspect, or elevation. Smaller scale factors, such as avalanche slides, bluff faces, or local topographic features, are modeled as spatially autocorrelated random processes ($\eta[x,y]$). All the other variables that could influence mean annual temperature are modeled as uncorrelated variability or measurement error ($\varepsilon[x,y]$; Haining 2003).

Stationarity, a spatial random process whose properties do not vary with location, is a desirable property of spatial data. A process is stationary if the mean, variance, and directionality are constant (Fortin et al. 2012). Stationarity is not a property of the pattern on the landscape but a property of the underlying process that generated the pattern. For example, the basal area of a pine (*Pinus* spp.) plantation would be more similar within the plantation than in neighboring patches of forest. Basal area is not stationary, but the underlying pattern that results in consistent basal area (i.e., the date trees were planted, consistent soil nutrients) is stationary. It is impossible to directly assess if a process is stationary or not because we cannot directly measure it. Instead, we have to explore this assumption by calculating mean and variance of a given variable within a moving window (or kernel) across the landscape (Haskard and Lark 2009; Fortin et al. 2012). In most landscapes, it is unrealistic to assume that spatial processes are stationary because the resulting pattern in our data is a function of multiple pattern-generating processes (i.e., locations of plants would be influenced by the spatial distribution of water, light intensity, and mineral content of the soil).

Model Classification

Spatial data are flexible and inherently hierarchical in modeling, which results in a wide variety of potential models. Determining the source of heterogeneity in one's data is critical in selecting the correct model. Whether the spatial heterogeneity is from an abiotic or a biotic source is an important consideration. Abiotic sources of spatial heterogeneity can come from variation in habitat quality, resource avail-ability, and dispersal boundaries, which influence ecological processes such as mortality rate, colonization, or dispersal probability. Biotic sources of heterogeneity can arise from many causes, such as animal behavior (territoriality, competition for resources, predation), local dispersal, or disease transmission.

Models can be represented as spatially explicit or spatially implicit (Fig. 13.1). Spatially explicit models utilize variables, covariates, or processes that have explicit spatial locations (i.e., latitude and longitude coordinates) because location influences the modeled process. Almost any model that directly incorporates spatial data is spatially explicit; examples include mark-recapture methods with spatial data (Efford 2004; Royle et al. 2014), home range estimation (e.g., fixed and adaptive kernels; Worton 1989), and land cover analysis/light detection and ranging (LiDAR) data, among a surfeit of others. Spatially implicit models incorporate implied assumptions of how wildlife interact with biotic features, but do not include geographic space. Island biogeography models (Wilson and MacArthur 1967) and resource se-

Space is implicit or explicitly modelled?	Implicit	Explicit	Explicit	Explicit
Scale?	Metapopulation	Individual	Population	Landscape
Dimensionality	1 (population or diversity)	1 (points in space)	2 (area, perimeter)	3 (latitude, longitude, height)
Discrete or continuous?	Not applicable	Discrete	Discrete	Continuous
Examples	Island Biogeography, Metapopulation models	Animal locations	Watersheds, protected areas, management units	Forest canopy heights via LiDAR

Fig. 13.1. Model classification diagram showing categories of spatial models with examples.

lection functions (Manly et al. 1993) are spatially implicit because spatial relations are implied, but the precise location of each habitat patch or island is not specified nor accounted for in the models.

Spatial data are often divided into the categories of continuous or discrete (Fig. 13.1). Continuous spatial data can be recorded as any possible value within a certain range and are often referenced by location coordinates. These data are represented by surfaces such as elevation, density, or canopy cover. Discrete data are typically nominal, ordinal, and ratio variables such as land use, soil classification, or forest type. In other words, continuous data are measured, and discrete data are counted. Similar to a-spatial data, spatial data types will influence data analysis.

Spatial heterogeneity is inherently scale dependent (Fig. 13.1). Mere acknowledgment of scale is insufficient—understanding what is causing heterogeneity at different scales is typically the crux of effective spatial analysis. Processes and relations that are clear at fine scales often dissipate or vanish at broader scales. Wildlife species use habitat at different scales (Johnson 1980); for example, in Japan Asiatic black bears (*Ursus thibetanus*) select areas with high hard mast at a small scale, but select areas with limited amounts of agriculture at large scale (Carr et al. 2002). Unless we understand the underlying scale and complexity of our data, we increase the risk of making spurious management decisions.

Dimensionality can also influence how models are classified (Fig. 13.1). The traditional spatial analysis focuses on one- and two-dimensional techniques. One-dimensional data are typically point datasets (e.g., locations of plots, animals, or other features). Two-dimensional data contain both perimeter and area (e.g., home range, territories, bodies of water). Three-dimensional data contain *x*, *y*, and *z* values; they are commonly triangulated irregular network (TIN) surfaces that use points with three dimensions to interpolate a surface (e.g., forest canopy surface derived from LiDAR or terrain surfaces derived from a network of points). Dimensionality can also take

less literal, but equally important definitions—other elements such as time or management scenarios are also dimensions. Comparisons across time or among management scenarios are valuable. Analyzing discrete time slices and creating change maps that compare different time periods or management plans can elucidate patterns on the landscape that are relevant to management and conservation.

Spatial Data Analysis

Spatial data analysis employs a wide array of metrics and tools to assess patterns in spatial data. Analysis typically falls into one of three major groups: determining spatial dependence, understanding point patterns, or making predictions. Nearest neighbor methods, Moran's *I*, Geary's *c* (Cliff and Ord 1981), or the Getis-Ord *G* statistic (Getis and Ord 1992) can be used to understand dependence (i.e., autocorrelation) in spatial data. Pattern analysis techniques such as Ripley's *K* or hot-spot analysis elucidate patterns in underlying point processes. Geostatistics (i.e., semivariograms, kriging, linear mixed models) are used to interpolate and make predictions of spatial point data.

Assessing Spatial Dependence

Geostatistics measure spatial dependence in data and enable researchers to avoid many of the issues that occur when analyzing autocorrelated data (Getis 1999). They are based on hypothesis tests of the null hypothesis—H_0: the data are randomly distributed as a realization of a random field (Plant 2012). Statistics of this nature are a special case of Gamma, also known as the Mantel statistic (Mantel 1967). Gamma is defined by:

$$\Gamma = \sum_{i=1}^{n} \sum_{j=1}^{n} w_{ij} c_{ij},$$

where w_{ij} is the *ij*th element in the *spatial weights matrix*—a matrix that represents the spatial structure in data—and represents the concordance between locations i and j; c_{ij} is the quantity that describes how much weight is applied to the pair ij (Cliff and Ord 1981; Plant 2012). This special case of Gamma occurs when a matrix representing possible pointwise, locational associations (i.e., the spatial weights matrix) among all points is multiplied by a matrix that represents some nonspatial association among the points (Mantel 1967). The nonspatial association can be any statistical relationship, such as an ecological, economic, or epidemiological one (Getis 1999). Mantel tests characterize dependence in data utilizing the same principle (Mantel 1967).

Gamma is a natural fit to assess spatial dependence in data. It describes spatial association based on covariances (Moran's I), subtraction (Geary's C), and addition (Getis-Ord G statistic; Getis and Ord 1992). All three of these statistics assess the degree of autocorrelation globally, which is to say that all pointwise measurements are taken into account simultaneously. If the spatial weights matrix becomes a column vector (i.e., the spatial dependence between one point and all others), Gamma becomes a local measure (I_i, C_i, and notably G_i). The local G_i statistic is commonly used to assess spatial autocorrelation at the local scale. Statistics like these are often used as precursors to other local analyses, such as kernel analysis or moving window analysis.

Finding the degree of autocorrelation in spatial data is a fundamental question in spatial statistics. Cliff and Ord (1973, 1981) demonstrated the issues of model misspecification from autocorrelation and how to test for spatial dependence in regression residuals. Their monograph (Cliff and Ord 1981) is the basis for most of the Cliff-Ord tools contained in GIS programs today. They extensively detailed the spatial weights matrix and how to conduct two major tests for *global* autocorrelation: Moran's I and Geary's C tests (Cliff and Ord 1973; Getis 1999). Global statistics measure the degree of autocorrelation in the entire dataset. Moran's I (Moran 1950) uses the covariance among dependent locations, and Geary's C (Geary 1954) utilizes the numerical

differences of dependent locations to assess spatial dependence; both are special cases of the Mantel statistic, Gamma (Mantel 1967). Moran's I is formulated as:

$$I = \frac{n}{S_0} \frac{\sum_i \sum_j w_{ij}(Y_i - \bar{Y})(Y_j - \bar{Y})}{\sum_i (Y_i - \bar{Y})^2}$$

where $S_0 = \frac{1}{4}i\ \frac{1}{4}jw_{ij}$, w_{ij} is the spatial weights matrix, n is the number of samples indexed by i and j, and Y is the variable of interest. We can reformulate this as a special case of Gamma, where C_{ij} (the similarity measure) is replaced with the latter half of the Moran's I equation:

$$\Gamma = \sum_{i=1}^{n} \sum_{j=1}^{n} w_{ij} \frac{n}{S_0} \frac{(Y_i - \bar{Y})(Y_j - \bar{Y})}{\sum_i (Y_i - \bar{Y})^2}.$$

Note that the C_{ij} substitution resembles the Pearson correlation coefficient (Plant 2012). Geary's C is very similar to Moran's I with a few differences:

$$C = \frac{n-1}{2S_0} \frac{\sum_i \sum_j w_{ij}(Y_i - \bar{Y})^2}{\sum_i (Y_i - \bar{Y})^2}$$

which can be formulated as a special case of Gamma:

$$\Gamma = \sum_{i=1}^{n} \sum_{j=1}^{n} w_{ij} \frac{n-1}{2S_0} \frac{(Y_i - \bar{Y})^2}{\sum_i (Y_i - \bar{Y})^2}.$$

The underlying similarity measurement in Geary's C is a squared difference, which is similar to the underlying calculations in a variogram (Plant 2012).

Moran's I and Geary's C can be interpreted similarly to the Pearson's correlation coefficient (r), but on different scales with some differences (Getis 1999). I values range from −1 to 1, just like r values; negative values indicate negative spatial autocorrelation, and positive values indicate positive spatial autocorrelation (Moran 1950; Getis 1999). C values differ from r because of the scaling; values <1 indicate positive spatial autocorrelation whereas values >1 indicate negative spatial autocorrelation (Geary 1954; Getis 1999).

Which global statistic should be used when? The statistical power of I is greater than that of C, as shown through Monte Carlo simulation (Cliff and Ord 1981:175), which suggests that I is the preferred statistic. Griffith et al. (1999) demonstrated that I is preferred over C; however, they advocated calculating both to assess if the statistics corroborate. When there are highly irregular outliers in the Y_i values, differences between I and C may occur (Griffith et al. 1999). Compute both statistics and compare them; if they indicate similar degrees of autocorrelation, there is no need for concern. If the measures are incongruent, it is important to understand why. Such differences may be important characteristics of a dataset and be worth exploring and reporting (Plant 2012).

It is also important to measure *local autocorrelation*—autocorrelation in small portions of the overall dataset. For example, there may be local hotspots of similar values that are not characteristic of the underlying pattern in a random field of data. In this regard, consider the population density of American pika (*Ochotona princeps*), a lagomorph typically found only in boulder fields at or above the timberline. This species is very sensitive to high temperatures and cannot easily migrate between boulder fields (Smith 1974). Consequently, their habitat consists of small isolated patches of habitat at high elevation. If pika density is our variable of interest, we would find that there are small patches on the landscape with a disproportionly higher density and much of the landscape would have lower, but consistent density values. Test of local autocorrelation would help understand the spatial autocorrelation and the structure of the spatial heterogeneity in these patches.

The first widely used metrics of local spatial autocorrelation were the Getis-Ord (Getis and Ord 1992) statistics called $Gi(d)$ and $Gi*(d)$. An area of interest is divided into i sections, $i = 1, 2, 3, n$; i is an index of each aforementioned section and d is the Euclidean distance from element i (Cliff and Ord 1992; Plant 2012). This statistic allows one to generate hypothesis tests about spatial concentrations (i.e., the degree of spatial autocorrelation in the data) of the "sum of Y values of the j points within d

distance units of the ith point" (Getis and Ord 1992:190). The statistic is formulated as:

$$G_i(d) = \frac{\sum_{j=1}^{n} w_{ij}(d)Y_j}{\sum_{j=1}^{n} Y_j}, \, j \neq i.$$

The $Gi*(d)$ statistic is related but allows for situations where $j = i$. $Gi(d)$ statistics measure the concentration (positive autocorrelation) or diffusion (negative autocorrelation) of the sum of values within a given section in an area of interest. If high values of Y are within d of point i, $Gi(d)$ will be high as well. One benefit to this statistic is that $Gi(d)$ is Z-distributed and makes for standardized hypothesis testing (Getis and Ord 1992).

Local Moran's I is another statistic that can indicate spatial clustering (Anselin 1995). The formula is very similar to the global Moran's I:

$$I_i = (Y_i - \overline{Y}) \sum_{j=1}^{n} w_{ij}(Y_j - \overline{Y}).$$

When summed across all regions, local Moran's I will equal the global Moran's I ($\sum_i I_i = I$) (Anselin 1995, Anselin 1999). This property allows us to determine which areas of the region of interest contribute most to the global autocorrelation statistic. Interpretation of the local statistic is the same as that of the global Moran's I statistic: positive values indicate clustering of similar values; negative values indicate dissimilar values (Anselin 1995). This tool is commonly used to delineate hot and cool spots—local areas of extreme high or low values, respectively—and to detect spatial outliers (Anselin 1995). Returning to our pika example, a hot spot could indicate rock outcrops with abundant pika, whereas a cool spot would be an open area with few pika.

Interpolation and Point Process Analysis

Geostatistics is a field of statistics used to understand how location influences patterns in nature. It uses spatial (and sometimes spatiotemporal) coordinates of the data to analyze and describe spatial patterns, relationships, and interpolations. Originally developed by the mining industry in the 1950s to predict the probability distribution of mineral ores, geostatistics were designed to interpolate values where samples were not taken or to understand the uncertainty within spatial data (Krige 1951; Cressie 1991). Since then, geostatistical techniques have developed beyond simple interpolation to assessing complex, multivariate problems that allow us to understand spatiotemporal phenomenon and make better predictions. Field measurements of landscape variables is costly and labor-intensive, but with geostatistics, researchers can sample a variable of interest, predict measurements for unsampled locations (with associated measures of uncertainty), and assess underlying spatial trend.

Kriging, one of the most common classes of geostatistical models, uses the distance or direction between sample points to quantify spatial correlation. This correlation can explain the variation in a predicted surface. It is a form of regression analysis using observed values of a given variable (e.g., rainfall, biomass, bunchgrass density) of surrounding data points, weighted by values of spatial dependence. The general formula is a weighted sum of the data locations:

$$\hat{Z}(S_0) = \sum_{i=1}^{N} \lambda_i Z(S_i),$$

where $Z(s_i)$ = the measure value of the ith location, λi = an unknown weight for the measured value at the ith location, S_0 = the prediction location, and N = the sample size (Cressie 1991). Spatial dependence influences the weights (λi); it depends on the fitted model's distance to measured points and prediction location and on spatial relationships of the measured locations. Nearly all interpolation algorithms predict the value of new locations as the weighted sum of data values at surrounding measured locations. Most assign weights according to mathematical functions (sometimes called "linear distance decay functions") that give a decreasing weight with increased distance from measured locations (Isaaks and Srivastava 1989).

Kriging is a valuable interpolation tool with many advantages but has disadvantages. Kriging helps mitigate the effects of spatial data clustering by treating points in a cluster more like single, isolated data points. In some cases, weights are assigned such that a cluster is treated more like a single point (Plant 2012). If points are dense and distributed uniformly across the area of interest, interpolated estimates will be consistent and similar among kriging techniques. Kriging provides an estimate of prediction error (kriging variance) in addition to an estimate of the variable itself. The error estimate enables one to evaluate precision in interpolated values; users should keep in mind that error maps are just scaled maps of distance to the nearest points. If the underlying kriging function does not match the spatial pattern, estimates and error maps are spurious. Further, if data locations are in sparse clusters with significant distance between them, the estimates will be unreliable, regardless of the technique chosen.

Wildlife landscape data are often continuous in nature (such as basal area, canopy cover, tree diameter, population density, etc.), so it is best to model these data as such (Plant 2012). One such way to characterize the continuous nature of spatial data is to use a *variogram*—a figure that displays the variance between field values at different locations across realizations of a random field (Cressie 1991). Variograms are used to visualize autocorrelation in the data. Random fields contain *spatial lag* (h), which is a variable that accounts for the average values of neighboring locations. For example, we can understand how similarly sized aspen (*Populous tremuloides*) trees are clustered by measuring the diameter of a sample of trees and their neighboring trees. The pairwise difference in spatial lag between neighboring locations allows researchers to measure and account for spatial autocorrelation (Cressie 1991). The underlying data in a variogram vector that has magnitude and direction of the spatial lag for each pair of points. Generally, we assume that the spatial lag is *isotropic* (independent of direction), but this assumption can be relaxed to include direction with a more general model (e.g., Isaaks and Srivastava 1989; Cressie 1991). If data are isotropic, the variogram depends only on the magnitude of spatial lag in the data. Cressie (1991:58) defined the variogram $\gamma(h)$ as a function of spatial lag:

$$\gamma(h) = \frac{1}{2} var[Y)x + h) - Y(h)],$$

where, x is a vector of spatial coordinates (x, y), h is the spatial lag, and Y is the measured quantity. Because there is a constant of ½ in the formula, the correct term for $\gamma(h)$ is *semivariogram* (Plant 2012). In this text, the terms "variogram" and "semivariogram" are used interchangeably.

Users can also fit the variogram graphically by plotting the *semivariance*, $\gamma(h)$, against the lag distance (h). Semivariance is a measure of how much overall variation there is in the dataset, after adjusting for correlations among closely -spaced locations. To interpret the variogram, the user needs to understand its three components: sill, range, and nugget. The *sill* is the semivariance value at which the curve levels off, sometimes called the amplitude of the curve (Fig. 13.2). The *range* is the lag distance at which the semivariogram curve reaches the sill, after which autocorrelation is negligible (Fig. 13.2). Theoretically, the semivariogram value at the origin of the graph should be 0 (i.e., measurements are taken without error and lags are negligible very close to zero (Plant 2012); if this is not the case, the semivariogram value very near zero is referred to as the "nugget." The *nugget* is the variability of the semivariogram curve at values smaller than typical spacing; nugget variability includes unexplained measurement error (Fig. 13.2).

Sometimes it is easier to work with the covariogram, derived from the correlation of the lag $\rho(h)$. Cressie (1991:53) defined this quantity as $C(h)$, where

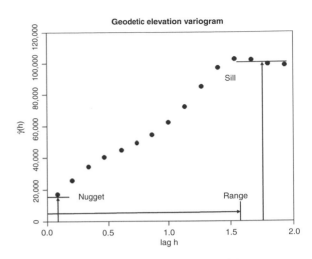

Fig. 13.2. Semivariogram of U.S. Geological Survey geodetic marker locations and elevations in West Virginia.

$$C(h) = \operatorname{cov}\left[\Upsilon(x), \Upsilon(x + h)\right],$$
$$\text{iff, } C(0) > 0, \text{ then } \rho(h) = \frac{C(h)}{C(0)}.$$

There are some assumptions about stationarity here, but they are unlikely to be violated with ecological data; readers should consult Banerjee et al. (2004) for a thorough discussion stationarity assumptions. The covariogram and the semivariogram are related because the $C(0)$ is equal to the sill of the semivariogram $\gamma(h)$ (Cressie 1991; Plant 2012). More generally, the covariogram $C(h)$ and the semivariogram $\gamma(h)$ are related: $\gamma(h) = C(0) - C(h)$. Both the covariogram and the semivariogram will give estimates of autocorrelation in point-based location data (Fig. 13.3). After the degree of autocorrelation is calculated, in-depth analysis can proceed. If autocorrelation is negligible, one can use traditional statistical methods that make assumptions about the independent nature of the observations (i.e, analysis of variance, linear models, etc.); if autocorrelation is considerable—which is likely in many ecological datasets—researchers should account for the autocorrelation explicitly in their analysis techniques, such as geographically weighted regression or generalized linear mixed models, or generalized additive models (Bivand et al. 2008). Failure to do

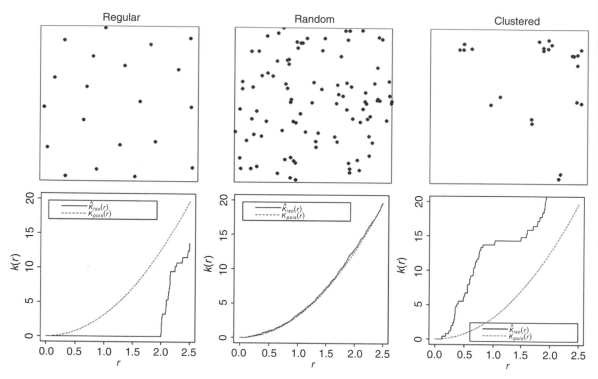

Fig. 13.3. Covariogram of crime data (residential vehicle thefts and burglaries per 1,000 household in the neighborhood) in Columbus, Ohio, from Anselin (1988). Dark circles indicate statistically significant covariances at the $\alpha = 0.05$ level. In this case, the first two bins show significant autocorrelation in the first four distance classes. No significant autocorrelation exists after these distance classes.

so can result in biased estimates, incorrect precision values, and spurious results.

Point pattern analysis summarizes point patterns, tests hypotheses about the patterns, and fits models to estimate parameters in the pattern (Ripley 1981). The Ripley's *K* statistic is a mathematical function used on completely mapped data (i.e., data that contain locations of all occurrences in a predefined study area or window). Users can fit bivariate or multivariate generalizations of Ripley's *K* to describe relationships among spatial point patterns. The general *K* function for a spatial point process is formulated as:

$$K(d) = \lambda^{-1} E[Z],$$

where λ is the density (number of occurrences per unit area) and *Z* is the number of expected occurrences under the null hypothesis of *complete spatial*

randomness (CSR)—a pattern where point events are arranged in a completely random manner (Ripley 1981). This function is frequently used to assess whether a spatial point process is regularly spaced, random, or clustered. The simplest form for *K(d)* is a homogenous Poisson process. This is formulated as $K(d) = \pi t^2$, the area of a circle with radius *t*. For a point process more regularly spaced than CSR, we expect fewer occurrences within distance *t* of a randomly chosen point; for a point process more clustered than CSR, we expect more events within distance *t* of a randomly chosen point. To estimate *K(d)*:

$$\hat{K}(d) = \hat{\lambda}^{-1} \frac{1}{N} \sum_i \sum_j \delta(dist(i,j) < d),$$

where $i \neq j$, $dist(i, j)$ is the Euclidean distance between occurrences, δ is the density of occurrences,

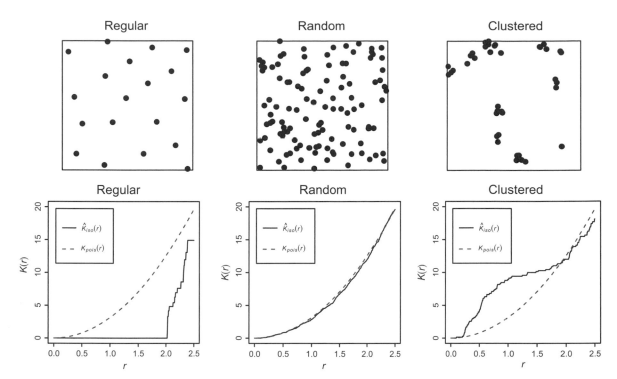

Fig. 13.4.   Three spatial point pattern processes (top three panels) and their associated Ripley's *K* functions (bottom three panels). The left column indicates a regularly spaced point pattern, the middle column indicates a complete spatial randomness process, and the right column indicates a clustered point process.

and $\delta(\text{dist}(i, j) < d) = 1$ if $\text{dist}(i, j) < d$ and 0 otherwise (Ripley 1981). Because this method requires a defined study region, often termed the *bounding box*, there is an alternate formulation that corrects for the edges. It precludes occurrences that occur outside the bounding box, but within *d*. It is defined as:

$$\hat{K}_{edge}(d) = \hat{\lambda}^{-1} \frac{1}{N} \sum_i \sum_j w_{ij} \delta(\text{dist}(i, j) < d),$$

where $w_{ij} = 1$ is the distance between *i* and *j* is less than the distance between occurrence *i* and the bounding box edge (Diggle 1979). We can plot this function to test if the point pattern has CSR (Fig. 13.4, top panels). Departures from the line (CSR) indicate clustering or regularity; deviations above the line indicate clustering, and deviations below the line indicate regularity (Fig. 13.4, bottom panels).

Prediction: Spatial Data Analysis with Linear Mixed Models

Spatial data are a special form of longitudinal data, which track the same sample at different points in time (e.g., individual animals, plant communities, ecosystems, etc.). Autocorrelation among samples can be accounted for using a linear mixed model. Interpolation can take a linear, mixed model form where the random effect that we fit is a correlated spatial process. The equation for this model is:

$$y = \beta X + Z\alpha + \epsilon,$$

where *y* is a vector of response data (i.e., rainfall, tree basal area, etc.), *X* and *Z* are matrices of explanatory variables, β represents a vector of fixed effects, and α and ϵ are random vectors (Wikle and Royle 2009). Columns of *Z* are indicator variables, such that every observation is associated with some element of α. These random effects carry an assumption of multivariate normality (Wikle and Royle 2009):

$$\begin{bmatrix} \alpha \\ \epsilon \end{bmatrix} \sim Gaussian\left(\begin{bmatrix} 0 \\ 0 \end{bmatrix}, \begin{bmatrix} \Sigma_\alpha & 0 \\ 0 & \Sigma_\epsilon \end{bmatrix} \right).$$

In this formulation, random errors are independent (i.e., $\Sigma_\epsilon = \sigma_\epsilon^2 I_{nxn}$, where I is an n-dimensional identity matrix). The parameter, σ_ϵ^2, is a measurement-error variance that includes any variance unexplained by the random effect. This is the "nugget" mentioned earlier.

Often, wildlife data consist of random variables that are not from a Gaussian distribution (e.g., count data, binomial data). As with traditional linear mixed models, we can formulate a spatial GLMM by specifying a link function (denoted g [parameter]) of the location parameter for the distribution or explicitly include a spatial process in our linear models (Wikle and Royle 2009:13).

Properties and Limitations of Spatial Data

Spatial data have inherent properties that are important to consider when conducting an analysis. At present, spatial data are available in a wide variety of formats, obtainable by more methods, produced by more agencies, and at a higher resolution than ever before. Much of these data are freely available to all but have underlying limitations that must be clearly understood. Spatial data can be differentiated in space, time, or thematic dimensions (Veregin 2005). For each dimension, data quality components vary in accuracy, precision, and consistency. The questions each user should ask are: (1) Are these data suitable to answer the question? (2) Where did the data come from? (3) Are they accurate and precise enough for the analysis? (4) How complete are these data? and (5) At what scale were these data collected?

No spatial data are error-free. When spatial layers are created, there is a certain amount of variation within the database. Those errors propagate with every operation and analysis, which can sometimes generate output so error-ridden that we make spurious conclusions (Haining and Arbia 1993). Further, few keep a record of the accuracy of intermediate results, which makes it impossible to evaluate the accuracy of the final product. Consider a simple quantitative model $A_i(x) = v_i(x) + U_i(x)$, where:

$A_i(x)$ = reality, the true value of the parameter;
$v_i(x)$ = the deterministic value of the parameter in the GIS map;
$U_i(x)$ = the stochastic error about the deterministic value (typically mean = 0, but non zero variance and spatial autocorrelation).

Rearranged, we can estimate the error—the difference between reality and the map representation of reality $(U_i(x) = A_i(x) - v_i(x))$—, and record the amount of error in each step of analysis.

There are three steps to analyzing error propagation in spatial data: definition, identification, and analysis. To properly define the model uncertainty (Ui), one can (1) generate a marginal probability distribution function at each location, (2) assess spatial correlation (part of the stationarity assumption) with a semivariogram or correlogram, (3) assess temporal correlation if the variables change through time, and (4) conduct cross-correlation analysis with other spatial layers with uncertainty (Cressie 1991:53; Pannatier 1996; Heuvelink 2005). The next step is to identify the error model through methods such as sampling (standard error or confidence intervals), ground-truthing (root mean square error difference between the spatial layer and reality), interpolation error using geostatistics for kriging, or kappa statistics to assess errors in classification methods (Heuvelink 2005). Once the error model is identified, then error propagation analysis can be conducted using a Taylor series method (Heuvelink 2005) or a Monte Carlo simulation (Hammersley and Handscomb 1979). In practice, model uncertainty is a major source of error and should be included in analysis. Ignoring the error would underestimate the uncertainty of the system and could cause spurious conclusions. Researchers can include model error by adding a term in the residual error of model equations, or by assigning errors to the model coefficients (Heuvelink 2005).

Scale (both temporal and spatial) is an important aspect of all spatial data that is often overlooked or is preempted by limited resources such as funding or logistics. All studies are affected by spatial and temporal scale. Researchers are limited by the logistics of data collection; resources and time are limited. Collecting data on an irrelevant scale can reduce strength of inference or result in spurious conclusions. A study conducted at one spatial scale may have a tight, parsimonious explanation for how a species selects habitat, but the relationship may not hold in the context of a broad, coarse scale. Habitat selection, for example, is inherently hierarchical: foraging may happen at a fine scale, whereas defending a territory or finding a mate occurs at a coarser scale (Johnson 1980). There are also differences in how abiotic and biotic factors operate at different scales of habitat selection. (Urban et al. 1987).

How does one choose appropriate spatial scales for a study? It is question-dependent and requires a thorough understanding of a species' natural history. Scales at which Gambel's quail (*Callipepla gambelii*) operate are much finer than those used by a mountain lion (*Puma concolor*). The range of spatial scales that are important to a species often depends on its mobility, behavior, and home range. Individual animals are not often influenced by scales larger than their home range, except in cases of dispersal and migration (Bissonette et al. 1997).

Errors, Errors, Errors: The Use and Misuse of Spatial Data

Spatial data require researchers to pay special attention to the underlying properties of their datasets.

Early concerns about spatial autocorrelation in ecological datasets focused on the inflated type I error rate in hypothesis tests from a lack of independence among samples (Legendre 1993). Spatial autocorrelation can influence the parameter estimates by altering the variance-covariance matrix of linear models (Anselin 1988; Anselin and Bera 1998). Coefficients may change when using spatial or nonspatial models; the magnitude or sign of such a change is unpredictable and unstable among methods. Model selection is not spared from effects of spatial autocorrelation; autocorrelation in model residuals causes Akiaike's Information Coefficient (AIC, Akaike 1985) to favor unstable and overfit models (Diniz-Filho, Bini, and Hawkins et. al 2003).

There is no right or wrong way to deal with spatial autocorrelation in ecological models. This is an active area of research in which new methods are emerging rapidly (Dormann et al. 2007; Beale et al. 2010). The general process is to (1) test for the degree of autocorrelation and (2) if spatial autocorrelation is significant, adjust or utilize it in some way. One could account for spatial autocorrelation by creating synthetic variables that capture spatial patterns to include as predictor variables in regression models (Diniz-Filho et al. 2003). One could also implicitly incorporate the spatial autocorrelation in geographically weighted regression (Fotheringham et al. 2002) or in a linear mixed model (Wikle and Royle 2009; Plant 2012). Autocorrelation can also be incorporated into spatially implicit models (Nielsen et al. 2002)

Spatial dependence may require users to alter their traditional sampling designs. Lack of independence in the data is essentially a reduction in the sample size; what remains is the effective sample size (ESS; Getis and Ord 2000). Spatial dependency is loss of information. If a dataset consists of n sample points, what is the ESS of the dataset? If the observations are independent, the answer is n. This is rarely the case because of Tobler's first law of geography (Tobler 1970). If all points are perfectly autocorrelated, then ESS is 1; if points have only some dependence, ESS is less than n (Vallejos and Osorio 2014). ESS is also dependent on the sampling design. Typically, ESS is overestimated when the sampling is regular (i.e., a typical, uniform grid of sample points), underestimated when the sampling is clustered, and least biased when the sampling grid is random (Vallejos and Osorio 2014). Pilot studies assist in determining underlying correlation of variables of interest. If the pilot study yields sufficient information to estimate a variogram or spatial autocorrelation, those data should be used to maximize the ESS (Vallejos and Osorio 2014). If a pilot study is not logistically feasible, simple stratified sampling (Gelfand et al. 2010) or general randomized tessellation stratification sampling (GRTS; Stevens and Olsen 2004) are good choices.

Much attention has been paid to spatial autocorrelation in wildlife datasets, but this is not the only challenge for concern. Problems with model misspecification can far outweigh effects of spatial autocorrelation. In some cases, the addition of several environmental variables in a linear model can decrease autocorrelation in model residuals to negligible levels (Diniz-Filho et Al. 2003). Misspecification occurs when models fail to include important variables, include irrelevant variables, or use an incorrect functions form (e.g., fail to use an interaction term). This problem has been discussed for decades (Larson and Bancroft 1963), and as with a-spatial statistics, model assumptions need to be checked for proper validation. One simple check is to plot the residuals against the model predictions and the explanatory variables; the relation between residuals and each of these model components should be random when plotted. If a pattern emerges, this often indicates model misspecification (Plant 2012).

Uncertainty is information, too. It guides wildlife monitoring and research by influencing decision-making. Understanding the uncertainty in our data facilitates making better decisions. When incorporating spatial data, geographic information systems (GIS) are commonly used to describe and display data. Often, there is a perception that GIS problems are a-statistical. Emphasis is placed on creating and displaying spatial data in maps, rather than digging into the statistical modeling of spatial data (Wikle and Royle 2009). Functionality for spatial modeling is limited or nonexistent in GIS programs, despite various "toolboxes" that allow users to conduct basic spatial statistical analysis. Users are seldom trained to utilize formal model fitting procedures and assessment techniques. It is critical that all users of spatial data (1) understand the nature of the data, (2) learn how to quantify the uncertainty of map predictions, and (3) appropriately apply uncertainty estimates in their analysis and conclusions.

Recent Advances and Future Trends

We live in geospatial times. We possess the tools to better understand the spatial nature of wildlife data and glean understanding by incorporating spatial data into our models. Just a few decades ago, computational processing power was expensive, and as a result, spatial statistics were limited in scope. We now have new ways to incorporate spatial data into analysis, and processing power has become exponentially cheaper. Further, we are generating more data than ever. It is now commonplace to collect spatial information whenever a measurement is taken. This is fortunate given the rise of big data, and there is an increased need to address large-scale ecological problems.

Geospatial information can play an integral role in understanding complex natural resource issues, but incorporating this information adds complexity to analyses. Classical spatial analysis techniques assume that spatial processes are stationary; without this assumption, parameter estimation is difficult or impossible (Wikle and Royle 2009). In reality, it is unlikely that an ecological spatial process is truly stationary. Spatial dependence is likely to vary in different portions of the study region. For example, wind, evapotranspiration, and temperature patterns influence spatial variability in annual rainfall patterns. Analysis of nonstationary point processes is an active area of research, but currently, analysts can model nonstationary covariance structures if they have sufficient samples or replicates (Wikle and Royle 2009). Transformation methods provide a way to assess nonstationarity with small problems; see Wikle and Royle (2009) for more detail about these methods. For larger problems or datasets, computational demands are nontrivial and require parallel processing methods to generate results in a reasonable time frame (Wikle and Royle 2009). Increased reliance on computationally intense statistical techniques requires exploration of parallel processing, cloud computing, and big data solutions, to efficiently answer complex questions.

Recent advances in geostatistics allow the incorporation of spatial data into traditional analysis techniques. For example, capture-mark-recapture methods (CMR; e.g., Williams et al. 2002) have greatly improved with the inclusion of spatial data. Spatially explicit capture-recapture methods (SECR, Efford 2004; SCR, Royle et al. 2014) are a profound development over traditional CMR methods. Regression has been forever changed with geographically weighted regression methods; we can now ask why something is occurring rather than just where it occurs (Fotheringham et al. 2002). Geographically weighted regression methods are underutilized in wildlife studies but their use is certain to increase in the future. Advances in Bayesian statistics have also increased the use of spatially explicit data. This flexible modeling method allows for the use of spatial data in hierarchical models to further our understanding of wildlife science. The field of spatial statistics is rapidly changing and it is difficult to foresee a world in which we do not incorporate spatial knowledge into our inferences. The old days of ignoring spatial autocorrelation, scale, and underlying spatial patterns because they were "noise" are over.

LITERATURE CITED

Akaike, H. 1985, Prediction and entropy. In A celebration of statistics, edited by A. C. Atkinson and S. E. Fienberg, pp. 1–24. Springer, New York.

Anselin, L. 1988. Spatial econometrics: Methods and models. Kluwer Academic, Dordrecht.

Anselin, L., 1995. Local indicators of spatial association—LISA. Geographical Analysis 27:93–115.

Anselin, L., 1999. The future of spatial analysis in the social sciences. Geographic Information Sciences 5:67–76.

Anselin, L. and A. K. Bera. 1998. Spatial dependence in linear regression models, with an introduction to spatial econometrics. Statistics Textbooks and Monographs, 155:237–290.

Banerjee, S, B. P. Carlin, and A. E. Gelfand. 2004. Hierarchical Modeling and Analysis for Spatial Data. Chapman and Hall–CRC Press. Boca Raton, FL.

Beale, C. M., J. J. Lennon, J. M. Yearsley, M. J. Brewer, and D. A. Elston. 2010. Regression analysis of spatial data. Ecology Letters 13:246–264.

Bissonette, J. A., D. J. Harrison, C. D. Hargis, and T. D. Chapin. 1997. The influence of spatial scale and scale-sensitive properties on habitat selection by American marten. In Wildlife and landscape ecology: Effects of pattern and scale, edited by J. A. Bissonette, pp. 368–385. Springer, New York.

Bivand, R. S., E. J. Pedesma, and V. Gomez-Rubio. 2008. Applied spatial data analysis with R. Springer, New York.

Burrough, P. A., and R. A. McDonnell. 1998. Principles of geographical information systems. Oxford University Press. Oxford.

Carr, M. M., J. Yoshizaki, F. T. VanManen, M. R. Pelton, O. C. Huygens, H. Hayashi, and M. Maekawa. 2002. A multi-scale assessment of habitat use by Asiatic black bears in central Japan. Ursus 13:1–9.

Cliff, A. D., and J. K. Ord. 1973. Spatial autocorrelation. Pios, London.

Cliff, A. D., and J. K. Ord. 1981. Spatial processes: Models & applications. Taylor & Francis. Milton Park, Didcot, United Kingdom.

Cressie, N. A. C. 1991. Statistics for spatial data. John Wiley & Sons, New York.

Diggle, P. J. 1979. On parameter estimation and goodness-of-fit testing for spatial point processes. Biometrics 35:87–101.

Diniz-Filho, J. A. F., L. M. Bini, and B. A. Hawkins. 2003. Spatial autocorrelation and red herrings in geographical ecology. Global Ecology and Biogeography, 12:53–64.

Dormann, F., J. M. McPherson, M. B. Araújo, R. Bivand, J. Bolliger, G. Carl, R. G. Davies, A. Hirzel, W. Jetz, W. D. Kissling, and I. Kühn. 2007. Methods to account for spatial autocorrelation in the analysis of species distributional data: A review. Ecography 30:609–628.

Efford, M. G. 2004. Density estimation in live-trapping studies. Oikos 106:598–610.

Fortin, M. J., P. M. James, A. MacKenzie, S. J. Melles, and B. Rayfield. 2012. Spatial statistics, spatial regression, and graph theory in ecology. Spatial Statistics 1:100–109.

Fotheringham, A. S., C. Brunsdon, and M. E. Charlton. 2002. Geographically weighted regression: The analysis of spatially varying relationships. Wiley, Hoboken, NJ.

Geary, R. C. 1954. The contiguity ratio and statistical mapping. Incorporated Statistician 5:115–146.

Gelfand, A., P. Digle, M. Fuentes, and P. Guttorp. 2010. Handbook of spatial statistics. Chapman and Hall/CRC, Boca Raton, FL.

Getis, A. 1999. Spatial statistics. Geographical Information Systems 1:239–251.

Getis, A., and J. K. Ord. 1992. The analysis of spatial association by use of distance statistics. Geographical Analysis 24:189–206.

Griffith, D.A., L. J. Layne, J. K. Ord, and A. Sone. 1999. A casebook for spatial statistical data analysis: A compilation of analyses of different thematic data sets. Oxford University Press, Oxford.

Haining, R. P. 2003. Spatial data analysis: Theory and practice. Cambridge University Press, Cambridge.

Haining R. P., and G. Arbia. 1993 Error propagation through map operations. Technometrics 35:293–305.

Hammersley, J. M., and D. C. Handscomb. 1979. Monte Carlo methods. Chapman and Hall, London.

Hanski, I. 1999. Metapopulation ecology. Oxford University Press, Oxford.

Haskard, K. A., and R. M. Lark. 2009. Modelling nonstationary variance of soil properties by tempering an empirical spectrum. Geoderma 153:18–28.

Heuvelink, G. M. B. 2005. Propagation of error in spatial modeling with GIS. In Geographical information systems: Principles, techniques, management and applications, 2nd ed., edited by P. A. Longley, M. F. Goodchild, D. J. Maguire, and D. W. Rhind, pp. 207–217. John Wiley and Sons, Hoboken, NJ.

Isaaks, E. H., and R. M. Srivastava. 1989. Applied geostatistics. Oxford University Press, New York.

Johnson, D. H. 1980. The comparison of usage and availability measurements for evaluating resource preference. Ecology 61:65–71.

Knegt, D., M. B. Coughenour, A. K. Skidmore, I. M. A. Heitkönig, N. M. Knox, R. Slotow, and H. H. T. Prins. 2010. Spatial autocorrelation and the scaling of species-environment relationships. Ecology 91: 2455–2465.

Krebs, C. J. 2005. Ecology: The experimental analysis of distribution and abundance. Benjamin Cummings, San Francisco, CA.

Krige, D. G. 1951. A statistical approach to some basic mine valuation problems on the Witwatersrand. Journal of the Chemical, Metal, and Mining Society of South Africa 52: 119–139.

Larson, H. J., and T. A. Bancroft. 1963. Biases in prediction by regression for certain incompletely specified models. Biometrika 50: 391–402.

Legendre, P. 1993. Spatial autocorrelation: Trouble or new paradigm?. Ecology 74:1659–1673.

Leopold, A. 1933. Game management. University of Wisconsin Press, Madison.

Manly, B. F. J., L. L. McDonald, and D. L. Thomas. 1993. Resource selection by animals: Statistical design and analysis for field studies. Chapman and Hall, London.

Mantel, N. 1967. The detection of disease clustering and a generalized regression approach. Cancer Research 27:209–220.

Matloff, N. S. 1991. Statistical hypothesis testing: Problems and alternatives. Environmental Entomology 20:1246–1250.

Moran, P. A. 1950. A test for the serial independence of residuals. Biometrika 37:178–181.

Nielsen, S. E., M. S. Boyce, G. B. Stenhouse, and R. H. M. Monro. 2002. Modeling grizzly bear habitats in the Yellowstone ecosystem of Alberta: Taking autocorrelation seriously. Ursus 13:45–56.

Pannatier, Y. 1996. VARIOWIN: software for spatial data analysis in 2D. Springer, New York.

Plant, R. E. 2012. Spatial data analysis in ecology and agriculture using R. CRC Press, Boca Raton, FL.

Ripley, B. D. 1981. Spatial statistics. Wiley, New York.

Royle, J. A., R. B. Chandler, R. Sollman, and B. Gardner. 2014. Spatial capture recapture. Academic Press, Waltham, MA.

Smith, A. T. 1974. The Distribution and dispersal of pikas: Influences of behavior and climate. Ecology 55:1368–1376.

Stevens, D. L., and A. R. Olsen. 2004. Spatially balanced sampling of natural resources. Journal of the American Statistical Association 99:262–278.

Tilman, D., and P. Kareiva. 1998. Spatial ecology: The role of space in population dynamics and interspecific interactions. Princeton University Press, Princeton, NJ.

Tobler, W. 1970. A computer movie simulating urban growth in the Detroit region. Economic Geography 46:234–240.

Urban, D. L., R. V. O'Neill, and H. H. Shugart Jr. 1987. A hierarchical perspective can help scientists understand spatial patterns. BioScience 37:119–127.

Vallejos, R., and F. Osorio. 2014. Effective sample
    size of spatial process models. Spatial Statistics
    9:66–92.

Veregin, H. 2005. Data quality parameters. In *New Develop-
    ments in Geographical Information Systems: Principles,
    Techniques, Management and Applications, edited by*
    M. F. G. P. A. Longley, D. J. Maguire, and D. W. Rhind,
    pp. 177–189. Wiley. Hoboken, NJ.

Wikle, C. K. and J. A. Royle. 2009. Spatial statistical
    modeling in biology. In Biometrics Volume II: Encyclo-
    pedia of life support systems, edited by S. R. Wilson, and

C. Burden, pp. 97–119. UNESCO, Eolss Publishers,
    Paris. http://www.eolss.net.

Williams, B., J. Nichols, and M. Conroy (eds). 2002. Analysis
    and management of animal populations. Academic Press,
    Cambridge, MA.

Wilson, E. O., and R. H. MacArthur. 1967. The theory of
    island biogeography. Princeton University Press, Prince-
    ton, NJ.

Worton, B. J. 1989. Kernel methods for estimating the
    utilization distribution in home-range studies. Ecology
    70:164–168.

PART V NUMERICAL METHODS

14 — Bayesian Analysis of Molecular Genetics Data

DAMON L. WILLIFORD AND
RANDY W. DEYOUNG

> Inferences from Bayesian statistical analyses are directly applicable to parameters
> that are of central interest to natural resource scientists.　　—Stauffer (2008:10)

Bayesian approaches to statistical inference are routinely used for the analysis of genetic and genomic data. Bayesian analyses allow us to evaluate population structuring, assign individuals to populations, estimate migration rates, make inferences about phylogeography, and detect selection (Table 14.1). Statistical analyses in a Bayesian framework offer many advantages over traditional frequentist analyses for the types of data and questions involved in the analysis of genetic and genomic data. Furthermore, advances in computing power and the development of user-friendly software packages (Table 14.2) have democratized Bayesian analytical tools, at the expense of reducing analyses to a push-button exercise, where data go into a Bayesian black box and results pour out the other side.

In this chapter, we give an overview and background on the use of Bayesian analysis of genetic and genomic data from the perspective of the end-users of Bayesian tools, not developers or statisticians. We will use the generic term "genetic data" to encompass all forms of genetic information. Regardless of whether the data are derived from starch-gel electrophoresis, Sanger sequencing, microsatellite genotypes, or next-generation sequences and single nucleotide polymorphisms (SNPs), the Bayesian theory underlying the analyses is the same. The Bayes-

ian approach to data analysis has some weaknesses and disadvantages—an important consideration in the context of statistical inference for genetic data. As one author simply but eloquently points out, it is important to understand that user-friendly programs that perform Bayesian analyses "are easy to use but easier to misuse" (Wang 2017:981). It is our hope that this review can demystify aspects of the canned Bayesian black box programs and enable common-sense use of Bayesian tools for the analysis of genetic data.

Statistical Inference: Frequentist versus Bayesian Approaches

Most ecologists are familiar with statistical inferences based on the frequentist paradigm, developed in the early twentieth century from a strange amalgam of the ideas of R. A. Fisher, J. Neyman, and E. Pearson (Cohen 1990; Huberty 1993; Cherry 1998). The frequentist approach is based on the creation of a null hypothesis, usually of no treatment effect, followed by an experiment or sampling of one or more populations of interest. The resulting data are then subjected to a statistical test, which results in a P-value and a decision as to whether to reject the null hypothesis. The P-value is the probability of obtaining

Table 14.1. Bayesian-based software for the analysis of molecular data.

Software	Function(s)	Citations
BAMBE	Computer program for phylogenetic inference from DNA sequence data.	Simon and Larget (1998)
BEST	Computer program for Bayesian hierarchical modeling to jointly estimate gene and species trees from multilocus DNA sequences.	Liu (2008)
BEAST, BEAST2	Software package containing several programs for phylogenetic inference, estimation of divergence dates, and estimation of changes in long-term effective population size from DNA sequence data.	Drummond et al. (2006); Dummond and Rambaut (2007); Drummond et al. (2012); Bouckaert et al. (2014)
MrBayes	Computer program phylogenetic inference and estimation of divergence times from DNA sequence data.	Huelsenbeck and Ronquist (2001); Ronquist and Huelsenbeck (2003); Huelsenbeck and Ronquist (2005); Ronquist et al. (2009a, b); Ronquist et al. (2012a)
PAML	Software package containing several computer programs including mcmctree, which estimates phylogenies from DNA sequence data.	Yang (1997, 2007)
PhyloBayes	Computer program for phylogenetic inference and estimation of divergence times from amino acid and DNA sequence data.	Lartillot et al. (2009, 2013)
STRUCTURE	Computer program for the inference of population structure from genotypic data via Markov Chain Monte Carlo.	Pritchard et al. (2000); Falush et al. (2003, 2007); Hubisz et al. (2009)
BAPS	Computer program for the inference of population structure from georeferenced genotypic data.	Corander et al. (2003, 2008)
GENELAND	R package for the inference of population structure from georeferenced genotypic data.	Guillot et al (2005, 2008)
TESS	R package for estimation diversification rates from molecular phylogenies.	Höhna et al. (2016)
BPEC	R package for the construction of haplotype networks based from DNA sequence data.	Manolopoulou et al. (2011); Manolopoulou and Emerson (2012)
ONeSAMP	Web-based program that estimates effective population size from microsatellites via Approximate Bayesian Computation.	Tallmon et al. (2008)
abc	R package for model choice and estimating model parameters based on output from other computer programs.	Csilléry et al. (2012)
ABCtoolbox	Software package for computing summary statistics, running simulations, estimating model parameters, and performing model choice from microsatellite and DNA sequence data via ABC. Interacts with other computer programs.	Wegmann et al. (2010)
DIYABC	Computer program for computing summary statistics, running simulations, estimating model parameters, and performing model choice from microsatellite, haplotypic, and single nucleotide polymorphism (SNP) data via ABC.	Cornuet et al. (2008, 2010, 2014a, b)
EasyABC	R package for performing simulations for ABC.	Jabot et al. (2013)
BATWING	Computer program for modeling population divergence, population size and growth, and mutation rates from DNA sequence data and microsatellites.	Wilson et al. (2002, 2003)
msBayes	ABC program for testing simultaneous divergence among co-distributed pairs of populations from DNA sequence data.	Hickerson et al. (2007)
BayesAss	Inference of recent migration rates between populations from genotypic data.	Wilson and Rannala (2003)
COLONY	Computer program for the inference of sibship and parentage among individuals and estimates of effective population size using genotypic data.	Jones and Wang (2010)
BayesScan	Detection of selection in genotypic data.	Foll and Gaggiotii (2008); Fischer et al. (2011)

Table 14.2. Applications of Bayesian inference to molecular ecology.

Application	Taxa
Higher phylogenetics	Black basses and sunfishes (Near et al. 2004; Bagley et al. 2011), skinks (Brandley et al. 2005, 2012), snakes (Pyron and Burbrink 2009), New World quails (Hosner et al. 2015), carnivorans (Flynn et al. 2005), deer (Gilbert et al. 2006)
Intraspecific phylogenetics	Cricket frogs (Gamble et al. 2008), snakes (Castoe et al. 2007; Burbrink et al. 2008; Guiher and Burbrink 2008; Krysko et al. 2016), brown creeper (Manthey et al. 2011), northern cardinal (Smith et al. 2011), spotted skunk (Ferguson et al. 2017)
Divergence times	Cycads (Condamine et al. 2015), angiosperms (Bell et al. 2010; Couvreur et al. 2010), Hymenoptera (Ronquist et al. 2012a), black basses and sunfishes (Near et al. 2003, 2005a), crocodiles (Oaks 2011), humpback whale (Jackson et al. 2014)
Bayesian skyline plots	Pacific herring (Grant et al. 2012), green anole (Tollis et al. 2012), snakes (Guiher and Burbrink 2008; Pyron and Burbrink 2009), northern bobwhite (Williford et al. 2014, 2016), steppe bison (Shapiro et al. 2004), roe deer (Baker and Hoelzel 2014), spotted skunk (Ferguson et al. 2017)
Bayesian clustering	Oaks (Cavender-Bares and Pahlich 2009; Neophytou 2014), salmon (Dionne et al. 2002), black basses (Lutz-Carrillo et al. 2006; Stepien et al. 2007; Bean et al. 2013), amphibians (Zamudio and Wieczorek 2007; Wang et al. 2009; Richmond et al. 2014), green anole (Manthey et al. 2016), cormorants (Barlow et al. 2011), prairie-grouse (Oyler-McCance et al. 2016), wild turkeys (Seidel et al. 2013), feral pigs (Hampton et al. 2004; Spencer and Hapton 2005), mule deer (Latch et al. 2014), wild cats (Lecis et al. 2006; O'Brien et al. 2009), wolves (Bohling et al. 2013)
Contemporary effective population size estimation	Frogs (Phillipsen et al. 2011), lesser prairie-chicken (Pruett et al. 2011), house sparrows (Baalsrud et al. 2014), bottlenose dolphins (Louis et al. 2014), brown bear (Skrbinšek et al. 2012)
ABC model selection	Ladybird beetles (Lombaert et al. 2011), fruit flies (Pascual et al. 2007), Coeur d'Alene salamander (Pelletier and Carstens 2014), fox snakes (Row et al. 2011), Tasmanian devil (Bruniche-Olson 2014), orangutans (Nater et al. 2015)
Sibship and parentage	Freshwater prawns (Karaket and Poompuang 2012), salmonids (Kanno et al. 2011; Lin et al. 2016; Ackerman et al. 2017), green turtle (Frey et al. 2013), northern bobwhite (Miller 2014), okapi (Stanton et al. 2015), gray wolf (Galaverni et al. 2012), Antarctic fur seal (Bonin et al. 2014)

a value of the test statistic as or more extreme if the experiment was repeated, assuming that the null hypothesis is true. Therefore, inference in the frequentist paradigm depends in part on long-run probabilities derived from sampling theory, as well as the arbitrary choice of a P-value to represent the threshold for a significant treatment effect (Anderson et al. 2000).

In manipulative experiments, the frequentist approach can provide a valid tool for statistical inference. However, many have argued that the use of P-values is more of a decision rule than an interpretation of the evidence about the null and alternative hypotheses. Because P-values are affected by sample size, a P-value says nothing about the magnitude of the treatment effect; supporting information must be used to decide whether an effect is biologically mean-

ingful, regardless of statistical significance or lack thereof. In addition, the creation of a valid null hypothesis is problematic, and "more extreme" data can be understood only in the context of sampling design, aspects of which are not often reported or even comparable among studies (Johnson 1999). For observational studies, the development of a valid null hypothesis and reliance on long-run probabilities are tenuous because the observer lacks the ability to repeat or replicate the study (Anderson et al. 2000).

The Bayesian paradigm for statistical inference stems from the ideas of Reverend Thomas Bayes (1702–1761), which were published posthumously (Bayes 1763). Bayes developed the mathematical basis for conditional probability, now termed *Bayes's theorem*; the application of conditional probability to statistical inference is termed *Bayesianism*. The

	Resident (R)	Immigrant (I)	
Male (M)	0.2	0.2	0.4
Female (F)	0.6	0.0	0.6
	0.8	0.2	

Joint probabilities
$P_{(M,R)} = 0.2$ $P_{(F,R)} = 0.6$
$P_{(M,I)} = 0.2$ $P_{(F,I)} = 0.0$

Marginal probabilities
$P_{(M)} = 0.4$ $P_{(R)} = 0.8$
$P_{(F)} = 0.6$ $P_{(I)} = 0.2$

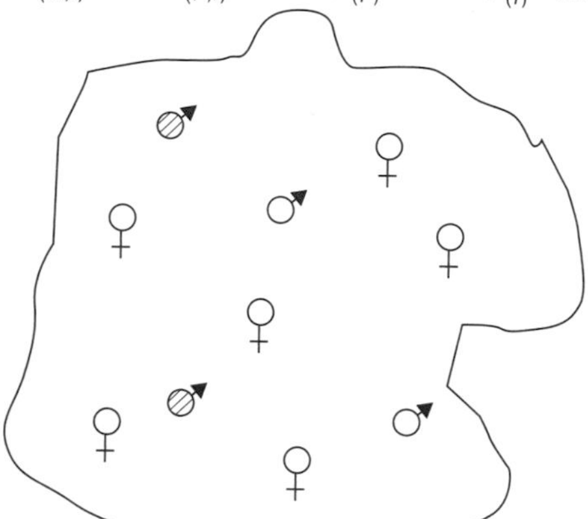

What is the probability P that an individual is resident (R), given it is a male (M)?

$$P(R \mid M) = \frac{P(R)\, P(M \mid R)}{P(M)}$$

$P(M) = P(M,R) + P(M,I)$ = marginal P of being male, computed by marginalizing over Residency status:

$$P(R \mid M) = \frac{P(M,R)}{P(M,R) + P(M,I)} \qquad P(M)$$

$$P(I \mid M) = \frac{P(M,I)}{P(M,R) + P(M,I)} \qquad P(M)$$

where : $P(R \mid M) + P(I \mid M) = \dfrac{P(M,R) + P(M,I)}{P(M,R) + P(M,I)} = 1.0$

thus :

$$P(R \mid M) = \frac{P(R)\, P(M \mid R)}{P(M,R) + P(M,I)} = \frac{P(R)\, P(M \mid R)}{P(R)\, P(M \mid R) + P(I)\, P(M \mid I)} = \frac{P(R)\, P(M \mid R)}{\sum_{\theta\{R,I\}} P(\theta)\, P(D \mid \theta)}$$

in generalized notation : $P(\theta \mid D) = \dfrac{P(D \mid \theta)\, P(\theta)}{\sum_{\theta} P(D \mid \theta)\, P(\theta)}$

where D = data and θ = hypotheses or estimated parameters
The denominator may be summarized as $P(D)$, the sum of the marginal probabilities of the data over all θ

Fig. 14.1.   In a given deme of a population of mammals, there are four males and six females, eight residents (open polygons) and two immigrants (shaded polygons). From these counts, we can tabulate the joint probabilities of sex and residence (cell values) and the marginal probabilities of sex or residence (row and column sums). This two-dimensional example illustrates how the "marginal" probability of being male can be computed by marginalizing over residency status. The resulting marginal probabilities are used to compute the quantity $P(D)$. *Source: Adapted from Lewis (2001); Ronquist et al. (2009a).*

Bayesian approach involves the specification of a hypothesis and a conjecture about the probability of the hypothesis prior to the collection of data or observations. After data or observations are available, one can update the probability based on the new information. The two probabilities are termed the *prior* and *posterior*, and Bayes's theorem describes how to update the prior probability as new information becomes available (Fig. 14.1).

In simple form, Bayes's theorem may be stated as:

$$P(H \mid D) = \frac{P(D \mid H)\, P(H)}{P(D)},$$

where the quantity $P(H \mid D)$ is the probability of hypothesis H, conditional on the observed data D, also termed the *posterior probability*; the quantity $P(D \mid H)$ is the probability of observing the data D conditional on hypothesis H, also termed the *likelihood of H*; the quantity $P(H)$ is the probability of hypothesis H before collection of data, also termed the *prior probability*; and the quantity $P(D)$ is the unconditional probability of observing the data D (Sober 2008). Note that the likelihood is a mathematical term that describes the probability of the data D under the hypothesis H; this quantity can have a different value than the unconditional probability of the data D, $P(D)$, or the posterior probability of the hypothesis H given the data D, $P(H \mid D)$. Based on the probability of the hypothesis H, one can compute the probability of "not H" as $1 - P(H)$ because probabilities must sum to 1.0. However, likelihoods are not required to sum to 1.0, so one cannot compute the likelihood of "*not H*" based on the likelihood of *H* (Sober 2008).

In most applications of Bayes's theorem to genetic data, the quantity of interest may not be a hypothesis, but the probability of a parameter or other non-observable information needed to make inferences from the data. For instance, allele frequencies are often considered "nuisance" parameters because an estimate of allele frequencies is needed to make inferences about populations or population processes, such as admixture (mixing of genomes of divergent parental populations or taxa; Buerkle and Lexer 2008) or migration. However, the allele frequencies themselves usually are not the main objective of the study (Shoemaker et al. 1999). Therefore, one may see Bayes's theorem written as the solution to $P(\theta|D)$, where $P(\theta)$ is the prior distribution of a parameter(s), including nuisance parameters such as allele frequencies, nucleotide substitution rates, and so on.

The prior information and a likelihood model enable the estimation of the probability of the hypothesis after the data are collected, which forms the Bayesian posterior probability. In contrast, the *P*-values from frequentist statistics provide the probability of observing more extreme data if the null hypothesis is true and the experiment was repeated many times. One strength of the Bayesian approach is that the outcome is not viewed as a decision rule, unlike the "$P < \alpha$" threshold between significant and nonsignificant results, where the predefined α is usually 0.05 (Shoemaker et al. 1999; Beaumont and Rannala 2004). Rather, the outcome of a Bayesian analysis is a statement about the degree of belief in a hypothesis, its probability, where the degree of belief may change as new information becomes available (Sober 2008).

The Bayesian approach is best suited to cases where there are discrete hypotheses, and one has some basis to estimate the probability of *H* and "*not H*" (Sober 2002). For instance, many phylogenetic analyses are variations of a straightforward set of alternatives: the probability that taxon 1 is more closely related to taxon 2 than either is to taxon 3. Without a clear set of definable alternatives, the probability of *H* and "*not H*" becomes difficult to justify. Consider the following: What is the probability that life began on the Earth? We have little basis for the estimation of a prior or a likelihood model because at this time, life is only known to exist on the Earth. Furthermore, the alternatives are not discreet. Either life began on Earth or it did not, so $P(\text{Earth}) + P(\text{not Earth}) = 1$. Perhaps life began on Mars, but why not elsewhere in our solar system—and why not an extrasolar option? In this case, $P(\text{Earth}) + P(\text{Mars}) + P(\text{Elsewhere}) + P(\text{Extrasolar}) = 1$. A combination of two or more alternatives is plausible because the mitochondria and chloroplasts of Eukaryotic organisms is the result of endosymbiosis between prokaryotic cells (Sagan 1967). Thus, $P(\text{Earth}) + P(\text{Mars}) + P(\text{Elsewhere}) + P(\text{Extrasolar}) + P(\text{Earth} + \text{Mars}) + P(\text{Earth} + \text{Elsewhere}) + P(\text{Earth} + \text{Extrasolar}) + P \ldots = 1$. The lack of a clear set of definable alternatives quickly becomes a "catch-all," and it is difficult to formulate unconditional probabilities or a valid likelihood model (Sober 2008). Fortunately, the genetic analyses confronted with Bayesian methods typically have discrete alternatives and the "catchall" situation is rarely an issue.

Bayesian Ideas Applied to Complex Problems

Consider the following example: the jaguarundi (*Puma yagouaroundi*) is a small Neotropical felid that ranges from South America to northern México. The jaguarundi historically occurred in the lower Rio Grande Valley of southern Texas, but was extirpated from Texas during the late 1900s; the last confirmed sighting was of a road-killed individual in 1984. Sightings of jaguarundi are frequently reported in the region, but rarely considered credible; the jaguarundi is similar in size and appearance to the domestic cat, and misidentification is common. A resident of the lower Rio Grande Valley obtains a photo of an unidentified felid using a remote camera; the image superficially resembles a jaguarundi, but is not conclusive. What is the probability that the photo is of an actual jaguarundi?

$$P(\text{jaguarundi} \mid \text{photo})$$
$$= \frac{P(\text{photo} \mid \text{jaguarundi}) \ P(\text{jaguarundi})}{P(\text{photo})}.$$

We can estimate the prior probability of a jaguarundi in the region, $P(\text{jaguarundi})$, based on the last confirmed sighting, which occurred 37 years ago: 1 confirmed occurrence in 37 years, or 1 in 13,505 days = 0.000074. Based on previous research, we can estimate the probability of obtaining a photo of a jaguarundi if a jaguarundi was in fact present, $P(\text{photo} \mid \text{jaguarundi})$, from the average number of photos per marked individual in previous studies, 47 photos per 365 days = 0.129. Finally, we can estimate the probability that a photo of a potential jaguarundi is reported from the region, $P(\text{photo})$, from the average number of reported photos that resemble jaguarundis in the lower Rio Grande Valley per year, 20 photos per 365 days = 0.055.

$$P(\text{jaguarundi} \mid \text{photo}) = \frac{(0.129)(0.000074)}{(0.055)}$$
$$= 0.00017.$$

Thus, a photo that resembles a jaguarundi is enough to raise the posterior probability of occurrence. However, the posterior probability of a jaguarundi in the region remains small because (1) the prior probability is low due to lack of confirmed sightings during the past 37 years; and (2) the occurrence of photos that resemble jaguarundi, false positives in this case, is 743 times more common than a confirmed sighting (0.055 / 0.000074).

The jaguarundi problem is a fairly simple one, in which we can use previous information to inform our choice of prior; in this case, we have an informative prior. The problem also has a discrete yes/no outcome: either there is a jaguarundi or there is not. In genetics problems, we often compute probabilities of

population membership or phylogenetic outcome in terms of distributions because the data are continuous. Each combination of allele frequencies or nucleotide substitution models can be viewed as a separate hypothesis, where the aggregate forms a continuum of probabilities (Chen et al. 2014). We can thus reformulate Bayes's theorem for a continuous distribution (Lewis 2017) as

$$f(\theta \mid D) = \frac{f(D \mid \theta) \ f(\theta)}{\int f(D \mid \theta) \ f(\theta) d\theta},$$

where $f(\theta \mid D)$ is the posterior probability density of parameter theta θ, given the data D. The other parts of the equation are replaced by distributions as well. The quantity $f(D \mid \theta)$ is the likelihood function, $f(\theta)$ is the prior probability distribution, and $f(D \mid \theta)$ is the marginal probability (Fig. 14.1) of the data integrated over all possible values of D and θ. Probabilities are now calculated for intervals rather than discrete data points by scaling the probability density function so that the total area under the curve is 1.0. The quantity $\int_\theta f(\theta) \ d\theta$ is the integral of the distributions $f(\theta)$ for each possible value of the data, $d\theta$. Priors can be stated in the form of distributions, where the prior is often represented by a distribution that is computationally efficient, yet allows one to incorporate the degree of prior knowledge, ranging from vague to informative (Shoemaker et al. 1999). For analysis of genetic data, common prior distributions include beta, exponential, gamma, and Dirichlet, a multivariate version of the beta distribution (Ronquist et al. 2009a, 2009b). If there is very little or no information, one can use a flat or uniform prior, where all values in the interval have the same probability (Fig. 14.2).

Simple examples are useful learning tools, yet real-world applications are much more complex. For instance, most applications of Bayesian methods to genetic data involve multiple parameters, which greatly complicate the calculation of marginal probabilities (Fig. 14.1). Consider a two-parameter Bayesian analysis (Lewis 2017):

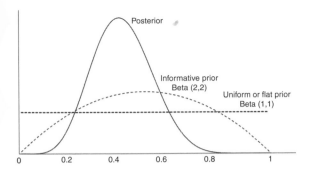

Fig. 14.2. Prior distributions may be represented by vague (uniform) priors or informative priors, depending on how much a priori information is available. In this case, both priors have the same mean, but the variance of the uniform prior is much greater. The posterior probability distribution, computed after data are collected, is usually much more informative than the prior. For small data sets, the prior can have a large influence on the posterior, but this is less of a problem for the analysis of genetic data using Markov Chain Monte Carlo (MCMC) methods, which sample extensively once the Markov chain has reached stationarity. *Source: Adapted from Lewis (2001).*

$$f(\theta,\delta\,|\,D) = \frac{f(D\,|\,\theta,\delta)\,f(\theta)\,f(\delta)}{\int_{\theta}\int_{\delta} f(D\,|\,\theta,\delta)\,f(\theta)\,f(\delta)\,d\theta\,d\delta}.$$

For analyses of population structure or migration rates, the number of alleles dictate the number of parameters. A simple data set with 10 microsatellite DNA loci and 10 alleles per locus has a minimum of 100 allele frequency parameters; the estimation of population clusters, admixture, and so on, adds more complexity. Although single nucleotide polymorphism (SNP) loci are bi-allelic, analyses based on SNP data often involve thousands of loci; a recent analysis of northern bobwhite (*Colinus virginianus*) in our lab used >19,000 SNP loci, which translates to >38,000 allele frequency parameters. Similarly, in phylogenetic analyses, parameters may include mutation rates, rate variation among sites, branch lengths, divergence times, and so on. As indicated earlier, nuisance parameters such as allele frequencies or rates of nucleotide substitution are needed to make inferences, but are not the goal of the analysis. Bayesian analyses offer a distinct advantage in that

one can incorporate uncertainty in estimates of the nuisance parameters, and even missing data, as well as the posterior probability. Nonetheless, as the number of parameters increases, it becomes first difficult, then impossible to compute the marginal probabilities over all possible parameter values.

Markov Chain Monte Carlo

For complex analyses, an approximate solution for the posterior distribution allows inference without the need to solve for the posterior, which becomes impossible as the number of parameters increases. Markov Chain Monte Carlo (MCMC) methods are a type of simulation well-suited to approximate posterior distributions for Bayesian analyses. Monte Carlo is a reference to the famous gaming casino in Monaco because the process incorporates random changes to the Markov chain to explore parameter space. The concept of MCMC first arose among scientists at the U.S. Los Alamos National Laboratory in the 1940s (Metropolis et al. 1953), when its use was limited because the procedure requires extensive computational effort. Desktop computers became sufficiently powerful in the 1990s to permit everyday users to apply MCMC, and the use of Bayesian methods in genetic analyses expanded dramatically. The ability to perform analyses in a reasonable amount of time on desktop computers with the aid of MCMC is a strong driving force behind the widespread use and development of Bayesian methods in genetic analyses (Beaumont and Rannala 2004; Shoemaker et al. 1999).

A Markov chain (Fig 14.3) will eventually iterate to an equilibrium in parameter space from any starting point, given enough repetitions. This equilibrium, or stationary distribution, is used to approximate the Bayesian posterior distribution. The most common approach for exploring parameter space is termed the *Metropolis-Hastings algorithm*, described by Metropolis et al. (1953) and later modified by Hastings (1970). The process begins at a random or arbitrary point in parameter space, φ_0, proposes a small change at random, φ_t, and then evaluates the posterior probability of the new value φ_t vis-à-vis the old value φ_{t-1}. If the ratio R of posterior probabilities P, depicted as $\dfrac{P(\varphi_t)}{P(\varphi_{t-1})}$, is >1, the proposal is accepted and the process repeats from the new value φ_t. If $\dfrac{P(\varphi_t)}{P(\varphi_{t-1})} < 1$, the proposal is either accepted with a probability proportional to R, or rejected. If the proposed move is accepted, the process repeats from the new value φ_t; if rejected, the chain remains at φ_{t-1}, proposes a new random value, and the process repeats (Rohnquist et al. 2009a).

Lewis (2001, 2012) proposed a useful analogy for MCMC, that of a robot climbing up "hills" in a two- or three-dimensional parameter space, where the hills are areas of high posterior probability. Moves that take the robot uphill or only slightly downhill are accepted because $R > 1$ or ~ 1; moves that plunge into steep "valleys" are usually rejected because R is

near 0 (Lewis 2017). If the Markov chain has enough repetitions, it will converge on a stationary distribution, and the number of sampled points in a given distribution of parameters will be the approximate posterior probability. Thus, if the posterior probability estimated from an analysis is 0.92, then 92% of points sampled were from that distribution or set of parameter values. Because the chain begins at a random or arbitrary point, it must sample parameter space to locate regions of high posterior probability. Therefore, it is common to discard the first few thousand(s) of points sampled to minimize the effect of the starting configuration; this period before stationarity is termed the *burn-in*. It is also common to practice *thinning* during MCMC runs that involve many hundreds of thousands or millions of iterations, where the algorithm will save the points visited at some interval; for instance 1 of every 100 points is saved (Ronquist et al. 2009a). Thinning can

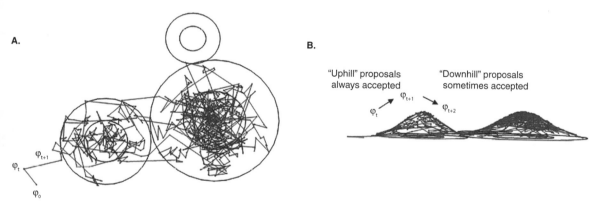

A.   **B.**

"Uphill" proposals always accepted   "Downhill" proposals sometimes accepted

φ_{t+1}

φ_t φ_{t+2}

φ_{t+1}

φ_t

φ_0

Fig. 14.3.   The basic Markov Chain Monte Carlo (MCMC) procedure:
i. Start at arbitrary or random point, φ_0.
ii. Propose a new value at random, φ_t.
iii. Calculate $P(D \mid \varphi_t)\, P(\varphi_t)$, compare to $P(D \mid \varphi_{t-1})\, P(\varphi_{t-1})$, where P is probability, D is the data.
iv. If $P(D \mid \varphi_t)\, P(\varphi_t) > P(D \mid \varphi_{t-1})\, P(\varphi_{t-1})$, accept the proposed value φ_t.
v. If $P(D \mid \varphi_t)\, P(\varphi_t) < P(D \mid \varphi_{t-1})\, P(\varphi_{t-1})$, accept with P proportional to the ratio of the 2 values.
vi. If φ_t is not accepted, remain at φ_{t-1}, propose a new value at random, φ_{t+1}.
vii. Repeat, record all points visited.
With sufficient repetitions, the chain will converge and most points occur in regions with high posterior probability. Because the probability of proposing a step from φ_t to φ_{t+1} is the same as the probability of φ_{t+1} to φ_t, the proportion of points in each region is proportional to the posterior probability. *Source: Adapted from Lewis (2001), Ronquist et al. (2009a).*
Hypothetical 1,000 iterations of MCMC in two dimensions (A) and three dimensions (B). Circles (A) and hills (B) represent regions with high posterior probability; sampled points are represented by dots and paths between points by solid lines.

reduce demand for memory and decrease the correlation among successive points.

The choice of proposed moves can affect how efficiently the Markov chain explores the parameter space, often termed its *mixing* behavior. For instance, if proposed moves are small, most are accepted, but the chain takes a long time to explore the parameter space and may not find regions of high posterior probability for many thousands of repetitions. If the proposed moves are large, most moves are rejected and the chain spends time "stuck" in a small portion of the parameter space (Fig. 14.3). The exploration of parameter space can be accomplished more efficiently using multiple Markov chains run simultaneously from different starting positions, or by sampling different parameters to propose moves of the

chain. In addition, >1 of the chains can be "heated," which allows the chain to move across "valleys" among hills of posterior probability; this approach is termed the *Metropolis-coupled MCMC* (Ronquist et al. 2009a; Lewis 2017). Exponentiation of the posterior probability to a value <1, termed *heating* the chain, essentially smooths the hills of posterior probability, and the chain can easily move across valleys between hills (Fig. 14.4). The heated chain(s) explores the exponentiated distribution, while "cold" chain(s) explore the non-exponentiated distribution. The non-exponentiated distribution is the target to make inferences from, while the heated distribution acts as a guide to help explore complex parameter spaces. The heated and cold chains are swapped periodically, but only points sampled by the cold chain

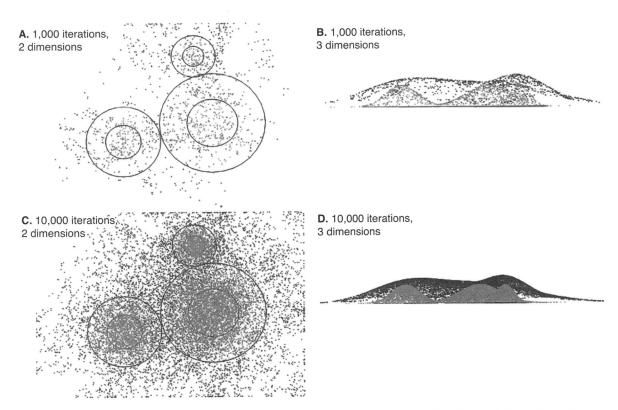

A. 1,000 iterations, 2 dimensions

B. 1,000 iterations, 3 dimensions

C. 10,000 iterations, 2 dimensions

D. 10,000 iterations, 3 dimensions

Fig. 14.4. Markov Chain Monte Carlo (MCMC), where the parameter space is sampled with two chains, a "heated" chain and a "cold" chain. Circles (A, C) and hills (B, D) represent regions of high posterior probability in two and three dimensions, respectively. The chain is heated by exponentiation of the posterior distribution to a power <1, allowing the heated chain to more fully explore the parameter space (A, C) by "flattening" valleys between "hills" of high posterior probability (B, D). The chains are swapped periodically; only points from the cold chain are retained for inference.

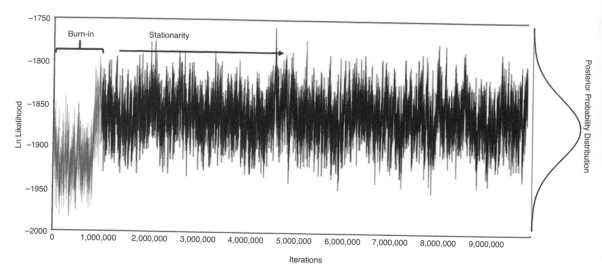

Fig. 14.5.   Trace plot of likelihood versus iteration of a Markov chain during Markov Chain Monte Carlo (MCMC) sampling of parameter space. The chain starts from a random or arbitrary point and must sample sufficiently to locate regions of high posterior probability. The initial iterations of the chain are influenced by the starting point and are usually discarded; this initial period is termed the "burn-in." As the chain approaches stationarity, or converges on the region of high posterior probability, the likelihood values level out. The rate that the chain samples regions of high posterior probability is termed "mixing." Poor mixing often stems from acceptance rate of movement proposals that is too high or too low, and the chain either takes too long to explore parameter space or remains "stuck" in the same region, respectively. Trace plots are one of several methods that should be used to assess convergence during MCMC. *Source: Adapted from Ronquist et al. (2009a).*

are retained to estimate the posterior probability. If successful, the process allows for thorough and efficient exploration of parameter space.

The end goal of the MCMC is to sample from a stationary distribution, where the Markov chain has converged on the region of highest posterior probability. In practice, it can be difficult to determine if the chain has converged and if the sampled region actually contains the highest posterior probability; convergence is largely a qualitative assessment and there are no formal diagnostics. Users are often advised to: (1) scrutinize the mixing behavior of runs using trace plots (Fig. 14.5) and other diagnostics; (2) compare results of different runs, where each run has a different starting point and/or different proposal move parameters; and (3) perform runs of different lengths, or number of repetitions. If all of the runs from steps 2 and 3 converge on the same solution, and the trace plots indicate good mixing, one can accept the results and use them to make inferences

from the data. If the runs differ markedly, the chain may need more iterations to fully explore the parameter space. Unfortunately, the assessment of convergence becomes time and computationally intensive for large or complex data sets, and still does not guarantee convergence.

Bayesian Analysis of Population Genetic Data

A new class of genetic marker, *DNA microsatellites*, came into widespread use in nonmodel organisms during the 1990s. Microsatellites are short, tandem repeats of 2–4 bases. These repetitive segments are common throughout the genome, and most are selectively neutral and variable within populations due to their high mutation rate (Avise 2004). A panel of 10–20 microsatellite loci has sufficient power to identify individuals and estimate relationships among individuals. Allelic variation at a locus is a function of the number of repeat units, so microsatellite loci can be easily assayed via the polymerase chain reaction (PCR) and electrophoresis. The availability of easily assayed, highly variable markers enabled ecologists to study gene flow, migration, dispersal, hybridization, parentage, and the effects of recent and past ecological, landscape, and anthropogenic influences on individuals and populations (DeYoung and Honeycutt 2005). Desktop computing power also became faster and more affordable during the same era. This perfect storm of technology and molecular biology created both motive and opportunity for more sophisticated analytical approaches, and Bayesian analyses proliferated. The use of SNP loci is beginning to supplant microsatellites for many population genetic studies, but Bayesian tools developed during the 1990s and early 2000s remain in widespread use.

Early analyses for the study of migration and dispersal, termed *assignment tests*, relied on maximum likelihood assignment of individuals to populations as a function of allele frequencies (Paetkau et al. 2004; Manel et al. 2005). Although useful, maximum likelihood approaches are limited by sampling requirements and required at least some a priori knowledge of populations. A suite of tools, collectively termed *Bayesian clustering algorithms*, were developed to address migration, dispersal, and population ancestry in a typically Bayesian manner (François and Waits 2016), by reformulating the question. Instead of asking "What is the probability that a sampled individual originated from one or more known populations?" the question became "Given the data, how many distinct genetic clusters are there, and which cluster(s) do the individuals belong to?"

The model-based clustering algorithm STRUCTURE (Pritchard et al. 2000) is arguably the most popular and widely used Bayesian analysis of population genetic data. The authors attempted to solve the problem of hidden population structure and admixture, which may produce spurious results during the analysis of case-control genetic data for human diseases (Noor 2013). The implications for conservation genetics of nonmodel species were immediately apparent, and the algorithm's popularity grew steadily; as of December 2017, the original paper had been cited >21,000 times. Subsequent modifications described in Falush et al. (2003, 2007), and Hubisz et al. (2009) made the program more user-friendly and modified or extended its capabilities. The algorithm attempts to fit sampled individuals into clusters (K) that minimize Hardy-Weinberg and linkage disequilibrium. The method is flexible and can accommodate situations where allele frequencies among clusters may be correlated. In addition, the algorithm can model admixture, where individuals may have ancestry to ≥2 distinct groups or populations; this is an important advantage for the analysis of many real-world population genetic questions, including dispersal and hybridization. Users can incorporate information on sampling locations into the analysis or run the analyses "blind" for some or all of the individuals. The end result is an estimate of the joint probability distribution of the genotype data and nuisance parameters, such as allele frequencies, and the cluster membership and admixture proportions for each individual (François and Durand 2010).

The basic procedure is for the user to model the data for $K = 1, 2, \ldots, j$ clusters and determine which K provides the best fit to the data. Pritchard et al. (2000) recommend the usual quality control for Bayesian analyses, including an assessment of burn-in and multiple runs of different lengths to ensure convergence. However, the choice of "best fit" to the data is qualitative, and the authors strongly emphasize interpretation of K using biological information (Pritchard et al. 2000). The authors selected K based on the estimated log $P(D|K)$ values among runs for different values of K, where D is the genotype data, defined as $P(X|K)$ in the notation of Pritchard et al. (2000). Later studies revealed that the values of log $P(D|K)$ tended to plateau as K increased, and did not always present a clear choice for best fit. An ad hoc solution based on the rate at which the log $P(D|K)$ values change between consecutive values of K, termed the *delta K method*, appears to provide an indication of the main structure in the data (Evanno et al. 2005) and has been widely used.

Users are advised to perform 5–10 or more replicate runs for each K because variation among runs for each K increases as the algorithm attempts to fit individuals into the hypothesized number of clusters when there is insufficient signal in the data on which to base the cluster. However, the delta K can assess only $K > 2$, and users should scrutinize $K = 1$ in comparison to values of $K \geq 2$. Another sign of over-fitting is the tendency to parcel out individual ancestry into equal portions among the hypothesized values of K. For instance, if the actual $K = 1$ and the user models for $K = 3$, the algorithm will force assignment into three clusters. Therefore, individuals may display assignment to each of the K in equal parts; in this case, each individual will have one-third of its ancestry parceled into each of the three clusters. Several recent algorithms can aid with the assessment and interpretation of clustering solutions derived from STRUCTURE runs, including CLUMPAK (Kopelman et al. 2015) and STRUCTURE HARVESTER (Earl and vonHoldt 2012).

Nonetheless, there is no perfect method for choosing K, and all approaches are subjective to some degree. Although the phrase "true number of K" often appears in the Results or Discussion sections of the literature, Pritchard et al. (2000) clearly indicate that the STRUCTURE algorithm can provide only an ad hoc approximation. Users should remember that "it usually makes sense to focus on values of K that capture most of the structure in the data and that seem biologically sensible" (Pritchard et al. 2010:15). Furthermore, users should have some clear hypotheses and alternative hypotheses before running the analysis. Otherwise, one might succumb to the temptation to over-interpret the output of Bayesian clustering algorithms, which can be the population genetics equivalent of Rorschach blots. Genetic data are nearly always the result of observational rather than manipulative studies; thus, inferential power is already compromised before a pipet has been lifted. If there are no clear a priori hypotheses, it is remark-

ably easy to make up a story to explain the results of Bayesian clustering! Users should treat such analyses as exploratory and take precaution to reduce the chances of spurious results, which are especially likely to arise during analysis of data sets with high dimensionality (Anderson et al. 2001), which include all data submitted to Bayesian clustering analyses. In cases such as these, it may be better to treat the exploratory analysis as a hypothesis-generating exercise rather than as hypothesis-testing or inferential.

Spatially Explicit Clustering

The STRUCTURE algorithm can incorporate prior knowledge about populations, but only in a qualitative manner. This is relatively inefficient for the study of fine-scale population structuring or the effect of landscape features on population differentiation, especially when sampling is uneven. The inclusion of spatial coordinates for samples enables the consideration of population clines and isolation by distance processes, in addition to the identification of genetic clusters (François and Durand 2010). Popular algorithms for spatially explicit Bayesian clustering include GENELAND (Guillot et al. 2005), TESS (Chen et al. 2007), and BAPS (Corander et al. 2008). Each has a different approach to the detection of clusters and can be run using models with or without admixture. In general, simulation studies and real-world data indicate that models that incorporate admixture usually perform better than models without admixture (François and Durand 2010).

The algorithms in TESS and BAPS employ a method termed *Voroni tessellation* to incorporate spatial information about individuals. The tessellation procedure encloses each sample point by a cell, where the boundaries of the cell are nearer to the sample point than to any other sample. Neighboring points have cells that share at least one edge, and neighbors have a greater chance of membership in the same cluster. The algorithms minimize departure from Hardy-Weinberg expectations through the introduction of correlation or spatial dependency (Chen et al. 2007;

Corander et al. 2008; François and Waits 2016). The expected genetic similarity among individuals is modeled assuming an isolation by distance process, where autocorrelation among points declines with increasing spatial distance between points. The GENELAND algorithm also uses Voroni tessellation, but constructs cells around groups of sample locations, as opposed to single points, and models boundaries that separate K random-mating subpopulations (François and Waits 2016).

The TESS and GENELAND algorithms use MCMC, while BAPS employs a numerical optimization procedure. Each algorithm relies on a different procedure to choose K: TESS uses a deviance information criterion, BAPS uses a split and merge process, and GENELAND uses a reversible jump algorithm (François and Waits 2016). All three appear to perform reasonably well within the confines of adequate sampling design and the presence of genetic signal in the data. Detection of barriers also depends on the number of generations that the barrier has existed. Overall, Bayesian clustering performs better for the detection of barriers than other methods, though a strong signal of isolation by distance can cause spurious clusters (Safner et al. 2011; Blair et al. 2012). The GENELAND algorithm appears to be a good choice for the detection of linear barriers, whereas TESS can accommodate complex and nonlinear barriers as well as complex demographic histories (François and Waits 2016).

Sampling Considerations and the Limits of Bayesian Clustering Algorithms

Sampling design and scale are important considerations for any study, but become critical for inference of population structure in relation to landscape or population history. Population genetic studies often rely on convenience sampling and can produce spurious clusters when sampling is uneven, especially in the presence of isolation by distance (Frantz et al. 2009; Schwartz and McKelvey 2009). This may occur when a continuous population is sampled at discrete intervals; the main driver of genetic differentiation is geographic distance among points because dispersal is finite, but the sampling creates an artefact of distinct clusters. A recent study showed that STRUCTURE clustering solutions can be biased low, indicating fewer clusters than actually present, when the sample size from each cluster is uneven (Puechmaille 2016). No other clustering methods were evaluated, but Puechmaille (2016) suggested that other Bayesian clustering algorithms were likely prone to the same bias. The bias could be alleviated with the use of a supervised clustering approach and subsampling (Puechmaille 2016), or by allowing the default ancestry prior, α, to vary from a small initial value combined with the model assuming uncorrelated allele frequencies (Wang 2017).

All clustering methods perform better with increasing genetic differentiation among clusters, and typically do well for $F_{ST} \geq 0.05-0.10$; the estimation of clusters at low levels of differentiation can be tenuous, especially when $F_{ST} < 0.03$ (Latch et al. 2006). Recent advances in SNP discovery and genotyping techniques (Peterson et al. 2012) have enabled clustering analyses for weakly differentiated populations due to the sheer power of many thousands or tens of thousands of SNP loci. Nonetheless, preliminary data analyses based on summary statistics, such as F_{ST} and F_{IS}, in combination with an assessment of isolation by distance or spatial autocorrelation, can be a valuable indicator of the type and strength of signal present in the data. Users should interpret K with caution when F_{ST} and F_{IS} are low, and/or there is evidence of spatial autocorrelation. Furthermore, the inner workings of the STRUCTURE algorithm, and all Bayesian clustering algorithms, can be obscured by the user-friendly graphical interface; the underlying algorithms are highly complex and can be sensitive to changes in parameter values (Wang 2017). Users should explore or understand how changes in parameter values affect the results before drawing firm conclusions based on the results.

What if after all of the long, complicated analyses and multiple runs, there is no structure, or $K = 1$? There is a natural temptation to keep looking for population structure using different methods until structure is found. However, it is far easier to obtain a result than an answer, and the risk of spurious associations rises dramatically during repeated analysis and re-analysis. The goal for population genetic studies should be to describe the biological basis for observed patterns of genetic differentiation. Sometimes the population(s) is simply weakly structured as a result of high dispersal. While not a novel or exciting result, such information remains useful for management (e.g., DeYoung et al. 2009). In practice, many users plan a priori to conduct separate analyses on the same data, and select methods that make fewer or different assumptions, such as a different Bayesian clustering algorithm, or a multivariate analysis, such as principal components (Blair et al. 2012; François and Waits 2016). Results that are recoverable with different analytical approaches are more likely to represent actual signal as opposed to spurious correlations, and are easier to justify. Users also must consider the ability to replicate or repeat the analysis. Many journals now require that authors archive the data that underpin the analyses in published work. A recent study by Gilbert et al. (2012) found that attempts to replicate published results on the archived data did not obtain the same result one third of the time. The authors stressed the importance of archiving the same data set or subsets used in analyses, long burn-in and run lengths to ensure convergence, replicate runs for each K and assessment of ancestry of individuals among replicates, and full description of the criteria for choice of K and the information the choice was based on, such as the Evanno et al. (2005) delta K, values of $\log P (D|K)$, and bar plots to compare clustering solutions among different values of K (Gilbert et al. 2012).

Bayesian Inference of Phylogeny

"Systematics," "classification," "taxonomy," and "phylogenetics" are often treated as synonymous but each has distinct definition within biology. *Systematics*

can be defined as the study of biological diversity, including the description of organisms, life history and phylogeny, and the evolutionary mechanisms that produce biological diversity, as well as the use of such knowledge to classify organisms (Simpson 1961; Sokal 1963; Mayr 1982). *Classification* is the ordering of organisms into groups on the basis of their relationships by common descent, whereas *taxonomy* is the science of classification. A phylogeny is the evolutionary history of a group of taxa or populations from a common ancestor (Futuyma 1998; Mayr 2001). *Phylogenetics* is, therefore, the science of inferring and studying phylogenies (Brinkman and Leipe 2001; Choudhuri 2014) and is a subdiscipline of systematics. The main tool of phylogenetics and systematics is the *phylogenetic tree*, a concept first introduced to biology by Charles Darwin (1859). A phylogenetic tree is a branching, tree-like diagram that illustrates the evolutionary relationships of a group of organisms (Choudhuri 2014). Throughout the late nineteenth and much of the twentieth century, taxonomic classification and the inference of phylogenies were based on the interpretation of morphological and other phenotypic evidence by a handful of experts in various fields. The apparent subjectivity of this situation spurred the search for more quantitative approaches to classification and eventually led to the development of two different schools of systematics: *numerical taxonomy*, or *phenetics* (Michener and Sokal 1957; Sneath and Sokal 1962; Sokal and Sneath 1963), and *cladistics*, or *phylogenetic systematics* (Hennig 1950, 1965, 1966). Numerical taxonomists infer phylogenetic relationships and classified organisms based on indexes of similarity derived from all available characters (i.e., traits), whereas cladists argue that it was necessary to differentiate between ancestral and derived character states and that phylogenies and classifications should be based only on shared derived character states. Throughout the 1960s, 1970s, and 1980s adherents of numerical taxonomy and cladism waged a long feud with one another and with practitioners of traditional (evolutionary) taxonomy. Cladists ultimately prevailed and their philosophical approach and much of their terminology are the standard for systematics and taxonomy; however, numerical taxonomists have left their imprint on phylogenetics. Numerical taxonomy was the first attempt to design a numerical inference of phylogenies, and phylogenetics is essentially a numerical and statistical discipline today (see Avise [2004] and Felsenstein [2004] for detailed history of the development and rivalry among numerical taxonomy and cladism).

Numerical taxonomy was hampered by its focus on similarity. Similarity is not enough to infer close relationships because organisms may be similar to one another due to shared derived character states, retention of ancestral character states, or homoplasy (similarity due to convergent evolution; Futuyma 1998). However, retention of ancestral traits and homoplasy plagued cladism as well. How does one determine objectively and conclusively whether the similarity of morphological character in two taxa is derived, ancestral, or convergent? The development of quantitative approaches to the inference of phylogenies in the 1960s and 1970s occurred in conjunction with improvements in computer technology and the development of the first molecular markers (gene frequencies and amino acid sequences; Avise [2004]; Felsenstein [2004]), the introduction of the concept of the molecular clock (Zuckerkandl and Pauling 1962, 1965), and the neutral theory of molecular evolution (Kimura 1968; King and Jukes 1969). Phylogenetics was further revolutionized by the invention of the polymerase chain reaction (PCR; Mullis et al. 1986; Mullis and Faloona 1987), which allowed scientists to create multiple copies of the same gene or genetic locus, and DNA sequencing (Sanger et al. 1977), providing the means of directly observing and comparing nucleotide composition of genes. Phylogenetics of extant organisms is now largely *molecular phylogenetics*. Phenotypic data continue to play a role in "total evidence" approaches, which infer phylogenies from molecular data and morphological characters from extant and fossil taxa (Kluge 1989; Springer et al. 2001; Nylander et al. 2004).

Today, four approaches are used for inference of phylogenies: *distance-based methods*, *maximum parsimony*, *maximum likelihood*, and *Bayesian inference* (Felsenstein 2004; Yang and Rannala 2012). Distance-based methods were first introduced by Fitch and Margoliash (1967) and Jukes and Cantor (1969), but the most widely used version today is neighbor-joining (Saitou and Nei 1987). Phylogenetic inference via distance-based methods start with the creation of a matrix of pairwise genetic distances. *Neighbor-joining*, a cluster algorithm, constructs a star-shaped tree from the distance matrix. Next, successive pairs of taxa are chosen to join together in such a way as to minimize the length of the tree, and the distance matrix is updated with the joined taxa, replacing the original taxa. The advantage of neighbor-joining is that it is very fast because it is not comparing many trees. However, neighbor-joining and other distance-based methods do not perform well on highly divergent sequences due to the large sampling errors associated with large distances; they also do not account for high variances of large distances, and perform poorly with data sets that contain highly divergent sequences or data sets that include gaps in the alignment (Bruno et al. 2000).

Maximum parsimony, maximum likelihood, and Bayesian inference are character-state methods in which all sequences in an alignment are compared simultaneously (Yang and Rannala 2012). The characters examined in these methods are nucleotides, and each one is examined individually to calculate an overall score for each tree that is constructed. A major advantage of character-state methods over distance-based methods is that the former construct and examine all possible trees or, at least, many likely trees. With large data sets, it is impractical and, in many cases, impossible to examine every conceivable tree, which has led to the development of optimization schemes to increase the speed at which maximum parsimony and maximum likelihood approaches find high-scoring trees (Takahashi and Nei 2000; Yang and Rannala 2012). Parsimony approaches were first developed by Edwards and Cavalli-Sforza (1964), Cavalli-Sforza and Edwards (1967), Camin and Sokal (1965), and Eck and Dayhoff (1966). The development of efficient algorithms for reconstructing changes on a tree, searching among trees for the most parsimonious tree, and evaluating the number of changes of state on a tree (Kluge and Farris 1969; Farris 1970; Fitch 1971) made maximum parsimony the premier method for phylogenetic reconstruction in the 1970s, 1980s, and early 1990s. Under maximum parsimony a score is calculated for each tree, and the tree with the lowest score, which is also the tree that minimizes the number of changes, is selected as the best (true) tree. The advantages of maximum parsimony include an extensive examination of possible trees and the fact that there is no chance of model misspecification because this approach does not make any assumptions about mutation rates, invariant sites, or saturation. These same advantages are also its biggest drawbacks. Extensive searches in tree space are computationally expensive, and large data set may require several days to complete. Multiple substitutions at the same site will result in long-branch attraction, in which long (fast-evolving) branches will attract one another during phylogenetic reconstruction (Felsenstein 1978; Nei and Kumar 2000; Sullivan and Joyce 2005; Yang and Rannala 2012).

Although likelihood approaches for inferring phylogenetic relationships were introduced in the 1960s and 1970s (Edwards and Cavalli-Sforza 1964; Cavalli-Sforza and Edwards 1967; Neyman 1971; Kashyap and Subas 1974), this method did not begin gaining popularity until the 1980s and 1990s, once an approach to making likelihood computations for DNA sequence data was developed (Felsenstein 1981). Modern phylogenetic inference via maximum likelihood involves the specification of a model of molecular evolution, optimization of branch lengths, calculation of maximum likelihood scores, and a search of the tree space for the maximum likelihood

tree. A contributing factor to the popularity of maximum likelihood is that it incorporates data regarding molecular evolution, making it useful for studying sequence evolution (Yang and Rannala 2012). As with maximum parsimony, maximum likelihood can be computationally expensive and slow if one is dealing with a large data set, although several algorithms have been developed to economize and speed up the process (Felsenstein 2004). Incorrect phylogenetic inference may result if the wrong model of molecular evolution is applied. A final concern for both maximum likelihood and maximum parsimony is that the tree with the best score is not necessarily the true tree and may perform poorly on large data sets with a small number of nucleotides (Nei et al. 1998; Takahashi and Nei 2000).

Phylogenetic reconstruction via Bayesian inference began in the 1990s, although Bayesian arguments had been used prior to that time (Edwards 1970; Farris 1973; Harper 1979; Smouse and Li 1987; Sinsheimer et al. 1996; for detailed history see Brown 2003; Felsenstein 2004; and Huelsenbeck et al. 2002). Rannala and Yang (1996) attempted the first full Bayesian inference of phylogeny, but computation difficulties hampered advancement of the field. Calculating the posterior distribution of a phylogenetic tree involves summing over all possible topologies and integrating all possible combinations of substation model parameters and branch lengths, which is not possible to do analytically except for very small, simple data sets (Felsenstein 2004; Huelsenbeck et al. 2002). The late 1990s and early 2000s saw the more frequent use of Bayesian methods due to advances in computer technology and the development and release of computer programs based on Markov chain Monte Carlo (MCMC), such as BAMBE (Simon and Larget 1998), MrBayes (Huelsenbeck and Ronquist 2001), and BEAST (Drummond et al. 2005). Bayesian phylogenetic inference via MCMC algorithm is accomplished by approximating the posterior probability distribution of phylogenetic trees (Metropolis et al. 1953; Hastings 1970; Green 1995;

Huelsenbeck and Ronquist 2001; Drummond and Rambaut 2007). The basic steps of the process involve (1) start with a tree selected at random, 2) propose a new tree, (3) calculate the probability of accepting the new tree, 4) draw a uniform, random value, and (5) replace the original tree with the new tree if the random value is less than the acceptance probability, or, if not, then reject the new tree and retain the original (Huelsenbeck and Ronquist 2001).

For phylogenetic analysis of multigene and multilocus data, Bayesian approaches have become as frequently used as, and in some areas of phylogenetic research superseded, neighbor-joining, maximum parsimony, and maximum likelihood methods. Reasons for this include the facts that results that are more easily interpreted, a priori information can be incorporated, and Bayesian methods are less computationally intense (Lemey et al. 2009; Blair and Murphy 2011; Yang and Rannala 2012). MrBayes and BEAST are the two most commonly used computer programs for Bayesian inference of phylogeny and divergence times. MrBayes is a simple computer program that uses a command-line interface. BEAST, in contrast, is a console application that is offered as part of a software package that contains several other computer programs, including BEAUti (for the creation of input XML files for BEAST), LogCombiner (merges independent log files of MCMC runs generated by BEAST), and TreeAnnotator (summarizes a sample of trees). Current versions of both programs are based on Metropolis-Hastings MCMC, allow the user to partition molecular data so that different models of molecular evolution can be applied to each gene, and have the ability to estimate divergence times (Ronquist et al. 2012b; Bouckaert et al. 2014). Although MrBayes and BEAST differ in some of their capabilities, the programs are similar in terms of phylogenetic reconstruction and a priori information that researchers must consider, including the application of a substitution model and rate, clock model, and calibration of tree nodes.

Length and Number of MCMC Runs and Heating

The user of any Bayesian-based computer program will need to set the length of the MCMC run (chain) and determine the number of MCMC runs necessary to achieve stationarity. BEAST and MrBayes require the user to determine the number of generations, or the length of the MCMC chain. Chain length should be proportional to the square of the number of taxa/individuals, but the researcher will also have to consider the size of the data, complexity of the model, and quality of the sample desired (Drummond 2007). Researchers may need to experiment with variations in chain length to achieve an effective sample size for parameters of interest (Drummond 2007). For each execution of BEAST, the program produces a single MCMC run. The effective sample size of parameters can also be increased by using the computer program LogCombiner to combine log files from multiple runs (Drummond 2007). LogCombiner can also be used to combine separate output files of trees generated from separate runs. The computer programs TRACER (Drummond and Rambaut 2007) and AWTY (Nylander et al. 2008) can be used to assess the stationarity of independent and combined runs and effective sample size of various parameters. MrBayes differs from BEAST by running two separate MCMC analyses for each execution (Ronquist et al. 2011). Users can examine the effective samples size as well as the average standard deviation of split frequencies (ASDSF; Lakner et al. 2008) and the potential scale reduction factor (PSRF; Gelman and Rubin 1992) computed by MrBayes to assess whether the two runs have converged on stationarity. The ASDSF, calculated by comparing clade frequencies from across multiple MCMC runs, should approach 0 as the two tree samples become more similar (Ronquist et al. 2012b). The PSRF, calculated for branch lengths, node times, and substitution model parameters, compares the variance within and between runs and should approach 1 as the runs converge (Ronquist et al. 2012b).

MrBayes version 3.2 implements both the standard MCMC algorithm and a variation referred to as *Metropolis-coupled MCMC*, or *(MC)3* (Geyer 1991). An $(MC)^3$ approach involves n chains, of which $n-1$ are heated. The current version of MrBayes provides the researcher the option of setting the number of chains to one, in which no heating is used. The default setting for MrBayes is four chains, which involves three "heated" chains and one "cold" chain. Heating, or increasing the "temperature" of the Markov chain, refers to the more drastic alteration of parameter values at each proposed step in a chain (Altekar et al. 2004; Randle et al. 2005). An $(MC)^3$ analysis attempts to swap the states of the heated and cold chains. The heated chain sees the tree space as flattened compared with the cold chains, which allows heated chains to move across valleys in the tree space. A cold chain may occasionally swap states with a heated chain that may be exploring a different peak, which allows the cold chain to span a deep valley in a single jump. Is heating necessary? The $(MC)^3$ improves mixing and convergence (Geyer 1991; Gilks and Roberts 1996; Huelsenbeck et al. 2001; Kuhner 2009), but increases the execution time of an analysis. Using more than three heated chains may help in analyzing large data sets or difficult data sets, but heating may not be necessary for all data sets (Ronquist et al. 2011). The researcher will need to experiment running the analysis without heating and with various numbers of heated chains to determine if heating is necessary. The current version of MrBayes implements a parallel Metropolis-coupled MCMC, which improves the speed of $(MC)^3$ computations (Altekar 2004). BEAST 2 now has the capability of performing $(MC)^3$ (Bouckaert et al. 2014).

Molecular Evolution Model and Rate

The rate of nucleotide substitution varies among genes, among the gene and loci across taxa, and among nucleotide sites within a gene (Nei and Kumar 2000). In addition to difference in substation rates,

any attempt to reconstruct a phylogeny must account for back mutation ("multiple hits"). Back mutation is the main source of homoplasy, the evolution of identical DNA or amino acid sequences due to independent mutations rather than common descent (Estoup et al. 2002). The extent to which homoplasy obscures or erases the phylogenetic signal depends on the length of time since two lineages diverged and the rapidity of molecular evolution of the gene being utilized. Properly accounting for variation in substitution rates and homoplasy is necessary to avoid incorrect phylogenetic inference or correct but poorly supported topologies. Substitution rate is the rate at which mutations become fixed in a population, as opposed to mutation rate, or the rate at which nucleotide changes occur in the genome (Ho and Larson 2006).

Many models of nucleotide substitution have been developed to account for both variation in the rate of molecular evolution and homoplasy. These are also many options for selection of the most appropriate substitution model; we refer the reader to Sullivan and Joyce (2005) for a detailed review of this topic. If the substitution rate varies extensively within an evolutionary lineage, an overly simplistic model may produce an incorrect tree topology due to the long-branch attraction—long (fast-evolving) branches will attract one another during phylogenetic reconstruction (Felsenstein 1978; Nei and Kumar 2000; Sullivan and Joyce 2005). The apparent solution is to use a more complex model that provides a better fit to the data; however, this does not guarantee better results. Variance increases with model complexity and leads either to incorrect phylogenetic inference or correct, but poorly supported trees (Takahashi and Nei 2000; Nei and Kumar 2000; Sullivan and Joyce 2005). Although the use of more complex models has been recommended for Bayesian phylogenetic inference (Huelsenbeck and Rannala 2004), Rannala (2002) warned that slower convergence may result from overparameterization. Nei and Kumar (2000) provide thoughtful advice on selecting a model of nucleotide substitution. They recommend using simpler substitution models when the number of substitutions per site is low and the number of nucleotides is small as these generally yield better results. Most phylogenetic analyses in wildlife biology are still based on a single gene and are focused on the relationships among intraspecific populations or congeneric species. Such data sets will generally involve short DNA sequences (≤1,000 bp) characterized by a small number of substitutions, and the analysis of such data sets may not require a Bayesian approach. Neighbor-joining analysis based on the proportion of nucleotide differences produces good results when the number of nucleotides is small, regardless of the number of taxa included in the analysis (Nei and Kumar 2000).

More complex models will be required when the number of substitutions per site is high and the number of nucleotides examined is large, as is often the case when conducting phylogenetic analyses of more distantly related taxa (Takezaki and Gojobori 1999; Nei and Kumar 2000). The cost of extraction, amplification, and sequencing of DNA has decreased dramatically in the past 20 years, which has resulted in the increasing frequency of studies based on multigene or multilocus data sets or whole genomes. DNA sequences from multiple genes can be concatenated and examined as a unified unit with one substitution model applied to the entire data. This approach, however, is oversimplified as it fails to account for variation in substitution rates among genes and loci and underestimates the phylogenetic signal, which may lead to either incorrect topologies or excessive credibility of posterior probabilities (Nylander et al. 2004; Brandley et al. 2005; Castoe and Parkinson 2006; Brown and Lemmon 2007). Bayesian approaches are better suited than other methods for phylogenetic analyses involving multiple genes because software such as BEAST and MrBayes allows users to partition the data set by gene or codon and apply better-fitted models of substitutions to each gene or codon (Castoe and Parkinson 2006). For distantly related taxa, partitioning by gene and by codon for protein-coding genes generally improves

phylogenetic inference and statistical support for inferred relationships, especially deeper nodes of the tree (Brandley et al. 2005; Castoe and Parkinson 2006; Brown and Lemmon 2007; Ward et al. 2010). The user will need to consider whether a Bayesian approach is necessary, decide whether the data needs to be partitioned and how small those partitions need to be, sample appropriate taxa, and include genes that provide sufficient phylogenetic signal. Simulation studies have found that posterior probabilities tend to be more liberal than bootstrapping (Suzuki et al. 2002). Studies using real data suggest that bootstrapping and jackknifing are overly conservative and tend to underestimate support for inferred clades, whereas posterior probability support may be only slightly inflated (Simmons et al. 2004; Taylor and Piel 2004). As Nei and Kumar (2000) point out, researchers should not automatically trust the phylogenetic tree produced by a computer program. It is incumbent upon the user to carefully examine the tree in light of what is known about substitution rates among taxa and genes and prior knowledge about the evolutionary relationships of the taxa under study. Users can assess the accuracy of Bayesian phylogenetic inference by testing subsets of the full data set (Alfaro and Holder 2006) or by conducting a separate phylogenetic analysis using maximum parsimony, maximum likelihood, or neighbor-joining.

Clock Models

Often scientists want to know when particular clades in a phylogenetic tree diverged from their closest relatives. One way of determining when a lineage of organisms arose is to look at the fossil record. Unfortunately, the fossil record is imperfect. Some taxa, such as cervids and whales, have left an extensive fossil record, whereas we have few or no fossils for many other taxa. In many cases, the fossil record provides only crude estimates for the divergence of various taxa. Molecular data offer an opportunity to refine these divergence times and to estimate divergence times for taxa that have very poor fossil records.

To achieve this, one can use a Bayesian phylogenetic computer program such as BEAST or MrBayes to calibrate the tree using a known substitution rate or constrain the age of a node in accordance with the age of a fossil or a geological event.

Zuckerkandl and Pauling (1962) discovered a linear relationship between the accumulation of amino acid substitutions in hemoglobins and species ages based on the fossil record. Amino acid changes seemed to occur at a relatively constant, or clock-like, rate, which suggested that substitutions in genes could be used as a *molecular clock* to determine how long ago two lineages diverged from a common ancestor (Zuckerkandl and Pauling 1962). Additional research showed that the rate of the molecular clock varied among different proteins, with hemoglobin faster than cytochrome *c*, and cytochrome *c* faster than histones (Zuckerkandl and Pauling 1965). The neutral theory of molecular evolution (Kimura 1968; King and Jukes 1969) offered an explanation for the apparent constant rate of amino acid substitution in protein-coding genes. According to the neutral theory of molecular evolution, most mutations had no impact on the function of a protein, being essentially "neutral" in terms of fitness, whereas beneficial mutations were rare, and natural selection quickly removed harmful mutations from a population. If most mutations were neutral, then the number of mutations separating two lineages was the result of the length time that had passed since those lineages shared a common ancestor (Kimura and Ohta 1971). Early phylogenetic work based on molecular data assumed a constant molecular clock for a specific type of gene or protein (Kimura and Ohta 1974; Wilson et al. 1977). Additional research, however, showed that the molecular clock for a particular gene could vary extensively among and within lineages (Lessios 1979; Li and Tanimura 1987; Li et al. 1987; Pereira and Baker 2006; Alter and Palumbi 2009). The molecular clock could be best described as a "sloppy" clock—the sloppiness being caused by variation in DNA repair mechanisms, spatial and temporal variation in selection pressures, and differences in population size and reproductive rate (Bromham and Penny 2003).

Several relaxed clock models have been developed to compensate for the sloppiness of the molecular clock including clocks based on uncorrelated exponential and lognormal distributions (Drummond et al. 2006), Brownian motion model (Thorne and Kishino 2002), compound Poisson process (Huelsenbeck et al. 2000), and gamma distribution (Lepage et al. 2007). A user should first determine whether or not a constant, or strict, molecular clock fits the data. For a data set consisting of a single gene, Tajima's (1993) relative rate test or maximum likelihood test of the molecular clock (Takezaki et al. 1995) can be used to determine if a strict clock is appropriate. Testing various relaxed clock models can be accomplished using Bayes factors calculated from harmonic mean estimator (Newton and Raftery 1994), posterior simulation-based analog of Akaike's (1974) information criterion (AIC) through Markov chain Monte Carlo (AICM, Raftery et al. 2007), path sampling (Ogata 1989), or stepping-stone sampling (Xie et al. 2011). Harmonic mean estimator is the fastest method to calculate Bayes factors, but is less reliable than other methods, as it tends to overestimate marginal likelihoods (Baele et al. 2012, 2013). Path sampling and stepping-stone sampling are considered more accurate. MrBayes implements a stepping stone sampling function for testing the fit of relaxed clock models to the data. BEAST does not incorporate any model testing options; however, the BEAST wiki (http://www.beast.bio.ed.ac.uk/Model-selection) provides computer code devised by Baele et al. (2012, 2013) to create the XML file to conduct the model testing. The results can be examined in computer program Tracer. The drawback of using Bayes factors is that these can be computationally intense for data sets consisting of a large number of taxa, many independent loci, or both. Recently, a new method for clock model selection, ClockstaR, was developed for partitioned data sets (Duchêne and Ho 2014; Duchêne et al. 2014). ClockstaR, which is based on a clustering method and the R programming language, calculates a distance metric based on the branch lengths of all possible trees for each subset of data (Duchêne et al. 2014). ClockstaR appears to be as accurate as path or stepping-stone sampling, but is less computationally intensive (Duchêne and Ho 2014). Except for some rapidly evolving viruses, relaxed clock models should not be used for intraspecific data sets because such data sets usually possess few lineages and the rates of evolution will be essentially the same among lineages due to identical life history characteristics and DNA repair mechanisms (Drummond and Bouckaert 2015). A strict clock model is more appropriate for low-diversity intraspecific datasets with little variation in evolutionary rates among lineages (Brown and Yang 2011; Drummond and Bouckaert 2015).

Priors

The priors used in a Bayesian phylogenetic analysis are determined by the objectives of the analysis and the substitution and clock models selected. For example, selection of the Hasegawa-Kishino-Yano (HKY) substitution model, an uncorrelated lognormal relaxed clock, and the Yule branch process model in BEAST will require the user to set priors for birthRate (rate of birth of new lineages), kappa (transition:transversion ratio), ucld.mean (mean substitution rate under the uncorrelated lognormal relaxed molecular clock), and the ucld.Stdev (standard deviation of the uncorrelated lognormal relaxed molecular clock). If the objectives also include estimating divergence times among taxa, the priors will include at least one topological constraint and an estimate of when the lineages at that node last shared a common ancestor. BEAST and MrBayes allow the user to apply a variety of parametric distributions to each of the model parameters. With the exception of fossil-calibrated nodes for divergence estimation (see *Divergence Dating* below), many published studies are based on the default settings for the priors, although most do not explicitly state this.

Divergence Dating

Divergence times can be estimated by using a mutation rate (McFadden et al. 2008; Yasukochi et al. 2009) or calibrating one or more nodes (or tips) in the tree using dates based on fossils (Near et al. 2005a, 2005b; Guiher and Burbrink 2008), geological or climatic events (Fleischer et al. 1998; Genner et al. 2007; Ho et al. 2015), or sampling dates (for fast-evolving viruses or bacteria; Rieux and Balloux 2016). Divergence times inferred from molecular clocks often exceed divergence estimates derived from paleontological evidence by tens or hundreds of millions of years (Ho and Larson 2006; Pulquério and Nichols 2007). This is partly due to the fact that rates of molecular evolution for genealogical time scales (<1 million years) are many times faster than those measured over geological time scales (>1 million years; Ho et al. 2005; Ho and Larson 2006). The short-term mutation rate, the instantaneous rate at which nucleotide changes occur in the genome, transitions over time to a long-term substitution rate, the rate at which mutations are fixed in the population (Ho and Larson 2006). Most nucleotide changes that occur in a population will be eliminated by selection or genetic drift, which will result in fewer observed changes per unit time. Substitution rates, therefore, will be slower than mutation rates. Failure to distinguish between mutation and substitution rates and consider the relationship between the two is the likely source of conflicting divergence times based on molecular clocks and fossil data (Ho and Larson 2006).

Software packages such as BEAST provide a way of avoiding the misapplication of substitution and mutation rates by allowing users to calibrate a molecular clock by assigning an age or date derived from fossils or other nonmolecular sources to one or more nodes within the tree or to the branch tips. However, even fossil-calibrated molecular clocks often produce divergence estimates that are far more ancient than those based on paleontological data. For example, several attempts to estimate the divergence times for avian lineages suggest that most modern orders of birds originated in the Late Cretaceous (Crowe et al. 2006; Baker et al. 2007; Pereira et al. 2007; Brown et al. 2008; Wright et al. 2008); however, fossils of crown-group members of most extant avian orders are absent from Eocene or older deposits (≥56 million years ago; Mayr 2005, 2009). Some of the conflict between divergence times inferred from molecular clocks and the fossil record may be due to the incompleteness of the fossil record itself and the fact that a fossil represents the minimum age of a lineage (Springer 1995; Smith and Peterson 2002; Ho et al. 2007; Brocklehurst et al. 2014; Ksepka et al. 2014). Another source of error is the application of fossils to the wrong nodes in a phylogenetic tree or the use of fossils of uncertain age to calibrate molecular clocks (Graur and Martin 2004; Near et al. 2005b; Gandolfo et al. 2008; Mayr 2008; Ksepka 2009), which tends to result in older divergence estimates. More conservative selection of fossils in recent studies has yielded post-Cretaceous divergence estimates for most extant avian orders and families (Ericson et al. 2006; Cohen et al. 2012; Hosner et al. 2015; Prum et al. 2015).

Parham et al. (2012) provide a set of recommendations for the selection of fossils for the calibration of a molecular clock, including (1) list all museum numbers of specimens that demonstrate all relevant characters, (2) cite up-to-date phylogenetic analyses of all fossil taxa, (3) provide explicit statements about the reconciliation of morphological and molecular data sets, (4) provide locality and stratigraphic data of the fossils used, and (5) cite a published radioisotopic age and specify the numeric age selection of the calibration (see also Gandolfo et al. 2008; Sauquet et al. 2012; Hipsley and Müller 2014). Ho et al. (2015) provide advice on the use of geological and climatic events for calibrating molecular clocks. Estimates of divergence times inferred from molecular data in previous studies are termed secondary calibrations. Researchers should carefully scrutinize any secondary calibrations before using them to estimate divergence times. Secondary calibrations often have narrower confidence intervals than the original study, which results in a loss of variance and gives a false impression of precision (Schenk 2016). Well-justified fossil calibrations are now catalogued by the Palaeontologia Electronica Fossil Calibration Database (http://fossilcalibrations.org; Ksepka et al. 2015). The database is currently biased towards vertebrates and animals in general, but will become more inclusive as new fossils are approved by its advisory board (Ksepka et al. 2015). How many calibrations are needed to achieve accurate divergence estimates? Generally, the use of two or more calibrations will yield more accurate divergence times than a single calibration (Graur and Martin 2004; Near et al. 2003, 2005b; Fulton et al. 2010). Researchers may use internal (ingroup) or external (outgroup) calibrations or both. The reliance on a single external fossil calibration can lead to wildly inaccurate divergence times, especially for intraspecific data sets (Graur and Martin 2004; Ho et al. 2008). Internal calibrations are advantageous because they reduce statistical errors associated with large extrapolation of time but risk underestimating divergence times (van Tuinen and Hedges 2004). For taxa that have poor fossil records, it may be possible to estimate divergence times for the ingroup by increasing the number of outgroup taxa and using multiple external fossil calibrations (Schenk 2016). The accuracy of divergence estimates can be improved by using multiple calibration points, which will compensate for poorly supported divergence and changes in substitution rates (Graur and Martin 2004; Near et al. 2005b; Hugall et al. 2007; Sauquet et al. 2012).

Point calibrations, which assume that the provided age is absolute, were often used in early molecular clock studies. However, point calibrations are inappropriate for estimating divergence because a fossil can provide only the minimum age estimates for divergence of two clades as it postdates the clade of which it is a member. Point calibrations were presented as errorless values, leading to inflated confidence of posterior age estimates (Ho 2007). Bayesian relaxed clock methods allow users to specify priors on calibrated nodes in the form of parametric distributions (Drummond et al. 2006). Parametric distributions most often used for node calibration include uniform, normal, exponential (diminishing probability over time), lognormal, and gamma. The uniform distribution places an equal probability across an interval. It is useful for calibrating a node with a fossil from a single stratigraphic unit in which dates can be derived for the base and top but no other information is available. A normal distribution places equal diminishing probability, determined by the variance, on either side of the mean, and is perhaps best suited for fossil species that are known from the middle of a stratigraphic unit or for specific biogeographic events (Ho 2007). An exponential distribution results in decreasing probability over time as the discrepancy between the estimated nodal age and the age of the fossil increases. This distribution is most useful when the fossil record of the group under study is relatively complete and the age of the fossil is very close to the actual divergence event. An exponential distribution requires two parameters: (1) a "hard" lower bound that represents the age of the fossil and (2) a mean, which is selected so that 95% of the probability is contained with a hard lower bound and some "soft" maximum bound (Yang and Rannala 2012; Ho 2007). The lognormal distribution is viewed by some as the most appropriate distribution for fossil calibration (Hedges and Kumar 2004; Drummond et al. 2006). This distribution assumes that the actual divergence between two lineages occurred sometime before the earliest fossil evidence. As with the exponential distribution, mean age and standard deviation must be specified, and the mean can be selected so that 95% of the probability lies within a certain interval. The gamma distribution has two parameters, shape (α) and scale (β), and is more flexible compared to the other parametric distributions. The gamma distribution will collapse to an exponential distribution when $\alpha = 1$, but if $\alpha > 1$, the gamma distribution resembles a normal distribution. Ho (2007), Ho and Phillips (2009), and O'Reilly et al. (2015) provide reviews of parametric distributions and recommendations for their uses.

Total-evidence dating (or tip-dating; Ronquist et al. 2012a; O'Reilly et al. 2015) is an attempt to avoid the ad hoc hypotheses associated with node calibration. This approach integrates fossil species with their extant relatives using combined sets of morphological and molecular data (Heath et al. 2014; Gavryushkina et al. 2017; Ronquist et al. 2012a; O'Reilly et al. 2015; Zhang et al. 2016). Although several studies utilizing either total-evidence dating or the fossil birth-death process have yielded divergence times congruent with those obtained from the fossil record or molecular data, the morphological dating models appear to have a bias toward older divergence estimates (O'Reilly et al. 2015). For example, total-evidence dating indicated the crown group placental mammals originated 130–145 million years ago, which is far earlier than the estimate of 80–90 million years based on fossil or molecular data (Puttick et al. 2016). Puttick et al. (2016) suggested that the more ancient origin inferred from total-evidence dating may be due to the taxonomic breadth of the focal group itself, in which there may be widely varying rates of morphological change within different subclades or at different times. O'Reilly et al. (2015) and Zhang et al. (2016) provide recommendations for incorporating fossil and extant taxa to estimate divergence times.

The branching process prior is a model that determines how trees are generated and places a prior on how the trees grow (Condamine et al. 2015); it is necessary to specify a branching process prior if divergence times are to be estimated. The two most commonly used branching process priors are the birth-death (Kendall 1948; Yang and Rannala 1997) and Yule models (Yule 1924; Aldous 2001). The birth-death model forms trees with constant rates of speciation and extinction, whereas the Yule model forms trees with only constant rates of speciation. Both models are available in BEAST. MrBayes does not offer the Yule model but has options for the birth-death model as well as uniform and coalescent models.

The Yule model is the branching process prior used most frequently in molecular clock studies performed in BEAST, possibly because the manual recommends the Yule model (Condmaine et al. 2015). However, it is likely that each branching process prior may have differing effects on estimates of divergence times.

Few molecular clock studies have compared divergence times inferred from different branching process priors (Couvreur et al. 2010; Kergoat et al. 2014; Condamine et al. 2015). Condamine et al. (2015) found that the Yule model yielded ages for genus-crown groups of cycads (Cycadales) that were three times older than those obtained with the birth-death model. In contrast, Yule and birth-death models produced similar divergence estimates for genera and species within the mustard family (Brassicaceae, Couvreur et al. 2010). The extinction rate prior in the birth-death model tends to produce trees with a greater number of nodes toward the present because extinction has not had time to affect these nodes, an effect termed "pull to the present" (Gernhard 2008; Condamine et al. 2015). Extinction has had more time to shape the current diversity of ancient lineages, such as cycads, which first appeared during the mid-Permian Period (270 million years ago) and are characterized by extinction and loss of diversity (Hermsen et al. 2006; Condamine et al. 2015). In contrast, Brassicaceae evolved 38 million years ago and diversified rapidly during the Oligocene Epoch (23–33 million years ago; Couvreer et al. 2009). The branching process prior may have limited effect on the estimation of divergence times within a relatively recent lineage; however, researchers should provide explicit justification for the use of a specific branch process prior (Condamine et al. 2015). Alternatively, divergence times could be inferred with birth-death, Yule, or other branch process priors and results compared.

Bayesian Skyline Plots

Census population size is the total number of individuals of a species living within a specific area, whereas *effective population size* is the proportion of reproductively mature adults that contribute to the genetic variation of the next generation (Avise 2004). Population size affects genetic diversity, and this relationship impacts long-term viability of a species (Frankham 1996). A number of genetic-based methods have been developed to estimate census and effective sizes (Leberg 2005; Luikart et al. 2010; Hare et al. 2011). However, many biologists recognize that management and conservation of any species require knowledge of its long-term demographic history—specifically, the causes of population increases or decreases and the magnitude of those changes. Such data provide scientists with a better understanding of how a species has responded to past climate change and human activities. For DNA sequence data, there are several qualitative methods to infer whether a species has undergone past population growth. For example, high haplotype diversity and low nucleotide diversity, a star-shaped haplotype network, and unimodal distribution of pairwise differences are usually indicative of past demographic growth (Avise 2000). Tests for departure from selective neutrality such as Tajima's (1989) D, Fu and Li's (1993) D and F, Fu's (1997) F_S, and Ramos-Onsins and Rozas (2002) R_2 can also provide evidence of demographic growth. Although these methods are valuable, they do not provide actual estimates of population size or estimates of the timing of demographic events.

Historical changes in population size and the timing of demographic events can be estimated directly from DNA sequence data via approaches based on coalescent theory (Drummond et al. 2005; Ho and Shapiro 2011). *Coalescent theory* refers to the quantification of the relationship between line of descent of a gene and the demographic history of the population (Kingman 1982; Tavaré 1984; Hudson 1991). Coalescent-based methods require a demographic model, such as constant population size, exponential growth, logistic growth, or expansion growth. Unfortunately, it is usually not known in advance which model will best fit the data, and the use of the wrong model could lead to incorrect demographic inferences. Testing and comparing the fit of each demographic model was a time-consuming process. To account for ignorance regarding the best-fitting model, Pybus et al. (2000) developed the skyline plot, a piecewise-constant model that can fit a variety of demographic scenarios. However, the classic skyline plot produces a large amount of noise due to the stochasticity of the coalescent process. Strimmer and Pybus (2001) incorporated the AIC into a modified version, the generalized skyline plot, to reduce the noise. These methods, unfortunately, infer demographic history from an estimated genealogy rather than from the sequence data itself, and thus ignore the error associated with phylogenetic reconstruction (Drummond et al. 2003, 2005). Drummond et al. (2005) introduced the Bayesian skyline plot as a solution to the above problem.

The Bayesian skyline plot uses an MCMC procedure to estimate a posterior distribution of past changes in population sizes directly from DNA sequences via phylogenetic reconstruction. Credibility intervals representing error associated with phylogenetic reconstruction and coalescent uncertainty are produced for all time intervals all the way back to the most recent common ancestor of the DNA sequences under study. Drummond et al. (2005) showed that the Bayesian skyline plot could detect signatures of demographic decline or growth in empirical datasets that were missed by the generalized skyline plot. The extended Bayesian skyline plot was developed to handle multilocus data (Heled and Drummond 2008). Other MCMC approaches for demographic inference include the Bayesian multiple-change-point method (Opgen-Rhein et al. 2005), the Bayesian skyride (Minin et al. 2008), and the Bayesian birth-death skyline plot (Stadler et al. 2013). The multiple-change-point and skyride methods appear to be unable to detect rapid changes in population size (Ho and Shapiro 2011). The Bayesian birth-death skyline plot, a noncoalescent approach to modeling demographic changes in pathogens such as RNA viruses, was developed to address the inability of other methods to differenti-

ate between incidence and prevalence (Stadler et al. 2013). The birth-death model is based on a piecewise process in which each branching event corresponds to a "birth" (emergence of a new strain), each branch tip in the tree represents a sampling event, and "death" are unsampled recoveries or deaths (Stadler et al. 2013; Bouckaert et al. 2014). The current versions of the BEAST software package offer Bayesian skyline, extended Bayesian skyline plot, Bayesian skyride, and the Bayesian birth-death skyline plot (Bouckaert et al. 2014; Suchard et al. 2018). Classic and generalized skyline plots are available in the computer software program GENIE (Pybus and Rambaut 2002) and in the R package APE (Paradis et al. 2004); the multiple-change-point approach is implemented in APE and another R package, rbrothers (Irvahn et al. 2013).

We will focus mostly on the Bayesian skyline plot implemented in BEAST as this will be of interest to most wildlife biologists. The Bayesian skyline plot requires that the user specify models of molecular evolution and a molecular clock, set priors for the analysis, and attach a real-time scale to the branch lengths of the genealogical tree (Ho and Shapiro 2011). To calibrate the genealogical tree, it is possible to use a known mutation rate derived from previous studies. If the mutation rate is unknown, then a node in the tree will need to be calibrated with a fossil or biogeographic event. Genealogies of viruses and other rapidly evolving pathogens can be calibrated using sampling dates. For most species, Bayesian skyline plots are sensitive to both the data quantity (number of loci examined) and quality (Ho and Shapiro 2011). Errors in demographic inference can be reduced by increasing the number of loci examined (Heled and Drummond 2008). Multiple loci should be treated as unlinked because (1) doing so ensures that each has a mutual but independent demographic history, and (2) current Bayesian skyline plot methods are unable to account for partial linkage (Ho and Shapiro 2011). Sequencing errors and post-mortem DNA damage, even at low levels, can produce spurious demographic inferences (Heled and Drummond 2008; Axelsson et al. 2009). High-throughput technology can reduce sequencing errors. Undetected miscoding lesions in ancient DNA sequences results in overestimation of rates of molecular evolution and transition-transversion bias, as well as the false impression of rapid population growth (Axelsson et al. 2009; Ho and Shapiro 2011). One possible solution is to eliminate damaged sequences from the data set if the number of affected sequences is small. This is not an option if post-mortem damage is present in a large number of sequences or some of the affected sequences are integral to the analysis. Eliminating damaged sites risks reducing the phylogenetic and demographic signals (Rambaut et al. 2009). An alternative strategy is to use phylogenetic reconstruction to detect post-mortem damage by identifying misplaced sequences within a tree—older sequences should occupy basal positions in the tree rather than being located at the tips (Ho et al. 2007; Rambaut et al. 2009; Ho and Shapiro 2011; Rieux and Balloux 2016). Once identified, miscoding lesions in misplaced sequences can be corrected.

Bayesian skyline plots are easy to interpret. A constant flat curve is indicative of demographic stasis. A flat curve preceding an upward deflection may represent a long period of demographic stasis followed by rapid population growth. A skyline plot with a stepped profile may indicate repeated expansion events. A skyline plot that declines as it approaches the present may be indicative of a recent population bottleneck. These interpretations, however, need to be made in light of several possible confounding factors, including sample size, number of polymorphisms, and population structure (Grant 2015). Flat skylines may be caused by a small sample size as well as demographic stability due to the fact that small sample sizes underestimate genetic variation and the magnitude of demographic expansions. DNA sequences with few polymorphisms will lack the genetic variation necessary to detect past demographic expansions. For this reason, it is nearly always preferable to use longer sequences of genes rather than shorter ones for demographic inference.

The coalescent model assumes panmixia (Donnelly and Tavaré 1995); violations of the model will lead to spurious results in Bayesian skyline plots (Heller et al. 2013). For example, based on a stepped Bayesian skyline plot, Kitchen et al. (2008) hypothesized that humans had experienced a long period of demographic stability between the initial colonization of Beringia and subsequent expansion into North America. However, Fagundes et al. (2008) found that the original dataset contained non–Native American DNA sequences and that the stepped profile disappeared once those sequences were removed. The new skyline plot based only on Native American DNA sequences suggested a long population bottleneck after the initial colonization of Beringia (Fagundes et al. 2008). Similarly, a stepped Bayesian skyline plot was observed when samples from three genetically divergent populations of Pacific herring (*Clupea pallasii*) were pooled, but the stepped profile was absent in skyline plots when each population was analyzed separately (Grant et al. 2012). Prior to a Bayesian skyline plot, relationships among DNA sequences in an intraspecific data set should be analyzed by constructing phylogenetic trees or haplotype networks to determine if genetically divergent lineages exist. If genetically distinct lineages are present, the best solution may be to construct Bayesian skyline plots for each lineage as well as the entire data set. Finally, the flat curve preceding an upward deflection may reflect not a stable population but the loss of demographic signal due to repeated loss of genetic diversity as a result of Pleistocene climatic oscillations preceding the Last Glacial Maximum (Grant and Cheng 2012; Grant et al. 2012; Grant 2015).

Approximate Bayesian Computation

Statistical models are used to make inferences about systems from collected data (White 2001). To make inferences, the model must be fitted to the data. This process is called *parameter estimation*. Parameters are numerical quantities of a system that cannot be measured directly and, therefore, must be estimated from the data (as opposed to variables, which are numerical quantities of a system that can be measured directly; Altman and Bland 1999; White 2001). One method of parameter estimation, that of least squares, was independently developed by Johann Carl Friedrich Gauss and Adrien-Marie Legendre during the late sixteenth and early seventeenth centuries (Plackett 1972; Stigler 1981). The least squares method, implemented in many regression software packages, is best suited to systems in which a model's deviation from the data is or can be safely assumed to be normally distributed (White 2001). Maximum likelihood, introduced by R. A. Fisher (1922, 1925), is a more flexible approach to parameter estimation and can be used in situations in which the data are not normally distributed. Parameter estimation via maximum likelihood requires the calculation of the likelihood function, or the probability of the observed data viewed as a function of the parameters (Held and Sabanés Bové 2014). The goal is to find the vector of parameter values that maximizes the likelihood function (White 2001).

Bayesian approaches to parameter estimation have been used less frequently in the biological sciences than maximum likelihood although this has changed in the last few decades as computing power has increased. Bayesian inference, unlike other methods of parameter estimation, is attractive because it makes explicit use of prior information. Traditional approaches to Bayesian inference require the calculation of marginal likelihood. This is possible if the model(s) under consideration is relatively simple; however, real-world data and the models needed to explain observed biological patterns are usually highly complex. Calculation of marginal likelihoods becomes more computationally expensive, and possibly intractable, as the complexity of the models increase (Beaumont 2010; Sunnåker et al. 2013; Green et al. 2015; this is also a problem for

maximum likelihood and the calculation of the likelihood function). An obvious solution would be to use a model selection approach that bypasses the likelihood calculation, which brings us to approximate Bayesian computation. *Approximate Bayesian computation* (ABC) is a likelihood-free approach for estimating the posterior distribution of parameters and model selection based on the comparison of summary statistics calculated from observed data and simulations of the data (Beaumont et al. 2002; Beaumont 2010; Sunnåker et al. 2013). Originally developed by population geneticists (Tavaré et al. 1997; Pritchard et al. 1999; Beaumont et al. 2002), ABC has become a standard procedure in phylogeography, landscape genetics, ecology, epidemiology, and systems biology (Beaumont 2010; Csilléry et al. 2010; Segelbacher et al. 2010). Recently, researchers in other scientific disciplines, such as archaeology (Tsutaya and Yoneda 2013; Crema et al. 2014), meteorology and climatology (Olson and Kleiber 2017), and cosmology (Cameron and Pettitt 2012; Weyant et al. 2013), have also found ABC useful for complex problems and phenomena.

Francis Galton was probably the first scientist to develop a rejection sampling scheme for the simulation of the posterior distribution (Stigler 2010). The philosophical and conceptual foundations of modern ABC, however, are rooted in a paper written by Donald Rubin (1984), in which he described a hypothetical sampling method and model rejection scheme that draws samples from the posterior distribution. Rubin (1984) argued that the advantage of a Bayesian approach to model selection was the ability to (1) determine the extent of scientific uncertainty by quantifying the sensitivity of conclusions drawn from observed data to model parameters, and (2) help reveal the solutions (i.e., more data, new theory, better insight into existing data and theory) for resolving the uncertainty. Diggle and Gratton (1984) introduced a simulation method for parameter estimation when the analytical approach to likelihood calculation was intractable. Diggle and Gratton's approach simulated the likelihood rather than the posterior distribution as in modern ABC methods. The first true implementation of an ABC model rejection scheme was developed by Tavaré et al. (1997), who devised a simulation method for the inference of coalescence times in which they replaced a full data set of human Y chromosome DNA sequences with the number of segregating sites. Pritchard et al. (1999) derived a model rejection scheme from the algorithm of Tavaré et al. (1997) for assessing human population changes via human Y chromosome microsatellite data. The term "approximate Bayesian computation" was coined by Beaumont et al. (2002) when they detailed further improvements to the methods developed by Tavaré et al. (1997) and Pritchard et al. (1999).

Estimation of the posterior distribution of a model parameter quantifies the remaining uncertainty in the parameter after observation (Liepe et al. 2014). The classic ABC rejection approach to parameter estimation begins by computing the summary statistics for the observed data. Next, the data are simulated several times under a specific model, and the summary statistics are computed for each simulation. The distance between the observed value for each summary statistic and the corresponding values from each simulation are calculated. A simulation is accepted as an approximate sample from the posterior distribution if the distances between observed and simulated summary statistics are smaller or equal to a previously specified tolerance level (ε); otherwise it is rejected. The posterior distribution of the parameter (θ) of interest is computed from the parameter points of the accepted simulations for each summary statistic. If a summary statistic is sufficient and if the data are sufficiently informative, the posterior distribution should have a small probability in a region around the true value of θ.

Methods for model selection include the likelihood ratio test (Neyman and Pearson 1933), coefficient of determination (adjusted R^2; Theil 1961), AIC, corrected AIC (Hurvich and Tsai 1989), Bayesian information criterion (BIC; Schwarz 1978), and

deviance information criterion (DIC; Spiegelhalter et al. 2002). The ABC rejection sampling scheme is the newest member of the model selection family. Model comparison and selection begin with sampling a particular model from the prior distribution for the models under consideration. Next, the parameters for that model are sampled from the model's prior distribution. Simulations are then performed as in single-model ABC parameter estimation, and the resulting relative acceptance frequencies for each model approximate posterior distribution for models. The plausibility of two models can be compared by computing the posterior odds ratio, which is related to Bayes factors (ratio of marginal likelihoods). The classic rejection approach is usually modified by using linear or nonlinear regression to account for dimensionality and correct for discrepancies between the observed data and simulations (Beaumont et al. 2002; Blum 2010; Leuenberger and Wegmann 2010). A drawback in using regression is that error increases with distance from the centroid, and this will be exacerbated by nonlinear relationships between the summary statistics and the parameters (Beaumont 2010). Additionally, the rejection method can be slow if the posterior distribution is very narrow in relation to the prior (Beaumont 2010). Others (Templeton 2009, 2010; Csilléry et al. 2010) have argued that more vigorous adjustment of model complexity in ABC is necessary and have advocated the use of Akaike's, Bayesian (Schwarz 1978), or deviance information criteria (Spiegelhalter et al. 2002) for model selection (see Beaumont et al. [2010] for an opposing view).

Two popular modifications of the original ABC methods include ABC Markov chain Monte Carlo (ABC-MCMC; Marjoram et al. 2003; Wegmann et al. 2009) and sequential Monte Carlo ABC (SMC-ABC; Sisson et al. 2007, 2009; Beaumont et al. 2009; Toni et al. 2009). ABC-MCMC more fully explores the parameter space and uses the distance between observed and simulated summary statistics to update parameter values. Parameter values that produce simulations close to the observed data are visited preferentially. As in the rejection method, the accuracy of ABC-MCMC can be improved through regression adjustments. Accuracy of ABC-MCMC depends on sufficient mixing and convergence of chains, which can be difficult to verify (Marjoram and Tavaré 2006; Beaumont 2010). Mixing tends to be slowest in the tails and can result in poor mixing overall, especially if the starting point for the MCMC is at the far end of a tail. SMC-ABC approximates the posterior distribution by using a large set of randomly chosen parameter values, called "particles" (Sisson et al. 2007). The particles are propagated over time by simple sampling mechanisms or rejected if generated data match the observation poorly. Repeated sampling from a gradually improving approximation of the posterior can produce a distribution of summary statistics closer to the posterior predictive distribution. However, repeated sampling is unlikely to improve the approximation if a large proportion of the simulated points lie within the tolerance regions. As tolerance decreases, the proportion of points accepted becomes very small without improvement in the approximation of the posterior.

Many ABC-based softwares have been developed for the analysis of genetic data. Computer programs such as DIYABC (Cornuet et al. 2008, 2014b), Pop-ABC (Lopes et al. 2009), msABC (Pavlidis et al. 2010), ABCtoolbox (Wegmann et al. 2010), and ONeSAMP (Tallmon et al. 2008) infer demographic histories. ABC-based software has also been developed to estimate parental contribution to admixture (2BAD; Bray et al. 2010), estimate F statistics of dominant genetic markers (ABC4F; Foll et al. 2008), and assist phylogeographic inference (ms-Bayes; Hickerson et al. 2007). The expansion and increased use of ABC in various fields have led to attempts to make the method more user-friendly and increase computational speed; however, those who choose to use ABC should be aware of its potential pitfalls and limitations. In addition to concerns regarding dimensionality in model selection, there is a pressing need for research to determine the best approach to selecting tolerance levels and identifying and selecting sufficient summary statistics (Csilléry et al. 2010; Sunnåker et al. 2013; Robert 2016). Little guidance or research exists on the selection of tolerance levels and summary statistics, which is unfortunate as these affect the estimation of posterior distributions (Toni et al. 2009; Turner and Van Zandt 2012; Calvet and Czeller 2015). Setting tolerance to zero or using very small values yields exact results but is computationally expensive and possible only with infinite computing power (Turner and Van Zandt 2012; Sunnåker et al. 2013; Robert 2016). Tolerance, therefore, has to be set $\varepsilon > 0$. Non-zero tolerance reduces variance but also increases bias due to uncorrected departures from additivity and linearity (Beaumont et al. 2002). The solution would seem, then, to be to set the tolerance as close to zero as possible, but strict tolerance levels tend to increase the frequency of Type I error (Stocks et al. 2014). Perhaps the better approach is to experiment with the type and number of summary statistics selected. Stocks et al. (2014) found that tolerance decreased in importance as the summary statistics captured more information.

One of the most difficult tasks in ABC is selecting sufficiently informative summary statistics. Most applications of ABC require the user to select and compute summary statistics; in contrast the newest version of DIYABC provides the user with a choice of summary statistics computed by the program itself (Cornuet et al. 2014a). For example, DIYABC can calculate, across microsatellite loci, for a single sample the mean number of alleles, mean gene diversity (Nei 1987), mean allele size variance, and the mean M index (Garza and Williamson 2001; Excoffier et al. 2005). The question remains, however, whether these statistics are sufficient. Insufficient summary statistics result in inflated credibility intervals and a loss of information (Robert et al. 2011; Sunnåker et al. 2013; Robert 2016). Furthermore, summary statistics that are sufficient in one situation may be insufficient in others (Robert et al. 2011; Robert 2016). For example, summary statistics that can separate constant population size from population expansion will not be suitable for separating structured and unstructured populations (Nunes and Balding 2010; Blum et al. 2013; Stocks et al. 2014). Marin ct al. (2014) argued that summary statistics used for parameter estimation should be used for model comparison and these should not be combined into a single summary statistic. Methods to identify informative summary statistics for parameter estimation include weighting schemes (Hamilton et al. 2005), partial least squares transformation of summary statistics (Wegmann et al. 2009), selecting summary statistics based on the relationship between simulated parameter values and data (Fearnhead and Prangle 2012), machine learning (or boosting) techniques (Aeschbacher et al. 2012), correlation (Hickerson et al. 2006; Pavlidis et al. 2010; Clotault et al. 2012), and principal components analysis (Cornuet et al. 2010; Stocks et al. 2014). Recently, procedures have been developed to identify informative summary statistics for model selection including linear discriminant analysis transformation of summary statistics before the regression step (Estoup et al. 2012), using regression with the model as a response variable to determine how summary statistics should be weighted (Prangle et al. 2014), and using a chi-square test to evaluate the expectations of summary statistics after simulation under each model is considered (Marin et al. 2014). Sousa et al. (2009) found that in situations such as admixture, whcrc informative summary statistics are difficult to identify, it is possible to use the full allelic distribution to obtain approximate posterior distributions. One promising development is piecewise ABC (PW-ABC), which avoids the use of summary statistics by using factorization to estimate the posterior density of the parameter (White et al. 2015). All approaches, including ABC, used for the inference of population structure and its causes via genetic data are sensitive to a variety of factors including the specific idiosyncrasies of the genetic markers used, ploidy type (haploid or diploid), mode of inheritance, extent of recombination, level of genetic variation, substitution rates, and deviations from sc

lective neutrality due to selection or hitchhiking. Simulation studies have indicated that low levels of genetic variation and small sample sizes may limit the ability of ABC to detect weak bottlenecks (Stocks et al. 2014). Shafer et al. (2015) used simulated data that mimicked genotyping-by-sequencing (GBS) data and empirical GBS data to assess the ability of ABC to estimate migration rate and split time, population size, and detect bottlenecks. Split time and migration rates could be reliably inferred with 1,000 loci. In contrast, population size was difficult to estimate even with 50,000 loci and only recent bottleneck scenarios without recovery could be reliably detected. All estimates, however, were improved by including rare alleles (Shafer et al. 2015).

Another potential pitfall of ABC is the use of a small number of models that are not representative of the full range of possibilities and may exclude the true model (Templeton 2009; Csilléry et al. 2010; Sunnåker et al. 2013). The results of ABC model choice are directly influenced by the number and type of models that are included in the analysis (Hickerson 2014; Pelletier and Carstens 2014). Currently, there is no validated, commonly accepted method for determining how many models to include in ABC model choice, which is unfortunate because two of the major advantages of ABC is that it is relatively flexible and can incorporate a large number of models (Csilléry et al. 2010). Recent studies by Pelletier and Carstens (2014) and Nater et al. (2015) may serve as good examples of how to conduct ABC model choice, especially regarding the number of models that should be examined. Both studies begin the ABC model selection process by first comparing a small number of general models and then conducting more intensive comparison using a larger set of submodels derived from the general models.

LITERATURE CITED

Ackerman, M. W., B. K. Hand, R. K. Waples, G. Luikart, R. S. Waples, C. A. Steele, B. A. Garner, J. McCance, and M. R. Campbell. 2017. Effective number of breeders from sibship reconstruction: Empirical evaluations using hatchery steelhead. Evolutionary Applications10:146–160.

Aeschbacher, S., M. A. Beaumont, and A. Futschik. 2012. A novel approach for choosing summary statistics in approximate Bayesian computation. Genetics 192:1027–1047.

Akaike, H. 1974. A new look at the statistical model identification. IEEE Transactions on Automatic Control 19:716–723.

Aldous, D. J. 2001. Stochastic models and descriptive statistics for phylogenetic trees, from Yule to today. Statistical Science 16:23–34.

Alfaro, M. E., and M. T. Holder. 2006. The posterior and the prior in Bayesian phylogenetics. Annual Reviews in Ecology, Evolution, and Systematics 37:19–42.

Altekar, G., S. Dwarkadas, J. P. Huelsenbeck, and F. Ronquist. 2004. Parallel Metropolis coupled Markov chain Monte Carlo for Bayesian phylogenetic inference. Bioinformatics 20:407–415.

Alter, S. E., and S. R. Palumbi. 2009. Comparing evolutionary patterns and variability in the mitochondrial control region and cytochrome *b* in three species of baleen whales. Journal of Molecular Evolution 68:97–111.

Altman, D. G., and J. M. Bland. 1999. Variables and parameters. The BMJ 318:1667.

Anderson, D. R., K. P. Burnham, W. R. Gould, and S. Cherry. 2001. Concerns about finding effects that are actually spurious. Wildlife Society Bulletin 29:311–316.

Anderson, D. R., K. P. Burnham, and W. L. Thompson. 2000. Null hypothesis testing: Problems, prevalence, and an alternative. Journal of Wildlife Management 64:912–923.

Avise, J. C. 2000. Phylogeography: The history and formation of species. Harvard University Press, Cambridge, MA.

Avise, J. C. 2004. Molecular markers, natural history, and evolution, 2nd. ed. Sinauer Associates, Sunderland, MA.

Axelsson, E., E. Willerslev, M. T. P. Gilbert, and R. Nielsen. 2009. The effect of ancient DNA damage on inferences of demographic histories. Molecular Biology and Evolution 25:2181–2187.

Baalsrud, H. T., B.-E.Sæther, I. J. Hagen, A. M. Myhre, T. H. Ringsby, H. Pärn, and H. Jensen. 2014. Effects of population characteristics and structure on estimates of effective population size in a house sparrow metapopulation. Molecular Ecology 23:2653–2668.

Baele, G., P. Lemey, T. Bedford, A. Rambaut, M. A. Suchard, and A. V. Alekseyenko. 2012. Improving the accuracy of demographic and molecular clock model comparison while accommodating phylogenetic uncertainty. Molecular Biology and Evolution 29:2157–2167.

Baele, G., W. L. S. Li, A. J. Drummond, M. A. Suchard, and P. Lemey. 2013. Accurate model selection of relaxed molecular clocks in Bayesian phylogenetics. Molecular Biology and Evolution 30:239–243.

Bagley, J. C., R. L. Mayden, K. J. Roe, W. Holznagel, and P. M. Harris. 2011. Congeneric phylogeographical sampling reveals polyphyly and novel biodiversity within black basses (Centrarchidae: *Micropterus*). Biological Journal of the Linnean Society 104:346–363.

Baker, A. J., S. L. Pereira, and T. A. Paton. 2007. Phylogenetic relationships and divergence times of Charadriiformes genera: Multigene evidence for the Cretaceous origin of at least 14 clades of shorebirds. Biology Letters 3:205–209.

Baker, K. H., and A. R. Hoelzel. 2014. Influence of Holocene environmental change and anthropogenic impact on the diversity and distribution of roe deer. Heredity 112:607–615.

Barlow, E. J., F. Daunt, S. Wanless, D. Álvarez, J. M. Reid, and S. Cavers. 2011. Weak large-scale population genetic structure in a philopatric seabird, the European shag *Phalacrocorax aristotelis*. Ibis 153:768–778.

Bayes, T. 1763. An essay toward solving a problem in the doctrine of chances. Philosophical Transactions of the Royal Society of London 53:370–418.

Bean, P. T., D. J. Lutz-Carrillo, and T. H. Bonner. 2013. Rangewide survey of the introgressive status of Guadalupe bass: Implications for conservation and management. Transactions of the American Fisheries Society 142:681–689.

Beaumont, M. A. 2010. Approximate Bayesian computation in evolution and ecology. Annual Reviews in Ecology, Evolution, and Systematics 41:379–406.

Beaumont, M. A., J.-M. Cornuet, J.-M. Marin, and C. P. Robert. 2009. Adaptive approximate Bayesian computation. Biometrika 96:983–990.

Beaumont, M. A., R. Nielsen, C. Robert, J. Hey, O. Gaggiotti, L. Knoweles, A. Estoup, M. Panchal, J. Corander, M. Hickerson, S. A. Sisson, N. Fagundes, L. Chikhi, P. Beerli, R. Vitalis, J.-M. Cornuet, J. Huelsenbeck, M. Foll, Z. Yang, F. Rousset, D. Balding, and L. Excoffier. 2010. In defence of model-based inference in phylogeography. Molecular Ecology 19:436–446.

Beaumont, M. A., and B. Rannala. 2004. The Bayesian revolution in genetics. Nature Reviews Genetics 5:251–261.

Beaumont, M. A., W. Zhang, and D. J. Balding. 2002. Approximate Bayesian computation in population genetics. Genetics 162:2052–2035.

Bell, C. D., D. E. Soltis, and P. S. Soltis. 2010. The age and diversification of the angiosperms re-revisited. American Journal of Botany 97:1296–1303.

Blair, C., and R. W. Murphy. 2011. Recent trends in molecular phylogenetic analysis: Where to next? Journal of Heredity 102:130–138.

Blair, C, D. E. Weigel, M. Balazik, A. T. H. Keeley, F. M. Walker, E. Landguth, S. Cushman, M. Murphy, L. Waits, and N. Balkenhol. 2012. A simulation-based evaluation of methods for inferring linear barriers to gene flow. Molecular Ecology Resources 12:822–833.

Blum, M. G. B. 2010. Approximate Bayesian computation: A nonparametric perspective. Journal of the American Statistical Association 105:1178–1187.

Blum, M. G. B., M. A. Nunes, D. Prangle, and S. A. Sisson. 2013. A comparative review of dimension reduction methods in approximate Bayesian computation. Statistical Science 28:189–208.

Bohling, J. H., J. R. Adams, and L. P. Watts. 2013. Evaluating the ability of Bayesian clustering methods to detect hybridization and introgression using an empirical red wolf data set. Molecular Ecology 22:74–86.

Bonin, C. A., M. E. Goebel, J. I. Hoffman, and R. S. Burton. 2014. High male reproductive success in a low-density Antarctic fur seal (Arctocephalus gazella) breeding colony. Behavioral Ecology and Sociobiology 68:597–604.

Bouckaert, R., J. Heled, D. Kühnert, T. Vaughan, C.-H. Wu, D. Xie, M. A. Suchard, A. Rambaut, and A. J. Drummond. 2014. BEAST 2: A software platform for Bayesian evolutionary analysis. PLoS Computer Biology 10(4):e1003537.

Brandley, M., H. Ota, T. Hikida, A. N. Montes de Oca, M. Fería-Ortíz, X. Guo, and Y. Wang. 2012. The phylogenetic systematics of blue-tailed skinks (Pleistodon) and the family Scincidae. Zoological Journal of the Linnaean Society 165:163–189.

Brandley, M. C., A. Schmitz, and T. W. Reeder. 2005. Partitioned Bayesian analyses, partition choice, and the phylogenetic relationships of scincid lizards. Systematic Biology 54:373–390.

Bray, T. C., V. C. Sousa, B. Parreira, M. W. Bruford, and L. Chikhi. 2010. 2BAD: An application to estimate the parental contributions during two independent admixture events. Molecular Ecology Resources 10:538–541.

Brinkman, F. S. L., and D. D. Leipe. 2001. Phylogenetics analysis. In Bioinformatics: A practical guide to the analysis of genes and proteins, 2nd ed., edited by A. D. Baxevanis and B. F. F. Ouellette, editors, pp. 323–358. John Wiley & Sons, Hoboken, NJ.

Brocklehurst, N., P. Upchurch, P. D. Mannion, and J. O'Connor. 2014. The completeness of the fossil record of Mesozoic birds: Implications for early avian evolution. PLoS ONE 7(6):e39056.

Bromham, L., and D. Penny. 2003. The modern molecular clock. Nature Reviews Genetics 4:216–224.

Brown, J. W. 2003. The state of Bayesian phylogenetics: Bayes for the uninitiated. Queen's University, Kingston, Ontario, Canada.

Brown, J. M., and A. R. Lemmon. 2007. The importance of data partitioning and the utility of Bayes factors in Bayesian phylogenetics. Systematic Biology 56:643–655.

Brown, J. W., J. S. Rest, J. García-Moreno, M. D. Sorenson, and D. P. Mindell. 2008. Strong mitochondrial DNA support for a Cretaceous origin of modern avian lineages. BMC Biology 2008 6:6.

Brown, R. P., and Z. Yang. 2011. Rate variation and estimation of divergence times using strict and relaxed clocks. BMC Evolutionary Biology 11:271.

Brüniche-Olsen, A., M. E. Jones, J. J. Austin, C. P. Burridge, and B. R. Holland. 2014. Extensive population decline in the Tasmanian devil predates European settlement and devil facial tumour disease. Biology Letters 10:20140619.

Bruno, W. J., N. D. Socci, and A. L. Halpern. 2000. Weighted neighbor joining: A likelihood-based approach to distance-based phylogeny reconstruction. Molecular Biology and Evolution 17:189–197.

Buerkle, A., and C. Lexer. 2008. Admixture as the basis for genetic mapping. Trends in Ecology & Evolution 23:686–694.

Burbrink, F. T., F. Fontanella, R. A. Pyron, T. J. Guiher, and C. Jimenez. 2008. Phylogeography across the continent: The evolutionary and demographic history of the North American racer (Serpentes: Colubridae: Coluber constrictor). Molecular Phylogenetics and Evolution 47:274–288.

Calvet, L. E., and V. Czellar. 2015. Accurate methods for approximate Bayesian computation filtering. Journal of Financial Econometrics 13:798–838.

Cameron, E., and A. N. Pettitt. 2012. Approximate Bayesian computation for astronomical model analysis: A case study in galaxy demographics and morphological transformation at high redshift. Monthly Notices of the Royal Astronomical Society 425:44–65.

Camin, J. H., and R. R. Sokal. 1965. A method for deducing branching sequences in phylogeny. Evolution 19:311–326.

Castoe, T. A., and C. L. Parkinson. 2006. Bayesian mixed models and the phylogeny of pitvipers (Viperidae: Serpentes). Molecular Phylogenetics and Evolution 39:91–110.

Castoe, T. A., C. L. Spencer, and C. L. Parkinson. 2007. Phylogeographic structure and historical demography of the western diamondback rattlesnake (Crotalus atrox): A perspective on North American desert biogeography. Molecular Phylogenetics and Evolution 42:193–212.

Cavalli-Sforza, L. L., and A. W. F. Edwards. 1967. Phylogenetic analysis: Models and estimation procedures. Evolution 21:550–570.

Cavender-Bares, J., and A. Pahlich. 2009. Molecular, morphological, and ecological niche differentiation of sympatric sister oak species, *Quercus virginiana* and *Q. geminata* (Fagaceae). American Journal of Botany 96:1690–1702.

Chen, C., E. Durand, F. Forbes, and O. François. 2007. Bayesian clustering algorithms ascertaining spatial population structure: A new computer program and a comparison study. Molecular Ecology Notes 7:747–756.

Chen, M.-H., L. Kuo, and P. O. Lewis, eds. 2014. Bayesian phylogenetics: Methods, algorithms, and applications. CRC Press, Boca Raton, FL.

Cherry, S. 1998. Statistical tests in publications of the Wildlife Society. Wildlife Society Bulletin 26:947–953.

Choudhuri, S. 2014. Bioinformatics for beginners: Genes, genomes, molecular evolution, databases and analytical tools. Elsevier, Amsterdam, Netherlands.

Clotault, J., A.-C. Thuillet, M. Buiron, S. De Mita, M. Couderc, B. I. G. Haussmann, C. Mariac, and Y. Vigouroux. 2012. Evolutionary history of pearl millet (*Pennisetum glaucum* [L.] R. Br.) and selection on flowering genes since its domestication. Molecular Biology and Evolution 29:1199–1212.

Cohen, C., J. L. Wakeling, T. G. Mandiwana—Neudani, E. Sande, C. Dranzoa, T. M. Crowe, and R. C. K. Bowie. 2012. Phylogenetic affinities of evolutionary enigmatic African galliforms: The Stone Partridge *Ptilopachus petrosus* and Nahan's Francolin *Francolinus nahani*, and support for their sister relationship with New World quails. Ibis 154:768–780.

Cohen, J. 1990. Things I have learned (so far). American Psychologist 45:1304–1312.

Condamine, F. L., N. S. Nagalingum, C. R. Marshall, and H. Morlon. 2015. Origin and diversification of living cycads: A cautionary tale on the impact of the branching process prior in Bayesian molecular dating. BMC Evolutionary Biology 15:65.

Corander, J., J. Sirén, and E. Arjas. 2008. Bayesian spatial modeling of genetic population structure. Computational Statistics 23:111–129.

Corander, J., P. Waldmann, and M. J. Sillanpaa. 2003. Bayesian analysis of genetic differentiation between populations. Genetics 163:367–374.

Cornuet, J. M., P. Pudlo, J. Veyssier, A. Dehne-Garcia, and A. Estoup. 2014a. DIYABC version 2.0: A user-friendly software for inferring population history through approximate Bayesian computations using microsatellite, DNA sequence, and SNP data. Institut National de la Recherche Agronomique, Campus International de Baillarguet, Saint-Gély-du Fesc Cedex, France.

Cornuet, J.-M., P. Pudlo, J. Veyssier, A. Dehne-Garcia, M. Gautier, R. Leblois, J.-M. Marin, and A. Estoup. 2014b. DIY ABC v2.0: A software to make approximate Bayesian computation inferences about population history using single nucleotide polymorphism, DNA sequence and microsatellite data. Bioinformatics 30:1187–1189.

Cornuet, J.-M., V. Ravigné, and A.Estoup. 2010. Inference on population history and model checking using DNA sequence and microsatellite data with the software DIY ABC (v1.0). BMC Bioinformatics 11:401.

Cornuet, J.-M., F. Santos, M. A. Beaumont, C. R. Robert, J.-M. Marin, D. J. Balding, T. Guillemaud, and A. Estoup. 2008. Inferring population history with DIY ABC: A user-friendly approach to approximate Bayesian computation. Bioinformatics 24:2713–2719.

Couvreur, T. L. P., A. Franzke, I. A. Al-Shehbaz, F. T. Bakker, M. A. Koch, and K. Mummenhoff. 2010. Molecular phylogenetics, temporal diversification, and principles of evolution in the mustard family (Brassicaceae). Molecular Biology and Evolution 27:55–71.

Crema, E. R., K. Edinborough, T. Kerig, and S. J. Shennan. 2014. An approximate Bayesian computation approach for inferring patterns of cultural evolutionary change. Journal of Archaeological Science 50:160–170.

Crowe, T. M., R. C. K. Bowie, P. Bloomer, T. G. Mandiwana, T. A. J. Hedderson, E. Randi, S. L. Pereira, and J. Wakeling. 2006. Phylogenetics, biogeography and classification of, and character evolution in, gamebirds (Aves: Galliforems): Effects of character exclusion, data partitioning and missing data. Cladistics 22:495–532.

Csilléry, K., M. G. B. Blum, O. E. Gaggiotti, and O. François. 2010. Approximate Bayesian computation (ABC) in practice. Trends in Ecology and Evolution 25:410–418.

Csilléry, K., O. François, and M. G. B. Blum. 2012. *abc*: an R package for approximate Bayesian computation (ABC). Methods in Ecology and Evolution 3:475–479.

Darwin, C. 1859. On the Origin of Species by Means of Natural Selection, or the Preservation of Favoured Races in the Struggle for Life. John Murray, London.

DeYoung, R. W., and R. L. Honeycutt. 2005. The molecular toolbox: Genetic techniques in wildlife ecology and management. Journal of Wildlife Management 69:1362–1384.

DeYoung, R. W., A. Zamorano, B. T. Mesenbrink, T. A. Campbell, B. R. Leland, G. M. Moore, R. L. Honeycutt, and J. J. Root. 2009. Landscape-genetic analysis of population structure in the Texas gray fox oral rabies vaccination zone. Journal of Wildlife Management 73:1292–1299.

Diggle, P. J., and R. J. Gratton. 1984. Monte Carlo methods of inference for implicit statistical models. Journal of the Royal Statistical Society B 46:193–227.

Dionne, M., F. Caron, J. Dodson, and L. Bernatchez. 2008. Landscape genetics and hierarchical genetic structure in Atlantic salmon: The interaction of gene flow and local adaptation. Molecular Ecology 17:2382–2396.

Donnelly, P., and S. Tavaré. 1995. Coalescents and genealogical structure under neutrality. Annual Reviews in Genetics 29:401–421.

Drummond, A., O. G. Pybus, and A. Rambaut. 2003. Inference of viral evolutionary rates from molecular sequences. Advances in Parasitology 54:331–358.

Drummond, A. J., and R. R. Bouckaert. 2015. Bayesian evolutionary analysis with BEAST. Cambridge University Press, Cambridge, UK.

Drummond, A. J., S. Y. W. Ho, M. J. Phillips, and A. Rambaut. 2006. Relaxed phylogenetics and dating with confidence. PLoS Biology 4(5):e88.

Drummond, A. J., S. Y. W. Ho, N. Rawlence, and A. Rambaut. 2007. A rough guide to BEAST 1.4. https://www.researchgate.net/profile/Nicolas_Rawlence.

Drummond, A. J., and A. Rambaut. 2007. BEAST: Bayesian evolutionary analysis by sampling trees. BMC Evolutionary Biology 7:214.

Drummond, A. J., A. Rambaut, B. Shapiro, and O. G. Pybus. 2005. Bayesian coalescent inference of past population dynamics from molecular sequences. Molecular Biology and Evolution 22:1185–1192.

Drummond, A. J., M. A. Suchard, D. Xie, and A. Rambaut. 2012. Bayesian phylogenetics with BEAUti and the BEAST 1.7. Molecular Biology and Evolution 29:1969–1973.

Duchêne, S., and S. Y. W. Ho. 2014. Using multiple relaxed-clock models to estimate evolutionary timescales from DNA sequence data. Molecular Phylogenetics and Evolution 77:65–70.

Duchêne, S., M. Molak, and S. Y. W. Ho. 2014. ClockstaR: Choosing the number of relaxed-clock models in molecular phylogenetic analysis. Bioinformatics 30:1017–1019.

Earl, D. A., and B. M. vonHoldt. 2012. STRUCTURE HARVESTER: a website and program for visualizing STRUCTURE output and implementing the Evanno method. Conservation Genetics Resources 4:359–361.

Eck, R. V., and M. O. Dayhoff. 1966. Atlas of protein sequence and structure 1966. National Biomedical Research Foundation, Silver Spring, MD.

Edwards, A. W. F. 1970. Estimation of the branch points of a branching diffusion process. Journal of the Royal Statistical Society B 32:155–174.

Edwards, A. W. F., and L. L. Cavalli-Sforza. 1964. Reconstruction of evolutionary trees. In Phenetic and phylogenetic classification, edited by V. H. Heywood and J. McNeill, pp. 67–76.. Publication number 6. Systematics Association, London.

Ericson, P. G. P., C. L. Anderson, T. Britton, A. Elzanowski, U. S. Johansson, M. Källersjö, J. I. Oholson, T. J. Parsons, D. Zuccon, and G. Mayr. 2006. Diversification of Neoaves: Integration of molecular sequence data and fossils. Biology Letters 2:543–547.

Estoup, A., P. Jarne, and J. M. Cornuet. 2002. Homoplasy and mutation model at microsatellite loci and their consequences for population genetic analysis. Molecular Ecology 11:1591–1604.

Estoup, A., E. Lombart, J.-M. Marin, T. Guillemaud, P. Pudlo, C. P. Robert, and J.-M. Cornuet. 2012. Estimation of demo-genetic model probabilities with approximate Bayesian computation using linear discriminant analysis on summary statistics. Molecular Ecology Resources 12:846–855.

Evanno, G., S. Regnaut, and J. Goudet. 2005. Detecting the number of clusters of individuals using the software STRUCTURE: A simulation study. Molecular Ecology 14:2611–2620.

Excoffier, L., A. Estoup, and J.-M. Cornuet. 2005. Bayesian analysis of an admixture model with mutations and arbitrary linked markers. Genetics 169:1727–1738.

Fagundes, N. J. R., R. Kanitz, and S. L. Bonatto. 2008. A reevaluation of the Native American mtDNA genome diversity and its bearing on the models of early colonization of Beringia. PLoS ONE 3(9):e3157.

Falush, D., M. Stephens, and J. K. Pritchard. 2003. Inference of population structure: Extensions to linked loci and correlated allele frequencies. Genetics 164:1567–1587.

Falush, D., M. Stephens, and J. K. Pritchard. 2007. Inference of population structure using multilocus genotype data: Dominant markers and null alleles. Molecular Ecology Notes 7:574–578.

Farris, J. S. 1970. Methods for computing Wagner trees. Systematic Zoology 19:83–92.

Farris, J. S. 1973. On comparing the shapes of taxonomic trees. Systematic Zoology 22:50–54.

Fearnhead, P., and D. Prangle. 2012. Constructing summary statistics for approximate Bayesian computation: Semi-automatic approximate Bayesian computation. Journal of the Royal Statistical Society B 74:419–474.

Felsenstein, J. 1978. Cases in which parsimony or compatibility methods will be positively misleading. Systematic Zoology 27:401–410.

Felsenstein, J. 1981. Evolutionary trees from DNA sequences: A maximum likelihood approach. Journal of Molecular Evolution 17:368–376.

Felsenstein, J. 2004. Inferring phylogenies. Sinauer Associates, Sunderland, MA.

Ferguson, A. W., M. M. McDonough, G. I. Guerra, M. Rheude, J. W. Dragoo, L. K. Ammerman, and R. C. Dowler. 2017. Phylogeography of a widespread small carnivore, the western spotted skunk (*Spilogale gracilis*), reveals temporally variable signatures of isolation across western North America. Ecology and Evolution 7:4229–4240.

Fischer, M. C., M. Foll, L. Excoffier, and G. Heckel. 2011. Enhanced AFLP genome scans detect local adaptation in high-altitude populations of a small rodent (*Mcirotus arvalis*). Molecular Ecology 20:1450–1462.

Fisher, R. A. 1922. On the mathematical foundation of theoretical statistics. Philosophical Transactions of the Royal Society of London A 222:309–368.

Fisher, R. A. 1925. Theory of statistical estimation. Proceedings of the Cambridge Philosophical Society 22: 700–725.

Fitch, W. M. 1971. Toward defining the course of evolution: Minimum change for a specific tree topology. Systematic Biology 20:406–416.

Fitch, W. M., and E. Margoliash. 1967. Construction of phylogenetic trees. Science 155:279–284.

Fleischer, R. C., C. E. McIntosh, and C. L. Tarr. 1998. Evolution on a volcanic conveyor belt: Using phylogeographic reconstructions and K-Ar-based ages of the Hawaiian Islands to estimate molecular evolutionary rates. Molecular Ecology 7:533–545.

Flynn, J. J., J. A. Finarelli, S. Zehr, J. Hsu, and M. A. Nedbal. 2005. Molecular phylogeny of the Carnivora (Mammalia): Assessing the impact of increased sampling on resolving enigmatic relationships. Systematic Biology 54:217–337.

Foll, M., M. A. Beaumont, and O. E. Gaggiotti. 2008. An approximate Bayesian computation approach to overcome biases that arise when using AFLP markers to study population structure. Genetics 179:927–939.

Foll, M., and O. Gaggiotti. 2008. A genome-scan method to identify selected loci appropriate for both dominant and codominant markers: A Bayesian perspective. Genetics 18:977–993.

François, O., and E. Durand. 2010. Spatially explicit Bayesian clustering models in population genetics. Molecular Ecology Resources 10:773–784.

François, O., and L. P. Waits. 2016. Clustering and assignment methods in landscape genetics. In Landscape genetics: Concepts, methods, applications, edited by N. Balkenhol, S. A. Cushman, A. T. Storfer, and L. P. Waits, pp. 114–128. John Wiley & Sons, Chichester, West Sussex, UK.

Frankham, R. 1996. Relationship of genetic variation to population size in wildlife. Conservation Biology 10:1500–1508.

Frantz, A. C., S. Cellina, A. Krier, L. Schley, and T. Burke. 2009. Using spatial Bayesian methods to determine the genetic structure of a continuously distributed population: Clusters or isolation by distance? Journal of Applied Ecology 46:493–505.

Frey, A., P. H. Dutton, and G. H. Balazs. 2013. Insights on the demography of cryptic nesting by green turtles (*Chelonia mydas*) in the main Hawaiian Islands from genetic relatedness analysis. Journal of Experimental Marine Biology and Ecology 442:80–87.

Fu, Y. X. 1997. Statistical tests of neutrality of mutations against population growth, hitchhiking and background selection. Genetics 147:915–925.

Fu, Y. X., and W. H. Li. 1993. Statistical tests of neutrality of mutations. Genetics 133:693–709.

Fulton, T. L., C. Strobeck, and A. Crame. 2010. Multiple fossil calibrations, nuclear loci and mitochondrial genomes provide new insight into biogeography and divergence timing for true seals (Phocidae, Pinnipedia). Journal of Biogeography 37:814–829.

Futuyma, D. J. 1998. Evolutionary biology, 3rd ed. Sinauer Associates, Sunderland, MA.

Galaverni, M., D. Palumbo, E. Fabbri, R. Caniglia, C. Greco, and E. Randi. 2012. Monitoring wolves (*Canis lupus*) by non-invasive genetics and camera trapping: A small-scale pilot study. European Journal of Widlife Research 58:47–48.

Gamble, T., P. B. Berendzen, H. B. Shaffer, D. E. Starkey, and A. M. Simons. 2008. Species limits and phylogeography of North American cricket frogs (*Acris*: Hylidae). Molecular Phylogenetics and Evolution 48:112–125.

Gandolfo, M. A., K. C. Nixon, and W. L. Crepet. 2008. Selection of fossils for calibration of molecular dating models. Annals of the Missouri Botanical Gardens 95:34–42.

Garza, J. C., and E. G. Williamson. 2001. Detection of reduction in population size using data from microsatellite loci. Molecular Ecology 10:305–318.

Gavryushkina, A., T. A. Heath, D. T. Ksepka, T. Stadler, D. Welch, and A. J. Drummond. 2017. Bayesian total-evidence dating reveals the crown radiation of penguins. Systematic Biology 66:57–73.

Gelman, A., and D. B. Rubin. 1992. Inference from iterative simulation using multiple sequences. Statistical Science 7:457–511.

Genner, M. J., O. Seehausen, D. H. Lunt, D. A. Joyce, P. W. Shaw, G. R. Carvalho, and G. F. Turner. 2007. Age of cichlids: New dates for ancient lake fish radiations. Molecular Biology and Evolution 24:1269–1282.

Gernhard, T. 2008. The conditioned reconstructed process. Journal of Theoretical Biology 253:769–778.

Geyer, C. J. 1991. Markov chain Monte Carlo maximum likelihood. In Computing science and statistics: Proceedings of the 23rd symposium of the interface, edited by E. M. Keramidas, pp. 156–163. Interface Foundation, Fairfax Station, VA.

Gilbert, C., A. Ropiquet, and A. Hassanin. 2006. Mitochondrial and nuclear phylogenies of Cervidae (Mammalia, Ruminatia): Systematics, morphology, and biogeography. Molecular Phylogenetics and Evolution 40:101–117.

Gilbert, K. J., R. L. Andrew, D. G. Bock, M. T. Franklin, N. C. Kane, J.-S. Moore, B. T. Moyers, S. Renaut, D. J. Rennison, T. Veen, and T. H. Vines. 2012. Recommendations for utilizing and reporting population genetic analyses: the reproducibility of genetic clustering using the program STRUCTURE. Molecular Ecology 21:4925–4930.

Gilks, W. R., and G. O. Roberts. 1996. Strategies for improving MCMC. In Markov chain Monte Carlo in practice, edited by W. R. Gilks, S. Richardson, and D. J. Spiegelhalter, pp. 89–114. Chapman & Hall, London.

Grant, W. S. 2015. Problems and cautions with sequence mismatch analysis and Bayesian skyline plots to infer historical demography. Journal of Heredity 106:333–346.

Grant, W. S., and W. Cheng. 2012. Incorporating deep and shallow components of genetic structure into the management of Alaskan red king crab. Evolutionary Applications 5:820–837.

Grant, W. S., M. Liu, TX. Gao, and T. Yanagimoto. 2012. Limits of Bayesian skyline plot analysis of mtDNA sequences to infer historical demographies in Pacific herring (and other species). Molecular Phylogenetics and Evolution 65:203–212.

Graur, D., and W. Martin. 2004. Reading the entrails of chickens: Molecular timescales of evolution and the illusion of precision. Trends in Genetics 20:80–86.

Green, P. J. 1995. Reversible jump Markov chain Monte Carlo computation and Bayesian model determination. Biometrika 82:711–732.

Green, P. J., K. Łatuszński, M. Pereyra, and C. P. Robert. 2015. Bayesian computation: A summary of the current states, and samples backwards and forwards. Statistical Computing 25:835–862.

Guiher, T. J., and F. T. Burbrink. 2008. Demographic and phylogeographic histories of two venomous North American snakes of the genus Agkistrodon. Molecular Phylogenetics and Evolution 48:543–553.

Guillot, G., F. Mortier, and A. Estoup. 2005. Geneland: A computer package for landscape genetics. Molecular Ecology Notes 5:712–715.

Guillot, G., F. Santos, and A. Estoup. 2008. Analysing georeferenced population genetics data with Geneland: A new algorithm to deal with null alleles and a friendly graphical user interface. Bioinformatics 24:1406–1407.

Hamilton, G., M. Currat, N. Ray, G. Heckel, M. Beaumont, and L. Excoffier. 2005. Bayesian estimation of recent migration rates after a spatial expansion. Genetics 170:409–417.

Hampton, J. O., P. B. S. Spencer, D. L. Alpers, L. E. Twigg, A. P. Woolnough, J. Doust, T. Higgs, and J. Pluske. 2004. Molecular techniques, wildlife management and the importance of genetic population structure and dispersal: A case study with feral pigs. Journal of Applied Ecology 41:735–743.

Hare, M. P., L. Nunney, M. K. Schwartz, D. E. Ruzzante, M. Burford, R. S. Waples, K. Ruegg, and F. Palstra. 2011. Understanding and estimating effective population size for practical application in marine species management. Conservation Biology 25:438–449.

Harper, C. W., Jr. 1979. A Bayesian probability view of phylogenetic systematics. Systematic Zoology 28:547–553.

Hastings, W. K. 1970. Monte Carlo sampling methods using Markov chains and their applications. Biometrika 57:97–109.

Heath, T. A., J. P. Huelsenbeck, and T. Stadler. 2014. The fossilized birth-death process for coherent calibration of divergence-time estimates. Proceedings of the National Academy of Science of the United States of America 111:E2957–E2966.

Held, L., and D. Sabanés Bové. 2014. Applied statistical inference. Springer-Verlag, Berlin.

Heled, J., and A. J. Drummond. 2008. Bayesian inference of population size history from multiple loci. BMC Evolutionary Biology 8:289.

Heller, R., L. Chikhi, and H. R. Siegismund. 2013. The confounding effect of population structure on Bayesian skyline plot inferences of demographic history. PLoS ONE 8(5):e62992.

Hennig, W. 1950. Grundzüge einer Theorie der phylogenetischen Systematik. Deutcher Zentralverlag, Berlin.

Hennig, W. 1965. Phylogenetic systematics. Annual Review of Entomology 10:97–110.

Hennig, W. 1966. Phylogenetic systematics. University of Illinois Press.

Hermsen, E. J., T. N. Taylor, E. L. Taylor, and D. Wm. Stevenson. 2006. Cataphylls of the Middle Triassic cycad Antarcticycas schopfii and new insights into cycad evolution. American Journal of Botany 93:724–738.

Hickerson, M. J. 2014. All models are wrong. Molecular Ecology 23:2887–2889.

Hickerson, M. J., G. Dolman, and C. Moritz. 2006. Comparative phylogeographic summary statistics for testing simultaneous vicariance. Molecular Ecology 15:209–233.

Hickerson, M. J., E. Stahl, and N. Takebayashi. 2007. msBayes: Pipeline for testing comparative phylogeographic histories using hierarchical approximate Bayesian computation. BMC Bioinformatics 8:268.

Hipsley, C. A., and J. Müller. 2014. Beyond fossil calibrations: Realities of molecular clock practices in evolutionary biology. Frontiers in Genetics 5(138):1–11.

Ho, S. Y. M. 2007. Calibrating molecular estimates of substitution rates and divergence times in birds. Journal of Avian Biology 38:409–414.

Ho, S. Y. W., T. H. Heupink, A. Rambaut, and B. Shapiro. 2007. Bayesian estimation of sequence damage in ancient DNA. Molecular Biology and Evolution 24:1416–1422.

Ho, S. Y. W., and G. Larson. 2006. Molecular clocks: When times are a-changin'. Trends in Genetics 22:79–83.

Ho, S. Y. W., and M. J. Phillips. 2009. Accounting for calibration uncertainty in phylogenetic estimation of evolutionary divergence times. Systematic Biology 58:367–380.

Ho, S. Y. W., M. J. Phillips, A. Cooper, and A. J. Drummond. 2005. Time dependency of molecular rate estimates and systematic overestimation of recent divergence times. Molecular Biology and Evolution 22:1561–1568.

Ho, S. Y. W., U. Saarma, R. Barnett, J. Haile, and B. Shapiro. 2008. The effect of inappropriate calibration: Three case studies in molecular ecology. PLoS ONE 3(2):e1615.

Ho, S. Y. W., and B. Shapiro. 2011. Skyline-plot methods for estimating demographic history from nucleotide sequences. Molecular Ecology Resources 11:423–434.

Ho, S. Y. W., K. J. Tong, C. S. P. Foster, A. M. Ritchie, N. Lo, and M. D. Crisp. 2015. Biogeographic calibrations for the molecular clock. Biology Letters 11:20150194.

Höhna, S., M. R. May, and B. R. Moore. 2016. TESS: An R package for efficiently simulating phylogenetic trees and performing Bayesian inference of lineage diversification rates. Bioinformatics 32:789–791.

Hosner, P. A., E. L. Braun, and R. T. Kimball. 2015. Land connectivity changes and global cooling shaped the colonization history and diversification of New World quail (Aves: Galliformes: Odontophoridae). Journal of Biogeography 42:1883–1895.

Huberty, C. J. 1993. Historical origins of statistical testing practices: The treatment of Fisher versus Neyman-Pearson views in textbooks. Journal of Experimental Education 61:317–333.

Hubisz, M., D. Falush, M. Stephens, and J. Pritchard. 2009. Inferring weak population structure with the assistance of sample group information. Molecular Ecology Resources 9:1322–1332.

Hudson, R. R. 1991. Gene genealogies and the coalescent process. Oxford Surveys in Evolutionary Biology 7:1–44.

Huelsenbeck, J. P., B. Larget, R. E. Miller, and F. Ronquist. 2002. Potential applications and pitfalls of Bayesian inference of phylogeny. Systematic Biology 51:673–688.

Huelsenbeck, J. P., B. Larget, and D. Swofford. 2000. A compound Poisson process for relaxing the molecular clock. Genetics 154:1879–1892.

Huelsenbeck, J. P., and B. Rannala. 2004. Frequentist properties of Bayesian posterior probabilities of phylogenetic trees under simple and complex substitution models. Systematic Biology 53:904–913.

Huelsenbeck, J. P., and F. Ronquist. 2001. MrBayes: Bayesian inference of phylogenetic trees. Bioinformatics 17:754–755.

Huelsenbeck, J. P., and F. Ronquist. 2005. Bayesian analysis of molecular evolution using MrBayes. In Statistical methods in molecular evolution, edited by R. Nielsen, pp. 183–226. Springer, New York.

Huelsenbeck, J. P., F. Ronquist, R. Nielsen, and J. P. Bollback. 2001. Bayesian inference of phylogeny and its impact on evolutionary biology. Science 294:673–688.

Hugall, A. F., R. Foster, and M. S. Y. Lee. 2007. Calibration choice, rate smoothing, and the pattern of tetrapod diversification according to the long nuclear gene RAG-1. Systematic Biology 56:543–563.

Hurvich, C. M., and C.-I. Tsai. 1989. Regression and time series model selection in small samples. Biometrika 76:297–307.

Irvahn, J., S. Chattopadhyay, E. V. Sokuernko, and V. N. Minin. 2013. rbrothers: R package for Bayesian multiple change-point recombination detection. Evolutionary Bioinformatics 9:235–238.

Jabot, F., T. Faure, and N. Dumoulin. 2013. EasyABC: Performing efficient approximate Bayesian computation sampling schemes using R. Methods in Ecology and Evolution 4:684–687.

Jackson, J. A., D. J. Steel, P. Beerli, B. C. Congdon, C. Olavarría, M. S. Leslie, C. Pomilla, H. Rosenbaum, and C. S. Baker. 2014. Global diversity and oceanic divergence of humpback whales (*Megaptera novaeangliae*). Proceedings of the Royal Society of London B 281:20133222.

Johnson, D. H. 1999. The insignificance of statistical significance testing. Journal of Wildlife Management 63:763–772.

Jones, O. R., and J. Wang. 2010. COLONY: A program for parentage and sibship inference from multilocus genotype data. Molecular Ecology Resources 10:551–555.

Jukes, T. H., and C. R. Cantor. 1969. Evolution of protein molecules. In Mammalian protein metabolism, vol. III, edited by M. N. Munro, pp. 21–132. Academic Press, New York.

Kanno, Y., J. C. Vokoun, and B. H. Letcher. 2011. Sibship reconstruction for inferring mating systems, dispersal and effective population size in headwater brook trout (*Salvenlinus fontinalis*) populations. Conservation Genetics 12:619–628.

Karaket, T., and S. Poompuang. 2012. CERVUS vs. COLONY for successful parentage and sibship determinations in freshwater prawn *Macrobrachium rosenbergii* de Man. Aquaculture 324–325:307–311.

Kashyap, R. L., and S. Subas. 1974. Statistical estimation of parameters in a phylogenetic tree using a dynamic model of the substitutional process. Journal of Theoretical Biology 47:75–101.

Kendall, D. G. 1948. On the generalized "birth-and-death" process. Annals of Mathematical Statistics 19:1–15.

Kergoat, G. J., P. Bouchard, A.-L. Clamens, J. L. Abbate, H. Jourdan, R. Jabbour-Zahab, G. Genson, L. Soldati, and F. L. Condamine. 2014. Cretaceous environmental changes led to high extinction rates in a hyperdiverse beetle family. BMC Evolutionary Biology 14:220.

Kimura, M. 1968. Evolutionary rate at the molecular level. Nature 217:624–626.

Kimura, M., and T. Ohta. 1971. On the rate of molecular evolution. Journal of Molecular Evolution 1:1–17.

Kimura, M., and T. Ohta. 1974. On some principles governing molecular evolution. Proceedings of the National Academy of Sciences of the United States of America 71:2848–2852.

King, J. L., and T. H. Jukes. 1969. Non-Darwinian evolution. Science 164:788–798.

Kingman, J. F. C. 1982. On the genealogy of large populations. Journal of Applied Probability 19:27–43.

Kitchen, A., M. M. Miyamoto, and C. J. Mulligan. 2008. A three-stage colonization model for the peopling of the Americas. PLoS ONE 3(2):e1596.

Kluge, A. G. 1989. A concern for evidence and a phylogenetic hypothesis of relationships among *Epicrates* (Boidae, Serpentes). Systematic Biology 38:7–25.

Kluge, A. G., and J. S. Farris. 1969. Quantitative phyletics and the evolution of anurans. Systematic Zoology 18:1–32.

Kopelman, N. M., J. Mayzel, M. Jakobsson, N. A. Rosenberg, and I. Mayrose. 2015. CLUMPAK: A program for identifying clustering modes and packaging population structure inferences across K. Molecular Ecology Resources 15:1179–1191.

Krysko, K. L., L. P. Nuñez, C. A. Lippi, D. J. Smith, and M. C. Granatosky. 2016. Pliocene-Pleistocene lineage diversifications in the eastern indigo snake (*Drymarchon couperi*) in the southeastern United States. Molecular Phylogenetics and Evolution 98:111–122.

Ksepka, D. T. 2009. Broken gears in the avian molecular clock: New phylogenetic analyses support stem galliform status for *Gallinuloides wyomingensis* and rallid affinities for *Amitabha urbsinterdictensis*. Cladistics 25:173–197.

Ksepka, D. T., J. F. Parham, J. F. Allman, M. J. Benton, M. T. Carrano, K. A. Cranston, P. C. J. Donoghue, J. J. Head, E. J. Hermsen, R. B. Irmis, W. G. Joyce, M. Kohli, K. D. Lamm, J. Leehr, J. L. Patané, P. D. Polly, M. J. Phillips, N. A. Smith, N. D. Smith, M. van Tuinen, J. L. Ware, and R. C. M. Warnock. 2015. The fossil calibration database—A new resource for divergence dating. Systematic Biology 64:853–859.

Ksepka, D. T., J. L. Ware, and K. S. Lamm. 2014. Flying rocks and flying clocks: Disparity in fossil and molecular dates for birds. Proceedings of the Royal Society of London B 281: 20140677.

Kuhner, M. K. 2009. Coalescent genealogy samplers: Windows into population history. Trends in Ecology & Evolution 24:86–93.

Lakner, C., P. van der Mark, J. P. Huelsenbeck, B. Larget, and F. Ronquist. 2008. Efficiency of Markov chain Monte Carlo tree proposals in Bayesian phylogenetics. Systematic Biology 57:86–103.

Lartillot, N., T. Lepage, and S. Blanquart. 2009. PhyloBayes 3: A Bayesian software package for phylogenetic reconstruction and molecular dating. Bioinformation 25:2286–2288.

Lartillot, N., N. Rodrigue, D. Stubbs, and J. Richer. 2013. PhyloBayes MPI: Phylogenetic reconstruction with infinite mixtures of profiles in a parallel environment. Systematic Biology 62:611–615.

Latch, E. K., G. Dharmarajan, J. C. Glaubitz, and O. E. Rhodes Jr. 2006. Relative performance of Bayesian clustering software for inferring population substructure and individual assignment at low levels of population differentiation. Conservation Genetics 7:295–302.

Latch, E. K., D. M. Reding, J. R. Heffelfinger, C. H. Alcalá-Galván, and O. E. Rhodes Jr. 2014. Range-wide analysis of genetic structure in a widespread, highly mobile species (*Odocoileus hemionus*) reveals the importance of historical biogeography. Molecular Ecology 23:3171–3190.

Leberg, P. 2005. Genetic approaches to estimating the effective size of populations. Journal of Wildlife Management 69:1385–1399.

Lecis, R., M. Pierpaoli, Z. S. Biró, L. Szemethy, B. Ragni, F. Vercillo, and E. Randi. 2006. Bayesian analyses of admixture in wild and domestic cats (*Felis silvestris*) using linked microsatellite loci. Molecular Ecology 15:119–131.

Lemey, P., A. Rambaut, A. J. Drummond, and M. A. Suchard. 2009. Bayesian phylogeography finds its roots. PLoS Computational Biology 5(9):e1000520.

Lepage, T., D. Bryant, H. Philippe, and N. Lartillot. 2007. A general comparison of relaxed molecular clock models. Molecular Biology and Evolution 24:2669–2680.

Lessios, H. A. 1979. Use of Panamanian sea urchins to test the molecular clock. Nature 280:599–601.

Leuenberger, C., and D. Wegmann. 2010. Bayesian computation and model selection without likelihoods. Genetics 184:243–252.

Lewis, P. O. 2001. Phylogenetic systematics turns over a new leaf. Trends in Ecology & Evolution 16:30–37.

Lewis, P. O. 2012. MCMC Robot: A Markov chain Monte Carlo teaching tool. http://www.mcmcrobot.org/.

Lewis, P. O. 2017. An introduction to Bayesian phylogenetics. Workshop on Molecular Evolution, Marine Biological Laboratory, Woods Hole, MA. https://molevol.mbl.edu/index.php/Paul_Lewis.

Li, W.-H., and M. Tanimura. 1987. The molecular clock runs more slowly in man than in apes and monkeys. Nature 326:93–96.

Li, W. H., M. Tanimura, and P. M. Sharp. 1987. An evaluation of the molecular clock hypothesis using mammalian DNA sequences. Journal of Molecular Evolution 25:330–342.

Liepe, J., P. Kirk, S. Filippi, T. Toni, C. P. Barnes, and M. P. H. Stumpf. 2014. A framework for parameter estimation and model selection from experimental data in systems biology using approximate Bayesian computation. Nature Protocols 9:439–456.

Lin, J. E., J. J. Hard, K. A. Naish, D. Peterson, R. Hillborn, and L. Hauser. 2016. It's a bear market: Evolutionary and ecological effects of predation on two wild sockeye salmon populations. Heredity 116:447–457.

Liu, L. 2008. BEST: Bayesian estimation of species trees under the coalescent model. Bioinformatics 24:2542–2543.Lombaert, E., T. Guillemaud, C. E. Thomas, L. J. Lawson Handley, J. Li, S. Wang, H. Pang, I. Goryacheva, I. A. Zakharov, E. Jousselin, R. L. Poland, A. Migeon, J. van Lentern, P. De Clercq, N. Berkvens, W. Jones, and A. Estoup. 2011. Inferring the origin of populations of introduced from a genetically structured native range by approximate Bayesian computation: Case study of the invasive ladybird Harmonia axyridis. Molecular Ecology 20:4654–4670.

Lopes, J. S., D. Balding, and M. A. Beaumont. 2009. PopABC: A program to infer historical demographic parameters. Bioinformatics 25:2747–2749.

Louis, M., A. Viricel, T. Lucas, H. Peltier, E. Alfonsi, S. Berrow, A. Brownlow, P. Covelo, W. Dabin, R. Deaville, R. de Stephanis, F. Gally, P. Gauffier, R. Penrose, M. A. Silva, C. Guinet, and B. Simon-Bouhet. 2014. Habitat-driven population structure of bottlenose dolphins, Tursiops truncatus, in the north-east Atlantic. Molecular Ecology 23:857–874.

Luikart, G., N. Ryman, D. A. Tallmon, M. K. Schwartz, and F. W. Allendorf. 2010. Estimation of census and effective population sizes: The increasing usefulness of DNA-based approaches. Conservation Genetics 11:355–373.

Lutz-Carrillo, D. J., C. C. Nice, T. H. Bonner, M. R. J. Forstner, and L. T. Fries. 2006. Admixture analysis of Florida largemouth bass and northern largemouth bass using microsatellite loci. Transactions of the American Fisheries Society 135:779–791.

Manel, S., O. E. Gaggiotti, and R. S. Waples. 2005. Assignment methods: Matching biological questions with appropriate techniques. Trends in Ecology & Evolution 20:136–142.

Manolopoulou, I., and B. C. Emerson. 2012. Phylogeographic ancestral inference using the coalescent model on haplotype trees. Journal of Computational Biology 19:745–755.

Manolopoulou, I., L. Legarreta, B. C. Emerson, S. Brooks, and S. Tavaré. 2011. A Bayesian approach to phylogeographic clustering. Interface Focus 1:909–921.

Manthey, J. D., J. Klicka, and G. M. Spellman. 2011. Cryptic diversity in a widespread North American songbird: Phylogeography of the brown creeper (Certhia americana). Molecular Phylogenetics and Evolution 58:502–512.

Manthey, J. D., M. Tollis, A. R. Lemmon, E. M. Lemmon, and S. Boissinot. 2016. Diversification in wild populations of the model organism Anolis carolinensis: A genome-wide phylogeographic investigation. Ecology and Evolution 6:8115–8125.

Marin, J.-M., N. S. Pillai, C. P. Robert, and J. Rousseau. 2014. Relevant statistics for Bayesian model choice. Journal of Royal Statistical Society B 76:833–859.

Marjoram, P., J. Molitor, V. Plagnol, and S. Tavaré. 2003. Markov chain Monte Carlo without likelihoods. Proceeding of the National Academy of Science of the United States of America 100:15324–15328.

Marjoram, P., and S. Tavaré. 2006. Modern computational approaches for analysing molecular genetic variation data. Nature Reviews Genetics 7:759–770.

Martin, A. P., and S. R. Palumbi. 1993. Body size, metabolic rate, generation time, and the molecular clock. Proceedings of the National Academy of Science of the United States of America 90:4087–4091.

Mayr, E. 1982. The growth of biological thought: Diversity, evolution, and inheritance. Harvard University Press, Cambridge, MA.

Mayr, E. 2001. What evolution is. Basic Books, New York.

Mayr, G. 2005. The Paleogene fossil record of birds in Europe. Biological Reviews 80:515–542.

Mayr, G. 2008. The fossil record of galliform birds: Comments on Crowe et al. (2006). Cladistics 24:74–76.

Mayr, G. 2009. Paleogene fossil birds. Springer, Berli.

McFadden, K. W., M. E. Gompper. D. G. Valenzeuela, and J. C. Morales. 2008. Evolutionary history of the critically endangered Cozumel dwarf carnivores inferred from mitochondrial DNA analyses. Journal of Zoology 276:176–186.

Metropolis, N., A. W. Rosenbluth, M. N. Rosenbluth, and A. H. Teller. 1953. Equation of state calculations by fast computing machines. Journal of Chemical Physics 21:1087–1092.

Michener, C. D., and R. R. Sokal. 1957. A quantitative approach to a problem in classification. Evolution 11:130–162.

Miller, K. S. 2014. Landscape genetics of northern bobwhite in Texas and the Great Plains. Ph.D. dissertation, Texas A&M University-Kingsville, Kingsville, Texas, USA.

Minin, V. N., E. W. Bloomquist, and M. A. Suchard. 2008. Smooth skyride through a rough skyline: Bayesian coalescent-based inference of population dynamics. Molecular Biology and Evolution 25:1459–1471.

Mullis, K., and F. Faloona. 1987. Specific synthesis of DNA in vitro via a polymerase catalyzed chain reaction. Methods in Enzymology 155:335–350.

Mullis, K., F. Faloona, S. Scharf, R. Saiki, G. Horn, and H. Erlich. 1986. Specific enzymatic amplification of DNA in vitro: The polymerase chain reaction. Cold Spring Harbor Symposium on Quantitative Biology 51:263–273.

Nater, A., M. P. Greminger, N. Arora, C. P. van Schaik, B. Goossens, I. Singleton, E. J. Verschoor, K. S. Warren, and M. Krützen. 2015. Reconstructing the demographic history of orang-utans using approximate Bayesian computation. Molecular Ecology 24:310–327.

Near, T. J., D. I. Bolnick, and P. C. Wainwright. 2004. Investigating phylogenetic relationships of sunfishes and black basses (Actinopterygii: Centrarchidae) using DNA sequences from mitochondrial and nuclear genes. Molecular Phylogenetics and Evolution 32:344–357.

Near, T. J., D. I. Bolnick, and P. C. Wainwright. 2005a. Fossil calibrations and molecular divergence time estimates in centrarchid fishes (Teleostei: Centrarchidae). Evolution 59:1768–1782.

Near, T. J., T. W. Kassler, J. B. Koppelman, C. B. Dillman, and D. P. Phillip. 2003. Speciation in North American black basses, Micropterus (Actinopterygii: Centrarchidae). Evolution 57:1610–1621.

Near, T. J., P. A. Meylan, and H. B. Shaffer. 2005b. Assessing concordance of fossil calibration points in molecular clock studies: An example using turtles. American Naturalist 165:137–146.

Nei, M. 1987. Molecular evolutionary genetics. Columbia University Press, New York.

Nei, M., and S. Kumar. 2000. Molecular evolution and phylogenetics. Oxford University Press, Oxford.

Nei, M., S. Kumar, and K. Takahashi. 1998. The optimization principle in phylogenetic analysis tends to give incorrect topologies when the number of nucleotides or amino acids used is small. Proceedings of the National Academy of Science of the United States of America 95:12390–12937.

Neophytou, C. 2014. Bayesian clustering analyses for genetic assignment and study of hybridization in oaks: Effects of asymmetric phylogenies and asymmetric sampling schemes. Trees Genetics & Genomes 10:273–285.

Newton, M. A., and A. E. Raftery. 1994. Approximate Bayesian inference with the weighted likelihood bootstrap. Journal of the Royal Statistical Society, Series B (Methodological) 56:3–48.

Neyman, J. 1971. Molecular studies of evolution: A source of novel statistical problems. In Statistical decision theory and related topics, edited by S. S. Gupta and J. Yackel, pp. 1–27. Academic Press, New York.

Neyman, J., and E. S. Pearson. 1933. On the problem of the most efficient tests of statistical hypotheses. Philosophical Transactions of the Royal Society of London A 231:289–337.

Noor, M. 2013. The 2013 Novitski prize: Jonathan Pritchard. Genetics 194:15–17.

Nunes, M. A., and D. J. Balding. 2010. On optimal selection of summary statistics for approximate Bayesian computation. Statistical Applications in Genetics and Molecular Biology 9:1544–6115.

Nylander, J. A. A., F. Ronquist, J. P. Huelsenbeck, and J. Nieves-Aldrey. 2004. Bayesian phylogenetic analysis of combined data. Systematic Biology 53:47–67.

Nylander, J. A. A., J. C. Wilgenbusch, D. L. Warren, and D. L. Swofford. 2008. AWTY (are we there yet?): A system for graphical exploration of MCMC convergence in Bayesian phylogenetics. Bioinformatics 24:581–583.

Oaks, J. R. 2011. A time-calibrated species tree of Crocodylia reveals recent radiation of the true crocodiles. Evolution 65:3285–3297.

O'Brien, J., S. Devillard, L. Say, H. Vanthomme, F. Léger, S. Ruette, and D. Pontier. 2009. Preserving genetic integrity in a hybridizing world: Are European wildcats (Felis silvestris silvesirls) in eastern France distinct from sympatric feral domestic cats? Biodiversity and Conservation 18:2351–2360.

Ogata, Y. 1989. A Monte Carlo method for high dimensional integration. Numerical Mathematics 55:137–157.

Olson, B., and W. Kleiber. 2017. Approximate Bayesian computation methods for daily spatiotemporal precipitation occurrence simulation. Water Resources Research 53:3352–3372.

Opgen-Rhein, R., L. Fahrmeir, and K. Strimmer. 2005. Inference of demographic history from genealogical trees using reversible jump Markov chain Monte Carlo. BMC Evolutionary Biology 5:6.

O'Reilly, J. E., M. dos Reis, and P. C. J. Donoghue. 2015. Dating tips for divergence time estimation. Trends in Genetics 31:637–650.

Oyler-McCance, S. J., R. W. DeYoung, J. A. Fike, C. A. Hagen, J. A. Johnson, L. C. Larsson, and M. A. Patten. 2016. Rangewide genetic analysis of lesser prairie-chicken reveals population structure, range expansion, and possible introgression. Conservation Genetics 17:643–660.

Paetkau, D., R. Slade, M. Burden, and A. Estoup. 2004. Genetic assignment methods for the direct, real-time estimation of migration rate: A simulation-based exploration of accuracy and power. Molecular Ecology 13:55–65.

Paradis, E., J. Claude, and K. Strimmer. 2004. APE: analyses of phylogenetics and evolution in R language. Bioinformatics 20:289–290.

Parham, J. F., P. C. J. Donoghue, C. J. Bell, T. D. Calway, J. J. Head, P. A. Holroyd, J. G. Inoue, R. B. Irmis, W. G. Joyce. D. T. Ksepka, J. S. L. Patané, N. D. Smith, J. E. Tarver, M. van Tuinen, Z. Yang, K. D. Angielczyk, J. M. Greenwood, C. A. Hipsley, L. Jacobs, P. J. Makovicky, J. Müller, K. T. Smith, J. M. Theodor, R. C. M. Warnock, and M. J. Benton. 2012. Best practices for justifying fossil calibrations. Systematic Biology 61:346–359.

Pascual, M., M. P. Chapuis, F. Mestre, J. Balanyá, R. B. Huey, G. W. Gilchrist, L. Serra, and A. Estoup. 2007. Introduction history of Drosophila subobscura in the New World: A microsatellite-based survey using ABC methods. Molecular Ecology 16:3069–3083.

Pavlidis, P., S. Laurent, and W. Stephan. 2010. msABC: A modification of Hudson's ms to facilitate multi-locus ABC analysis. Molecular Ecology Resources 10:723–727.

Pelletier, T. A., and B. C. Carstens. 2014. Model choice for phylogeographic inference using a large set of models. Molecular Ecology 23:3028–3043.

Pereira, S. L., and A. J. Baker. 2006. A mitogenomic timescale for birds detects variable phylogenetic rats of molecular evolution and refutes the standard molecular clock. Molecular Biology and Evolution 23:1731–1740.

Pereira, S. L., K. P. Johnson, D. H. Clayton, and A. J. Baker. 2007. Mitochondrial and nuclear DNA sequences support a Cretaceous origin of Columbiformes and a dispersal-driven radiation in the Paleogene. Systematic Biology 56:656–672.

Peterson, B. K., J. N. Weber, E. H. Kay, H. S. Fisher, and H. E. Hoekstra. 2012. Double digest RADseq: An inexpensive method for de novo SNP discovery and genotyping in model and non-model species. PLoS ONE 7(5): e37135. doi:10.1371/journal.pone.0037135.

Phillipsen, I. C., W. C. Funk, E. A. Hoffman, K. J. Monsen, and M. S. Blouin. 2011. Comparative analyses of effective population size within and among species: Ranid frogs as a case study. Evolution 65:2927–2945.

Plackett, R. L. 1972. Studies in the history of probability and statistics. XXIX: The discovery of the method of least squares. Biometrika 59:239–251.

Prangle, D., P. Fearnhead, M. Cox, P. Biggs, and N. French. 2014. Semi-automatic selection of summary statistics for ABC model choice. Statistical Applications in Genetics and Molecular Biology 13:67–82.

Pritchard, J. K., M. T. Seielstad, A. Perez-Lezaun, and M. W. Feldman. 1999. Population growth of human Y chromosomes: A study of Y chromosome microsatellites. Molecular Biology and Evolution 16:1791–1798.

Pritchard, J. K., M. Stephens, and P. Donnelly. 2000. Inference of population structure using multilocus genotype data. Genetics 155:945–959.

Pritchard, J. K., X. Wen, and D. Falush. 2010. Documentation for structure software: Version 2.3.

Pruett, C. L., J. A. Johnson, L. C. Larsson, D. H. Wolfe, and M. A. Patten. 2011. Low effective population size and survivorship in a grassland grouse. Conservation Genetics 12:1205–1214.

Prum, R. O., J. S. Berv, A. Dornburg, D. J. Field, J. P. Townsend, E. M. Lemmon, and A. R. Lemmon. 2015. A comprehensive phylogeny of birds (Aves) using targeted next-generation DNA sequencing. Nature 256:569–577.

Puechmaille, S. J. 2016. The program STRUCTURE does not reliably recover the correct population structure when sampling is uneven: Subsampling and new estimators alleviate the problem. Molecular Ecology Resources 16:608–627.

Pulquério, M. J. F., and R. A. Nichols. 2007. Dates from the molecular clock: How wrong can we be? Trends in Ecology and Evolution 22:180–184.

Puttick, M. N., G. H. Thomas, and M. J. Benton. 2016. Dating placentalia: Morphological clocks fail to close the molecular fossil gap. Evolution 70:873–886.

Pybus, O. G., and A. Rambaut. 2002. GENIE: Estimating demographic history from molecular phylogenies. Bioinformatics 18:1404–1405.

Pybus, O. G., A. Rambaut, and P. H. Harvey. 2000. An integrated framework for the inference of viral population history from reconstructed genealogies. Genetics 155:1429–1437.

Pyron, R. A., and F. T. Burbrink. 2009. Neogene diversification and taxonomic stability in the snake tribe Lampropeltini (Serpentes: Colubridae). Molecular Phylogenetics and Evolution 52:524–529.

Raftery, A., M. Newton, J. Satagopan, J. M. Satagopan, and P. N. Krivitsky. 2007. Posterior simulation using the harmonic mean identity. Bayesian Statistics 9:1–45.

Rambaut, A., S. Y. W. Ho, A. J. Drummond, and B. Shapiro. 2009. Accommodating the effect of ancient DNA damage on inferences of demographic histories. Molecular Biology and Evolution 26:245–248.

Ramos-Onsins, S. E., and J. Rozas. 2002. Statistical properties of new neutrality tests against population growth. Molecular Biology and Evolution 19:2092–2100.

Randle, C. P., M. E. Mort, and D. J. Crawford. 2005. Bayesian inference of phylogenetics revisited: Developments and concerns. Taxon 54:9–15.

Rannala, B. 2002. Identifiability of parameters in MCMC Bayesian inference of phylogeny. Systematic Biology 51:754–760.

Rannala, B., and Z. Yang. 1996. Probability distribution of molecular evolutionary trees: A new method of inference. Journal of Molecular Evolution 43:304–311.

Richmond, J. Q., A. R. Backlin, P. J. Tatarian, B. G. Solvesky, and R. N. Fisher. 2014. Population declines lead to replicate patterns of internal range structure at the tips of the distribution of the California red-legged frog (*Rana draytonii*). Biological Conservation 172:128–137.

Rieux, A., and F. Balloux. 2016. Inferences from tip-calibrated phylogenies: A review and a practical guide. Molecular Ecology 25:1911–1924.

Robert, C. P. 2016. Approximate Bayesian computation: A survey on recent results. In Monte Carlo and quasi–Monte Carlo methods, edited by R. Cools and D. Nuyens, pp. 185–205. Springer, Leuven, Belgium.

Robert, C. P., J.-M. Cornuet, J.-M. Marin, and N. S. Pillai. 2011. Lack of confidence in approximate Bayesian computation model choice. Proceedings of the National Academy of Science of the United States of America 108:15112–15117.

Ronquist, F., and J. P. Huelsenbeck. 2003. MrBayes 3, Bayesian phylogenetic inference under mixed models. Bioinformatics 19:1572–1574.

Ronquist, F., J. Huelsenbeck, and M. Teslenko. 2011. MrBayes version 3.2 manual: Tutorials and model summaries. https://www.mrbayes.sourceforge.net/mb3.2_manual.pdf.

Ronquist, F., S. Klopfstein, L. Vilhelmsen, S. Schulmeister, D. L. Murray, and A. R. Rasnitsyn. 2012a. A total-evidence approach to dating with fossils, applied to the early radiation of the Hymenoptera. Systematic Biology 61:973–999.

Ronquist, F., M. Teslenko, P. van der Mark, D. L. Ayres, A. Darling, S. Höhna, B. Larget, L. Liu, M. A. Suchard, and J. P. Huelsenbeck. 2012b. MrBayes 3.2: Efficient Bayesian phylogenetic inference and model choice across a large model space. Systematic Biology 61:539–542.

Ronquist, F., P. van der Mark, and J. P. Huelsenbeck. 2009a. Bayesian phylogenetic analysis using MrBayes: Theory. In The phylogenetic handbook: A practical approach to phylogenetic analysis and hypothesis testing, edited by P. Lemey, M. Salemi, and A.-M. Vandamme, pp. 210–236. Cambridge University Press, Cambridge, UK.

Ronquist, F., P. van der Mark, and J. P. Huelsenbeck. 2009b. Bayesian phylogenetic analysis using MrBayes: Practice. The phylogenetic handbook: A practical approach to phylogenetic analysis and hypothesis testing, edited by P. Lemey, M. Salemi, and A.-M. Vandamme, pp. 237–266. Cambridge University Press, Cambridge, UK.

Row, J. R., R. J. Brooks, C. A. MacKinnon, A. Lawson, B. I. Crother, M. White, and S. C. Lougheed. 2011. Approximate Bayesian computation reveals factors that influence genetic diversity and population structure of foxsnakes. Journal of Evolutionary Biology 24:2364–2377.

Rubin, D. B. 1984. Bayesianly justifiable and relevant frequency calculations for the applied statistician. Annals of Statistics 12:1151–1172.

Safner, T. M. P. Miller, B. H. McRae, M.-J. Fortin, and S. Manel. 2011. Comparison of Bayesian clustering and edge detection methods for inferring boundaries in landscape genetics. International Journal of Molecular Sciences 12:865–889.

Sagan, L. 1967. On the origin of mitosing cells. Journal of Theoretical Biology 14:225–274.

Saitou, N., and M. Nei. 1987. The neighbor-joining method: A new method for reconstructing phylogenetic trees. Molecular Biology and Evolution 4:406–425.

Sanger, F., S. Nicklen, and A. R. Coulson. 1977. DNA sequencing with chain-terminating inhibitors. Proceeding of National Academy of Science of the United States of America 74:5463–5467.

Sauquet, H., S. Y. W. Ho, M. A. Gandolfo, G. J. Jordan, P. Wilf, D. J. Cantrill, M. J. Bayly, L. Bromham, G. K. Brown, R. J. Carpenter, D. M. Lee, D. J. Murphy, J. M. K. Sniderman, and F. Udovicic. 2012. Testing the impact of calibration on molecular divergence times using a fossil-rich group: The case of *Nothofagus* (Fagales). Systematic Biology 61:289–313.

Schenk, J. J. 2016. Consequences of secondary calibrations on divergence time estimates. PLoS ONE 11(1):e0148228.

Schwarz, G. E. 1978. Estimating the dimension of a model. Annals of Statistics 6:461–464.

Schwartz, M. K., and K. S. McKelvey. 2009. Why sampling scheme matters: The effect of sampling scheme on landscape genetic results. Conservation Genetics 10:441–452.

Segelbacher, G., S. A. Cushman, B. K. Epperson, M.-J. Fortin, O. Francois, O. J. Hardy, R. Holderegger, P. Taberlet, L. P. Waits, and S. Manel. 2010. Applications of landscape genetics in conservation biology: concepts and challenges. Conservation Genetics 11:375–385.

Seidel, S. A., C. E. Comer, W. C. Conway, R. W. DeYoung, J. B. Hardin, and G. E. Calkins. 2013. Influence of translocations on eastern wild turkey population genetics. Journal of Wildlife Management 77:1221–1231.

Shafer, A. B. A., L. M. Gattepaille, R. E. A. Stewart, and J. B. W. Wolf. 2015. Demographic inferences using short-read genomic data in an approximate Bayesian computation framework: In silico evaluation of power, biases and proof of concept in Atlantic walrus. Molecular Ecology 24:328–345.

Shapiro, B., A. J. Drummond, A. Rambaut, M. C. Wilson, P. E. Matheus, A. V. Sher, O. G. Pybus, M. T. P. Gilbert, I. Barnes, J. Binladen, E. Willerslev, A. J. Hansen, G. F. Baryshnikov, J. A. Burns, S. Davydov, J. C. Driver, D. g. Froese, C. R. Harington, G. Keddie, P. Kosintsev, M. L. Kunz, L. D. Martin, R. O. Stephenson, J. Storer, S. Zimov, and A. Cooper. 2004. Rise and fall of the Beringian steppe bison. Science 306:1561–1565.

Shoemaker, J. S., I. S. Painter, and B. S. Weir. 1999. Bayesian statistics in genetics. Trends in Genetics 15:354–358.

Simon, D., and B. Larget. 1998. Bayesian analysis in molecular biology and evolution (BAMBE), version 1.01 beta. Department of Mathematics and Computer Science, Duquesne University, Pittsburgh, PA.

Simmons, M. P., K. M. Pickett, and M. Miya. 2004. How meaningful are Bayesian support values? Molecular Biology and Evolution 21:188–199.

Simpson, G. G. 1961. Principles of animal taxonomy. Columbia University Press, New York.

Sinsheimer, J. S., J. A. Lake, and R. J. Little. 1996. Bayesian hypothesis testing of four-taxon topologies using molecular sequence data. Biometrics 52:193–210.

Sisson, S. A., Y. Fan, and M. M. Tanaka. 2007. Sequential Monte Carlo without likelihoods. Proceedings of the National Academy of Science of the United States of America 104:1760–1765.

Sisson, S. A., Y. Fan, and M. M. Tanaka. 2009. Sequential Monte Carlo without likelihoods. Erratum. Proceedings of the National Academy of Science of the United States of America 106:16889.

Skrbinšek, T., M. Jelenčič, L. Waits, I. Kos, K. Jerina, and P. Trontelj. 2012. Monitoring the effective population size of a brown bear (*Ursus arctos*) population using new single-sample approaches. Molecular Ecology 21:862–875.

Smith, A. B., and K. J. Peterson. 2002. Dating time of origin of major clades: Molecular clocks and the fossil record. Annual Reviews in Earth and Planetary Sciences 30:65–88.

Smith, B. T., P. Escalante, B. E. Hernández Baños, A. G. Navarro-Sigüenza, S. Rohwer, and J. Klicka. 2011. The role of historical and contemporary processes on phylogeographic structure and genetic diversity in the northern cardinal, *Cardinalis cardinalis*. BMC Evolutionary Biology 11:136.

Smouse, P. E., and W.-H. Li. 1987. Likelihood analysis of mitochondrial restriction-cleavage patterns for human-chimpanzee-gorilla trichotomy. Evolution 41:1162–1176.

Sneath, P. H. A., and R. R. Sokal. 1962. Numerical taxonomy. Nature 193:855–860.

Sober, E. 2002. Bayesianism–Its scope and limits. Bayes's theorem. In Proceedings of the British Academy, volume 113, edited by R. Swinburne, pp. 21–38. Oxford University Press, Oxford.

Sober, E. 2008. Evidence and evolution: The logic behind the science. Cambridge University Press, Cambridge, UK.

Sokal, R. R. 1963. The principles and practice of numerical taxonomy. Taxon 12:190–199.

Sokal, R. R. 1986. Phenetic taxonomy: Theory and methods. Annual Reviews in Ecology and Systematics 17:423–442.

Sokal, R. R., and P. H. A. Sneath. 1963. Numerical taxonomy. W. H. Freeman, San Francisco.

Sousa, V. C., M. Fritz, M. A. Beaumont, and L. Chikhi. 2009. Approximate Bayesian computation without summary statistics: The case of admixture. Genetics 181:1507–1519.

Spencer, P. B. S., and J. O. Hampton. 2005. Illegal translocation and genetic structure of feral pigs in western Australia. Journal of Wildlife Management 69:7–384.

Spiegelhalter, D. J., N. G. Best, B. P. Carlin, and A. van der Linde. 2002. Bayesian measures of model complexity and fit. Journal of the Royal Statistical Society B 64:583–639.

Springer, M. S. 1995. Molecular clocks and the incompleteness of the fossil record. Journal of Molecular Evolution 41:531–538.

Springer, M. S., E. C. Teeling, O. Madsen, M. J. Stanhopes, and W. W. de Jong. 2001. Integrated fossil and molecular data reconstruct bat echolocation. Proceedings of the National Academy of Science of the United States of America 98:6241–6246.

Stadler, T, D. Kühnert, S. Bonhoeffer, and A. J. Drummond. 2013. Birth-death skyline plot reveals temporal changes of epidemic spread in HIV and hepatitis C virus (HCV). Proceedings of the National Academy of Sciences of the United States of America 110:228–233.

Stanton, D. W. G., J. Hart, N. F. Kümpel, A. Vosper, S. Nixon, M. W. Bruford, J. G. Ewen, and J. Wang. 2015. Enhancing knowledge of an endangered and elusive species, the okapi, using non-invasive genetic techniques. Journal of Zoology 295:233–242.

Stauffer, H.B. 2008. Contemporary Bayesian and frequentist research methods for natural resource scientists. Wiley, Hoboken, NJ.

Stepien, C. A., D. J. Murphy, and R. M. Strange. 2007. Broad- to fine-scale population genetic patterning in the smallmouth bass, Micropterus dolomieu, across the Laurentian Great Lakes and beyond: An interplay of behavior and geography. Molecular Ecology 16:1605–1624.

Stocks, M., M. Siol, M. Lascoux, and S. De Mita. 2014. Amount of information needed for model choice in approximate Bayesian computation. PLoS ONE 9(6):e99581.

Stigler, S. M. 1981. Gauss and the invention of least squares. Annals of Statistics 9:465–474.

Stigler, S. M. 2010. Darwin, Galton and the statistical enlightenment. Journal of the Royal Statistical Society A 173:469–482.

Strimmer, K., and O. G. Pybus. 2001. Exploring the demographic history of DNA sequences using the generalized skyline plot. Molecular Biology and Evolution 18:2298–2305.

Suchard, M. A., P. Lemey, G. Baele, D. L. Ayres, A. J. Drummond, and A. Rambaut. 2018. Bayesian phylogenetic and phylodynamic data integration using BEAST 1.10. Virus Evolution 4(1):vey016.

Sullivan, J., and P. Joyce. 2005. Model selection in phylogenetics. Annual Review in Ecology, Evolution, and Systematics 36:445–466.

Sunnåker, A. G. Busetto, E. Numminen, J. Corander, M. Foll, and C. Dessimoz. 2013. Approximate Bayesian computation. PLoS Computational Biology 9:e1002803.

Suzuki, Y., G. V. Glazko, and M. Nei. 2002. Overcredibility of molecular phylogenies obtained by Bayesian phylogenetics. Proceedings of the National Academy of Science of the United States of America 99:16138–16143.

Tajima, F. 1989. Statistical method for testing the neutral mutation hypothesis by DNA polymorphism. Genetics 123:597–601.

Tajima, F. 1993. Simple methods for testing molecular clock hypothesis. Genetics 135:599–607.

Takahashi, K., and M. Nei. 2000. Efficiencies of fast algorithms of phylogenetic inference under the criteria of maximum parsimony, minimum evolution, and maximum likelihood when a large number of sequences are used. Molecular Biology and Evolution 17:1251–1258.

Takezaki, N., and T. Gojobori. 1999. Correct and incorrect vertebrate phylogenies obtained by the entire mitochondrial DNA sequences. Molecular Biology and Evolution 16:590–601.

Takezaki, N., A. Rzhetsky, and M. Nei. 1995. Phylogenetic test of the molecular and linearized trees. Molecular Biology and Evolution 12:823–833.

Tallmon, D. A., A. Koyuk, G. Luikart, and M. A. Beaumont. 2008. ONeSAMP: A program to estimate effective population size using approximate Bayesian computation. Molecular Ecology Resources 8:299–301.

Tavaré, S. 1984. Line-of-descent and genealogical processes, and their applications in population genetics models. Theoretical Population Biology 26:119–164.

Tavaré, S., D. J. Balding, R. C. Griffiths, and P. Donnelly. 1997. Inferring coalescence times from DNA sequence data. Genetics 145:505–518.

Taylor, D. J., and W. H. Piel. 2004. An assessment of accuracy, error, and conflict with support values from genome-scale phylogenetic data. Molecular Biology and Evolution 21:1534–1537.

Templeton, A. R. 2009. Statistical hypothesis testing in intraspecific phylogeography: Nested clade phylogeographical analysis vs. approximate Bayesian computation. Molecular Ecology 18:319–331.

Templeton, A. R. 2010. Correcting approximate Bayesian computation. Trends in Ecology and Evolution 25:488–489.

Theil, H. 1961. Economic forecasts and policy. North-Holland Publishing, Amsterdam, Netherlands.

Thorne, J. L., and H. Kishino. 2002. Divergence time and evolutionary rate estimation with multilocus data. Systematic Biology 51:689–702.

Thornton, K., and P. Andolfatto. 2006. Approximate Bayesian inference reveals evidence for a recent, severe bottleneck in a Netherlands population of Drosophila melanogaster. Genetics 172:1607–1619.

Tollis, M., G. Ausubel, D. Ghimire, and S. Boissniot. 2012. Multi-locus phylogeographic and population genetic analysis of Anolis carolinensis: Historical demography of a genomic model species. PLoS ONE 7(6):e38474.

Toni, T., D. Welch, N. Strelkowa, A. Ipsen, and M. P. H. Stumpf. 2009. Approximate Bayesian computation scheme for parameter inference and model selection in dynamical systems. Journal of the Royal Society Interface 6:187–202.

Tsutaya, T., and M. Yoneda. 2013. Quantitative reconstruction of weaning ages in archaeological human populations using bone collagen nitrogen isotope ratios and approximate Bayesian computation. PLoS ONE 8(8):e72327.

Turner, B. M., and T. Van Zandt. 2012. A tutorial on approximate Bayesian computation. Journal of Mathematical Psychology 56:69–85.

van Tuinen, M., and S. B. Hedges. 2004. The effect of external and internal fossil calibrations on the avian evolutionary timescale. Journal of Paleontology 78:45–50.

Verardi, A., V. Lucchini, and E. Randhi. 2006. Detecting introgressive hybridization between free-ranging domestic dogs and wild wolves (Canis lupus) by admixture linkage disequilibrium analysis. Molecular Ecology 15:2845–2855.

Wang, I. J., W. K. Savage, and H. B. Shaffer. 2009. Landscape genetics and least-cost path analysis reveal unexpected dispersal routes in the California tiger salamander (Ambystoma californiense). Molecular Ecology 18:1365–1374.

Wang, J. 2017. The computer program structure for assigning individuals to populations: Easy to use but easier to misuse. Molecular Ecology Resources 17:981–990.

Ward, P. S., S. G. Brady, B. L. Fisher, and T. R. Schultz. 2010. Phylogeny and biogeography of dolichoderine ants: Effects of data partitioning and relict taxa on historical inference. Systematic Biology 59:342–362.

Wegmann, D., C. Leuenberger, and L. Excoffier. 2009. Efficient approximate Bayesian computation coupled with Markov chain Monte Carlo without likelihood. Genetics 182:1207–1218.

Wegmann, D., C. Leuenberger, S. Neuenschwander, and L. Excoffier. 2010. ABCtoolbox: A versatile toolkit for approximate Bayesian computations. BMC Bioinformatics 11:116.

Weyant, A., C. Schafer, and W. M. Wood-Vasey. 2013. Likelihood-free cosmological inference with Type Ia supernovae: Approximate Bayesian computation for complete treatment of uncertainty. Astrophysical Journal 764:116.

White, G. C. 2001. Statistical models: Keys to understanding the natural world. In Modeling in natural resource management: Development, interpretation, and application, edited by T. M. Shenk and A. B. Franklin, pp. 35–56. Island Press, Washington, DC.

White, S. R., T. Kypraios, and S. P. Preston. 2015. Piecewise approximate Bayesian computation: Fast inference for discretely observed Markov models using a factorised posterior distribution. Statistical Computing 25:289–301.

Williford, D., R. W. DeYoung, R. L. Honeycutt, L. A. Brennan, and F. Hernández. 2016. Phylogeography of the bobwhite quails (Colinus). Wildlife Monographs 193:1–49.

Williford, D., R. W. DeYoung, R. L. Honeycutt, L. A. Brennan, F. Hernández, E. M. Wehland, J. P. Sands, S. J. DeMaso, K. S. Miller, and R. M. Perez. 2014. Contemporary genetic structure of the northern bobwhite west of the Mississippi River. Journal of Wildlife Management 78:914–929.

Wilson, A. C., S. S. Carlson, and T. J. White. 1977. Biochemical evolution. Annual Review in Biochemistry 46:573–639.

Wilson, G. A., and B. Rannala. 2003. Bayesian inference of recent migration rates using multilocus genotypes. Genetics 163:1177–1191.

Wilson, I. J., D. Balding, and M. Weale. 2002. BATWING user guide. University of Aberdeen, King's College, Aberdeen, UK.

Wilson, I. J., M. E. Weale, and D. J. Balding. 2003. Inferences from DNA data: Population histories, evolutionary processes and forensic match probabilities. Journal of the Royal Statistical Society A 166:155–201.

Wright, T. F., E. E. Schirtzinger, T. Matsumoto, J. R. Eberhard, G. R. Graves, J. J. Sanchez, S. Capelli, H. Müller, J. Scharpegge, G. K. Chambers, and R. C. Fleischer. 2008. A multilocus molecular phylogeny of the parrots (Psittaciformes): Support for a Gondwanan origin during the Cretaceous. Molecular Biology and Evolution 25:2141–2156.

Xie, W., P. O. Lewis, Y. Fan, L. Kuo, and M.-H. Chen. 2011. Improving marginal likelihood estimation for Bayesian phylogenetic model selection. Systematic Biology 60:150–160.

Yang, Z. 1997. PAML: A program package for phylogenetic analysis by maximum likelihood. CABIOS Applications Note 13:555–556.

Yang, Z. 2007. PAML 4: Phylogenetic analysis by maximum likelihood. Molecular Biology and Evolution 24:1586–1591.

Yang, Z., and B. Rannala. 1997. Bayesian phylogenetic inference using DNA sequences: A Markov chain Monte Carlo method. Molecular Biology and Evolution 14:717–724.

Yang, Z., and B. Rannala. 2012. Molecular phylogenetics: Principles and practice. Nature Reviews 13:303–314.

Yasukochi, Y., S. Nishida, S.-H. Han, T. Kurosaki, M. Yoneda, and H. Koike. 2009. Genetic structure of the

Asiatic black bear in Japan using mitochondrial DNA analysis. Journal of Heredity 100:297–308.

Yule, G. U. 1924. A mathematical theory of evolution, based on the conclusions of Dr. J. C. Willis. Philosophical Transactions of the Royal Society of London B 213:21–87.

Zamudio, K. R., and A. M. Wieczorek. 2007. Fine-scale spatial genetic structure and dispersal among spotted salamander (*Ambystoma maculatum*) breeding populations. Molecular Ecology 16:257–274.

Zhang, C., T. Stadler, S. Klopfstein, T. A. Heath, and F. Ronquist. 2016. Total-evidence dating under the fossilized birth-death process. Systematic Biology 65:228–249.

Zuckerkandl, E., and L. Pauling. 1962. Molecular disease, evolution, and genic heterogeneity. In Horizons in biochemistry, edited by M. Kasha and B. Pullman, pp. 189–225. Academic Press, New York.

Zuckerkandl, E., and L. Pauling. 1965 Evolutionary divergence and convergence in proteins. In Evolving genes and proteins, edited by V. Bryson, and H. J. Vogel, pp. 97–166. Academic Press, New York.

15

JANE ELITH

Machine Learning, Random Forests, and Boosted Regression Trees

> The goals in statistics are to use data to predict and to get information about the underlying data mechanism. Nowhere is it written on a stone tablet what kind of model should be used to solve problems involving data. . . . the emphasis needs to be on the problem and on the data.   —Breiman (2001b:214)

Introduction

Machine learning is the part of computer science that deals with algorithms that can learn from and make predictions based on data. Some statisticians have adopted and developed methods from machine learning and refer to these as *statistical learning methods* (Hastie et al 2009). Whether we call it machine learning or statistical learning, we are concerned in this chapter with learning from data. The focus is on *supervised learning* problems—ones where there is an outcome or response variable that can be used to guide the learning. These contrast with unsupervised learning problems, where there is no response variable, and the goal is to learn structure in the data (e.g., through clustering algorithms). Even if you have had no previous exposure to machine learning methods, you will already be familiar with supervised learning problems. Chapter 2 in this book describes regression methods, which are essentially supervised learning methods. They deal with learning about the relationships between a response variable(s) and one or more predictor variables or covariates.

Sometimes the language of a new field can be challenging, so in this chapter I provide definitions and clarifications as necessary. Whether you are a sci-entist or practitioner, it is well worthwhile to at least read about and be open to machine learning methods, because they are likely to become more widely used as computational biology and ecology degrees teach these methods. Even if you doubt their importance to your work, it is worth understanding what they can and can't do. The paper "Statistical modeling: The two cultures" by Leo Breiman (Breiman 2001b), a distinguished statistician, provides an interesting place to start thinking about machine learning methods and how they contrast with traditional statistical models. Breiman presents compelling arguments for viewing machine learning methods as a necessary part of the toolbox of a statistician, because they enhance an analyst's capacity to use data to solve problems. For an accessible introduction to statistical learning methods written specifically for those outside the machine learning field, and including code and data for practice, see James et al. (2013).

Machine learning analyses can deal with many types of data. Here I will use the same notation as in Chapter 2 for the dependent (or response) variable (Y), and the independent (or predictor or explanatory) variables (X). Machine learning names for these terms include *labels* for the response and *features* for the predictor variables. Machine learning methods are available for a variety of response variables (e.g. continuous, ordered, or categorical), and hence can be used for regression and classification problems. Often the focus in machine learning applications is on prediction, but this does not mean that the model has to be treated as a black box. It is usually possible to extract summaries that promote understanding—for instance, regarding the main variables affecting the distribution of a species, or the shape of the species response to those variables. In other words, these models can be used for exploration of data and description of their structure; they can also be used for prediction to sites where the response has not been observed. In that way, they bear many similarities to statistical methods.

This chapter focuses on a small subset of the supervised learning methods, the *tree-based methods*, and in particular *random forests* and *boosted regression trees*, which are increasingly used in ecology. These are interesting in their own right, and are also useful as a vehicle for showing typical facets of a machine learning approach to analysis. First, though, an overview of model tuning and complexity is presented, because this is a key focus in machine learning.

Model Complexity, Training and Testing, and Cross-Validation

Many machine learning methods can fit very complex models—in fact, so complex that they can completely overfit the data, meaning that they fail to generalize well (Fig. 15.1). A model's complexity increases with the number of predictors, the complexity (or flexibility) of the permitted relationship between the response and a predictor, and the number of interactions (Chapter 2) between variables (Merow et al. 2014). Because the tree-based methods can easily handle many predictors and high-order interactions, and often fit complex responses, part of the art of using them is to control the model complexity. This is usually done by tuning a model so it predicts well to new data.

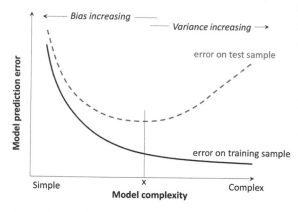

Fig. 15.1. Prediction error on test and training samples, as a function of model complexity.

The data available for modeling can be viewed as *training* data for fitting the model and *testing* (or validation or hold-out) data for evaluating the model. The train and test datasets are most commonly resampled subsets of one initial dataset, though in rare cases (e.g., Elith et al. 2006) they are from completely different surveys. James et al. (2013) provide an overview of resampling; here I will focus on the popular method of *cross-validation*. In k-fold cross-validation, the data set with N observations is first randomly split into k mutually exclusive groups (folds) of approximately equal size N/k. Typically k is in the range 4 to 10; if $k = N$, it is known as "leave one out cross-validation." For the first iteration one group is used as testing data and the remaining $k-1$ groups are used as the training sample. The process is then repeated until k iterations are completed, and over these, all data are used once for testing and $k-1$ times for training. The estimate of predictive performance on the testing data can be calculated at each iteration and then averaged, or calculated once on the saved combined set of predictions and observations. Calculating test statistics at each iteration has the advantage of enabling an assessment of the variability of the results (e.g., providing a mean and standard error).

Another common form of cross-validation is random repeated cross-validation. In this, for one iteration the data are split once into a training and a testing dataset, and the model is fitted and tested; this is repeated multiple times using a different random split each time, and average performance is estimated over all repeats. Variants of these two approaches allow nonrandom allocation to train and test datasets. For instance, for binary data, the proportions of the positive and negative case can be kept stable across the allocation ("prevalence-stratified"), or the test data might be kept separate spatially or temporally (see Roberts et al. [2017] for several ecological examples, and Valavi et al. [2019] for a relevant software package).

A further nuance with training and testing datasets is that machine learning practitioners often favor a three-way split of data if data are plentiful: two portions of data for training and validating the model (to enable thorough tuning of the model, with one portion used to fit a model under specific tuning parameters and the other to validate the performance of those parameters), then a third for an independent test of performance, using the finally selected optimal tuning of the model (Hastie et al. 2009). This three-way split keeps the evaluation of the final model completely independent of its tuning and is the best estimate of predictive performance to new sites, if sufficient data are available.

The idea in tuning a model is to allow enough complexity to fit the training data reasonably well, but not so much that it is very specific to that dataset, with no generality for predicting. Fig. 15.2 shows the effect of modeling a quadratic relationship $(y \sim x^2)$ too simply as a linear relationship, or with too much complexity, fitting the particular sample too tightly. The mapped predictions show the impact of these errors. This concept of fit versus generality is often described as the *bias-variance trade-off* (Fig. 15.1; and see James et al. [2013: chap. 2] for detail). If a modeling method has high variance, its estimate of the relationship between Y and X changes nontrivially each time a new training dataset is used. Flexible methods (ones that fit potentially complex response shapes) tend to have higher variance. Bias is the error introduced by using a model to approximate an observed phenomenon.

All models are biased to some extent, and simpler models, ones that estimate less flexible responses, tend to have higher bias than more complex ones. As the complexity of a model is increased, the bias decreases and the variance increases, but at different rates. When prediction error is measured on both training and test data we see that there is a point at which the model becomes too complex to predict well to new data—its variance has increased too much (to the right of the point x in Fig. 15.1). The goal of model tuning is to find the point (x in Fig. 15.1) at which the model is a reasonable representation of the data (compared with its tightest possible fit), and yet the model is simple enough to predict as well as it can to test data. Thinking about this trade-off, and using resampling methods to quantify it, is common for machine learning approaches. But the same idea of balancing bias and variance underpins approaches used commonly for model selection in statistical methods, including Akaike's Information Criterion (AIC; see Chapter 4 in this volume), and penalized regression methods, such as ridge regression and the lasso (Hastie et al. 2009). One word used commonly in relation to complexity of fit is *regularization*, which refers to tuning the model to an optimal level of complexity. Different regularization techniques are available, and often a specific one is implemented for a specific modeling method (Hastie et al. 2009).

Classification and Regression Trees

Decision trees are the basic unit in tree-based analyses. They are a broad class of models that can be used for decision analysis (where they may simply represent knowledge or thoughts) or for data analysis. Here we will consider *classification and regression trees* (e.g., CART, Breiman et al. 1984), which are decision tree algorithms applied to data, particularly for response variables that are categorical (classification) or numeric (regression).

To illustrate decision trees, I use data from Buston and Elith's (2011) analysis of factors determining

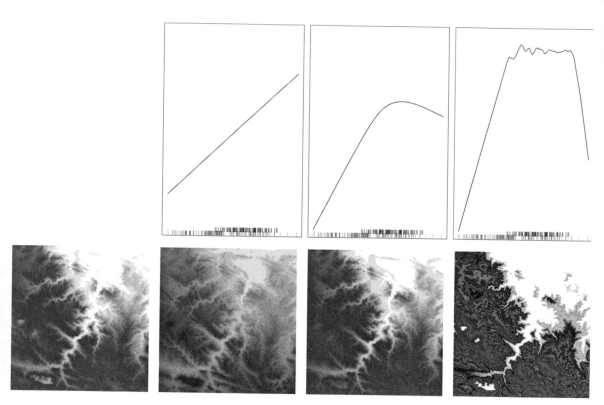

Fig. 15.2.   Simulated data (bottom left) for a simulated species with a quadratic relationship to standardized elevation (following Guillera-Arroita et al. 2015); values zero (white) to one (black). The top row shows generalized additive models with 1, 2, and 20 degrees of freedom in the smoothing splines fit to elevation, based on a sample of 250 presence/absence points. At the bottom of each, the two-row rug plots show the distribution of the sample for absences (grey, bottom row) and presences (black, top row). The y-axes show the predicted response on the scale of the linear predictor. Below each response function is the mapped prediction from that model.

clownfish reproductive success. Clownfish (*Amphiprion percula*) form close associations with the anemone *Heteractis magnifica*, living within an anemone as a single group comprised of a dominant breeding pair and zero to four nonbreeders. Within each group of fish there is a size-based dominance hierarchy, representing a queue for breeding positions. The female is largest, then the male, and nonbreeders are progressively smaller. If the breeding female dies, the male changes sex and assumes the dominant female position. The dataset was collected over one year in Madang Lagoon, Papua New Guinea (Buston 2002, 2004), with reproduction monitored every one to two days for 296 days. Here I use a subset of those data: namely, records for the number of eggs laid, for those fishes that laid. From the initial set of 31 covariates detailed in Buston and Elith (2011), here I use the six found to be important predictors of egg number. These are: the initial standard length (SL) for the dominant female (F.SL) and male (M.SL), change in mass over the year for the dominant female (F.d.mass) and change in length for the dominant male (M.d.SL), whether a given clutch of eggs was the pair's first, second, or n^{th} pair for the year (year clutch), and lunar month (with month 1 being the first month of monitoring). Buston and Elith (2011) discuss likely explanations for the importance of these variables from the viewpoint of the breeding capacity and parental involvement of the dominants. Over the 603 observations in this dataset, number of eggs laid was approximately normally distributed, with a mean of 361 and standard deviation of 187.

In the context of fitting a tree to data, a tree-based algorithm can be envisaged from two viewpoints. First, the *tree* (Fig. 15.3, right), with the root at the top and leaves at the bottom, is formed by repeatedly splitting the data, choosing a *split point*—a value of the predictor variable—that most effectively separates the response into mutually exclusive groups that are as homogenous as possible (De'ath and Fabricius 2000). For instance in Fig. 15.3 the first variable selected for splitting is M.d.SL and the split point is 1.6mm. All observations less than this go to one side (to the right in Fig. 15.3) and all greater than, to the other. There is only one split point each time a predictor is used in the tree, but any predictor can be used multiple times in a tree. Predictors can be numeric or categorical (i.e., factor) variables. For categorical variables, the levels (i.e., the categories) can be grouped in any combination to form two groups that enable the split; for numeric variables,

the split can be anywhere along the numeric scale. The first split forms the *root node*, splits within the tree are *internal nodes*, and the final groups are *terminal nodes*, or *leaves*.

The other way to think about this is from the viewpoint of the predictor space. Essentially, the repeated splitting is partitioning the predictor space into potentially high-dimensional rectangles or regions (e.g., Fig. 15.4). In the final model each region corresponds to the part of predictor space containing the leaves, and in each we make a prediction. Commonly that prediction is a constant—for instance, the mean of the observations.

The following discussion focuses on regression trees because regression analyses are common in ecology; classification will be addressed later. The number of clownfish eggs laid is normally distributed and suited to modeling with a regression tree. Say we have p predictor variables and n observations, so for the *ith* (response, predictor) observation (y_i, x_i), $x_i = (x_{i1}, x_{i2} \ldots x_{iP})$. As for all modeling approaches, a method is required for specifying or tuning the

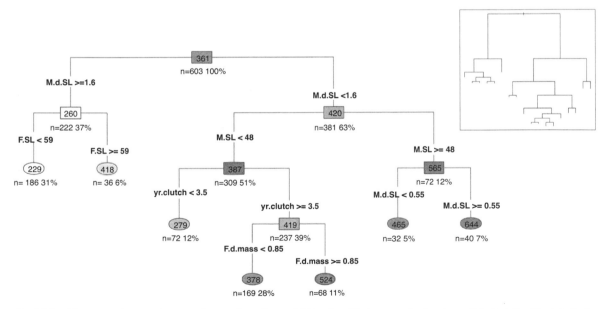

Fig. 15.3.   The pruned regression tree for number of eggs laid by *Amphiprion percula*, the clownfish (right) with detail of several of the earlier nodes (left). At left, rectangles are root and internal nodes; numbers represent the fitted number of eggs laid; terminal nodes are in ovals. Numbers below the nodes show number of observations (n =) per node and their proportion of the full dataset. Shading of nodes: darker with higher number of eggs laid. Variables and splits points shown in bold type.

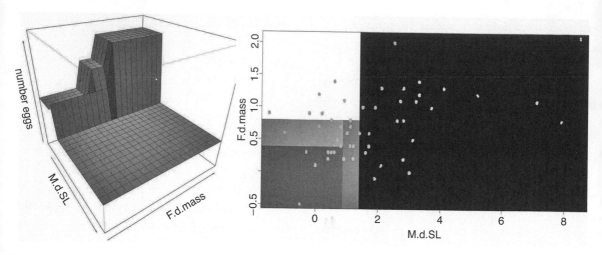

Fig. 15.4. Left: for the pruned regression tree of Fig. 15.3, a plot of the number of eggs laid (z axis) versus M.d.SL and F.d.mass, showing the response to these variables, with other variables held constant. The right image is the same data viewed from above (darker = lower number of eggs laid), with observations superimposed (pale gray circles).

model. Since the goal is to have a set of terminal nodes (say M regions: $R_1, R_2 \ldots R_M$) that are as homogeneous as possible, a measure of homogeneity (or conversely, of impurity) is needed. A common choice for regression trees is the residual sum of squares (RSS, equation 1 after James et al. [2013]), which should be minimized:

$$\sum_{m=1}^{M} \sum_{i \in R_m} (y_i - \hat{y}_{R_m})^2,$$

where $\hat{y}_{R_m}$ is the mean response for all observations in the training data within the mth region. As an aside, note that the response can be transformed before modeling—whether this is necessary depends on the selected measure of impurity and how the response varies in relation to mean values at the terminal nodes (De'ath and Fabricus 2000).

Model specification is a mix of *growing* and *pruning*. Growing is done with a top-down, greedy approach (James et al. 2013) known as *recursive binary partitioning*. It starts at the root node (i.e., at the "top" of the tree) and is greedy in the sense that at each step, the combination of predictor and split point that best minimizes node impurity is selected. This effectively looks only one step ahead—the next step

is optimized, without attempting to optimize the whole tree. It is *recursive* (as opposed to iterative) because the procedure repeatedly calls itself; i.e., the splitting process is always applied to the outputs of the splitting process. Each step involves evaluating all p predictors and all possible split points on each of the current set of terminal nodes, and choosing the set of predictor plus split point that results in the minimum impurity over the new set of terminal nodes. For instance, in Fig. 15.3, the first (predictor, split point) is (M.d.SL, 1.6mm) and the second is on the right branch of the first split, namely (M.SL, 48mm). The order of the splits—whether a node is the first, second, tenth, etc.—can be discovered from the fitted model object.

This is a fast procedure computationally, but when should it stop? A large tree might overfit the data, but a small tree might fail to explain important structure in the data. A common approach is to define, as a stopping criterion, the minimum number of observations allowed in any terminal node. The number can be set quite low. It is then likely that the tree T is too big, and pruning is used to reduce the tree's complexity. *Cost-complexity pruning* (Brieman et al. 1984) considers the nested set of subtrees with the same root as T, but produced by sequentially snipping off terminal subtrees (in other words, starting at the closest internal nodes to the leaves). Hastie et al. (2009) show how this works in detail; broadly speaking, the algorithm scans the set of subtrees and finds the one that minimizes the RSS for a selected penalty on complexity. To select the best-performing penalty, independent test data or cross-validation is used, then that selected penalty is used to dictate how T is pruned. In the clownfish example, the regression tree was fit and drawn using the *rpart, rpart.plot* and *plotmo* packages in R (https://www.r-project.org/). The final tree (right, Fig. 15.3) has 17 splits, pruned by cross-validation from the original 20 splits resulting from the model settings (minimum terminal node size of 20 observations). Fig. 15.5 shows the progression of training and cross-validation error for increasing tree size, and reinforces the notion that increases in complexity (here, number of nodes) can keep decreasing the error on the training data, but at some point the prediction to held-out data is compromised.

Classification trees are conceptually similar to regression trees in terms of structure, growing and pruning. However, the response is qualitative rather than quantitative; at terminal nodes the most commonly occurring class is predicted. Users are often interested, too, in the class probabilities at each terminal node; for these the class proportions are used. Different measures of impurity are available for analyzing the mix of classes in each node. Common criteria include the Gini index, cross-entropy (identical to the Shannon-Wiener index), or misclassification error (De'ath and Fabricius 2000; Hastie et al. 2009).

There are several advantages to decision trees. They are easily interpretable—the binary tree structure is a natural way both to think about how a response is affected by predictors and to present it to others. Interactions between variables do not have to be prespecified as in traditional regression models; the structure of a tree, where the predicted responses at the leaves depend on a sequence of splits on different predictors, is a natural way to fit an interaction. Fig. 15.4 (left panel) demonstrates an interaction. At high values of M.d.SL, F.d.mass has no effect on number of eggs laid. But at lower values, number of eggs laid increases with F.d.mass. Thus the level of M.d.SL affects the modeled relationship with F.d.mass. In decision trees, predictor variables need no transformation, because the splits are dependent only on rank order of the variable, so any monotonic transformation (such as a log or polynomial) makes no difference. Missing data in predictor variables are well handled. The regression methods described in Chapter 2 discard any sites for which predictors have missing data unless data imputation is used first. In decision trees, several methods exist for handling missing data (Hastie et al. 2009, De'ath and Fabricius 2000). The most general one involves creation of *surrogate* variables, which exploit correlations between predictors. First the "best" predictor and split point is found for all non-missing data. Then a list of surrogate predictors and corresponding split points is formed, with the performance of each being judged in relation to how well it mimics the best allocation of cases to its two resulting nodes. In this surrogate list the first variable is the

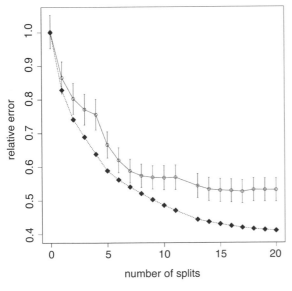

Fig. 15.5. Error (y axis) versus tree complexity (number of splits, x axis) as estimated on training data (filled diamonds and dotted lines) and 10-fold cross-validation (open circles and solid line, with standard errors). The optimal number of splits as estimated by cross-validation is 17.

most successful mimic and is used for all sites lacking data on the best predictor, provided that the data are not also missing for this first-listed surrogate. The list ensures a surrogate variable will be available for all observations. In the clownfish data there were 17 observations with missing data on two of the variables, yet the full dataset could be utilised in the model using this surrogate variables approach. This ability to use observations with missing predictor data is a valuable characteristic of trees, which guards against information loss.

A major disadvantage of trees is that they tend to be high variance methods, changing with each training dataset. They are not good at estimating smooth functions—this is mostly a weakness in regression settings, and leads to generally worse predictive performance than alternative regression methods. These and other problems (Hastie et al. 2009) led to exploration of their use in ensembles, as described in the following sections.

Ensembles of Trees

The idea of using ensembles of models—or collections of models—is not limited to machine learning approaches. Broadly speaking, ensembles are an approach for dealing with uncertainty about which is the true model and for overcoming bias and variance in component models. Examples are seen in weather forecasting and climate prediction. Intuitively, this is the "wisdom of the crowd": a group of people, especially if from diverse backgrounds, can overcome blind spots and biases in their knowledge if they come to a decision together. In machine learning the component models are viewed as *weak learners* or *base learners* or simple models, and the ensemble as a *strong learner* or *committee*. The ensemble can combine multiple models of the same algorithm (for instance, multiple classification trees) or multiple models from numerous different algorithms. *Weights* or *votes* need to be given to the component models: either each can have an equal "vote" in the committee, or some can be more influential. Ensembles are

less easily interpreted than their component simple models, but methods exist for summarizing important aspects of the ensemble. Ensembles can be used for both classification and regression, and numerous approaches exist for combining the base learners. The literature on these is large, with many nuances depending on the learning problem and the type of data that are available. See Dietterich (2000) and Witten et al (2016) for overviews. In the following sections two ensembling methods (bagging and boosting) are introduced, along with their specific use in the "random forests" and "boosted regression tree" algorithms now becoming well known in ecology. For these, I use as a case study data on the short-finned eel, *Anguilla australis*, in the rivers of New Zealand.

Within New Zealand, *A. australis* is a common freshwater species, frequenting lowland lakes, swamps, and sluggish streams and rivers in pastoral areas and forming a valuable traditional and commercial fishery. Short-finned eels take 10–20 years to mature then migrate to the sea to spawn, and the new generation reenters freshwaters as elvers (young eels) and migrate upstream. The full species dataset consisted of records of species caught from 13,369 sites spanning the major environmental gradients in New Zealand's rivers. *A. australis* was caught at approximately 20% of sites. In this chapter a random subsample from 1,000 of these sites is used, and the aim of analysis is to understand and map the distribution of *A. australis*. The explanatory variables are a set of 11 functionally relevant environmental predictors that summarize river conditions over several spatial scales: local (segment and reach-scale), upstream catchment-scale, and downstream to the sea. These help to define likely environmental correlates of suitable habitat and impediments to movements (such as very steep river sections, or dams). A twelfth variable describes fishing method. The data are available in the R package *dismo* (https://www.r-project.org/) and are described in a tutorial within *dismo* and in more detail in Elith et al. (2008).

Bagging and Random Forests

Ensembles can be constructed by creating many different samples of the training data and then fitting the base learners to each sample. A common approach is *bagging*, an acronym for *bootstrap aggregating* (Breiman 1996). On each iteration, from an original dataset with N observations we take a new sample *with replacement* of size N, called a *bootstrap sample*. On average a bootstrap sample contains 63.2% of the original training dataset, and has several observations repeated multiple times (Efron and Tibshirani 1993). The base learner is applied to each bootstrap sample. After B bootstrap samples (i.e., B iterations), B trees will have been built and used to predict an outcome at each observation in the sample. Then for each of the N observations the predictions are combined to make a single consensus prediction—e.g., by averaging for a numeric response or by taking the most commonly predicted class (the "majority vote") for classification. Bagging is effective for base learners that have high variance (i.e., are unstable) and low bias, like decision trees. It is less effective for—and in fact may degrade performance of—stable base learners. Breiman's paper introducing bagging (Breiman 1996)—an interesting read—concludes: "Bagging goes a ways toward making a silk purse out of a sow's ear, especially if the sow's ear is twitchy. It is a relatively easy way to improve an existing method, since all that needs adding is a loop in front that selects the bootstrap sample and sends it to the procedure and a back end that does the aggregation. What one loses, with the trees, is a simple and interpretable structure. What one gains is increased accuracy."

When taking a bootstrap sample, all the data *not* included in that sample (on average, 36.8% of the original dataset) can be identified and used as a holdout or "out-of-bag" (OOB) sample for estimating predictive error. The fitted model on any iteration is used to predict to the OOB sample. After the B iterations, the OOB predictions for each of the N observations are combined into a single consensus prediction as for the training samples. The error (e.g., the mean squared error for regression or classification error for classification) is then estimated. This use of OOB samples is an efficient and valid method for obtaining test samples, meaning that cross-validation is not required unless for other purposes. Bootstrap and OOB samples are used extensively in statistics and machine learning, well beyond bagging. Bagging is also used in many different ways (Witten et al. 2016).

Random forests (Breiman 2001) are a modification of bagging that perform well and are simple to tune. They have become very popular and are implemented in a range of software packages (Boulesteix et al. 2012). Building on the knowledge that bagging works best with dissimilar component models, the main idea in random forests is to build a set of de-correlated trees. For each bootstrapped dataset, a tree is grown, but not on the full set of p predictor variables. As each tree is grown, before *each split*, a subset z of the p predictor variables is randomly selected; z varies between 1 and p, but is typically small—e.g., the originators recommend $\sqrt{p}$ for classification and $p/3$ for regression (both rounded down to the nearest integer; Hastie et al. 2009). In fact, if $z = p$, the random forest would be identical to bagging. From the subset of z variables, the best variable and split point is chosen. Even though it might seem that ignoring many of the predictors is likely to increase variance in the final prediction, it doesn't. Using a subset of predictors is successful because bagging is most effective if the component trees are dissimilar. Within a set of predictor variables, one or two might be the most important predictors of the outcome—hence, if $z = p$ they will tend to be selected repeatedly across bootstrapped samples, and the set of trees will be highly correlated. By using $z \ll p$ and forcing other predictors to be considered at each split, the final set of trees are de-correlated, and the ensemble of them is more reliable. The trees are often grown deep (many splits) and are not pruned, because the aim is to produce trees with low bias. The bias of the random forest ensemble of trees is the same as the bias of the component trees (Hastie et al. 2009); the improved performance of the ensemble is entirely related to the reduction in variance compared with single trees.

In tuning random forests, the main decision is the value of z. While there are recommended defaults, it is best to test a range of values. For instance, when fitting a random forest model using the R package *randomForest* to the short-finned eel data where $p = 12$, I used the OOB error estimates to evaluate the effects of varying z between 1 and 9, over a range of ensemble sizes (e.g., Fig. 15.6). The results varied noticeably from run to run, but overall a midrange z reduced OOB error most consistently. As an aside, note here that I have used the OOB sample to tune the model (i.e., to find the best value for z). Following the point made in the section on training and testing data, if I want an estimate of predictive performance that is independent of the tuning, I should set up the analysis using a three-way split of the data, so that I have a completely held out test set for estimating predictive performance.

A more minor consideration in tuning a random forest model is how many bootstrap replications to use (in other words, how many trees to include in the forest). For both bagging and random forests, using too many replications does not increase overfitting (James et al. 2013)—its only effect is on the time taken to fit the model, which is barely a consideration with current fast computing. In practice, you need enough replications to stabilize the measure of interest—e.g., the test error or the estimate of variable importance. For instance, Fig. 15.6 demonstrates that OOB error for the eel data is quite variable for less than 300 trees and that using many trees (e.g., 600–1,000) reduces the variation without inflating the error. Boulesteix et al. (2012)—who work in bioinformatics and often deal with tens of thousands of predictors—suggest that the number of trees should increase with the number of candidate predictors so that each predictor has an opportunity for selection. Finally, the size of the tree can influence the outcome, and while deep trees are preferred, settings

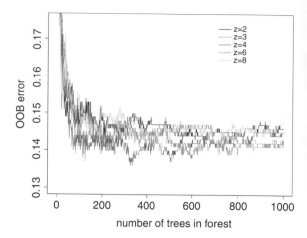

Fig. 15.6. Out of bag error versus number of trees for a random (classification) forest fitted to the data for *Anguilla australis*, the short-finned eel. Five values for number of predictors available at each split (*z*) are shown.

such as the minimum node size for splitting, the number of internal nodes, or a threshold value of the splitting criterion can be used to adjust tree depth (Boulesteix et al. 2012). However, Hastie et al. (2009) advise that using full-grown trees is rarely costly, and avoids the need for tuning this parameter. See James et al. (2013: section 8.3) for sample R code for fitting random forests.

Variable importance can be assessed for any sort of bagged model; here I describe options for random forests and show a worked example in Fig. 15.7. One measure accumulates—for each variable—data on the improvement in the homogeneity of nodes each time that variable is selected into any one of the B trees in the model. Variables that are often included and often improve the purity of splits will thus have higher importance than variables with little influence. Another variable importance measure focuses on prediction error rather than on the impurity measure and uses the OOB samples. For any given tree, the OOB prediction error is recorded. Then the values for each predictor variable in the OOB sample are, in turn, randomly permuted, and the prediction error is again recorded (Fig. 15.7). The increase in error resulting from this permutation is averaged over

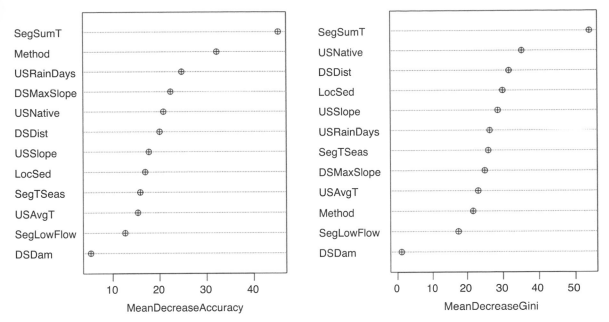

Fig. 15.7. Variable importance in the random forests model fitted to the data for *Anguilla australis,* the short-finned eel, using z = 4 and fitting 600 trees, as measured by prediction error (left) and node impurity as estimated by the Gini index (right). Note that the order of variable importance changes with the measure.

all *B* trees and used as a measure of the importance of the variable. Intuitively, this is relying on the idea that if a variable in its true (non-permuted) state decreases the error substantially compared to when it is permuted, then it is important for prediction. Boulesteix et al. (2012) discuss these two measures and detail situations in which one or the other may be preferred.

The model and its predictive performance can also be understood in other ways. For instance, response shapes can be plotted (Fig. 15.8a) and predictions mapped (Fig. 15.8b); predictions can also be assessed against observations, just as for any modeling method. For the latter, it is particularly important to test predictive performance on held out data rather than training data, since methods like this tend toward overfitting. A quick scan of the internet will turn up all sorts of additional methods for viewing the output, developing error estimates, and so on. The implementation of random forests most often used in ecology is that of the originators (Breiman 2001; and see Cutler et al. [2007] for an introduction

to classification by random forests for ecologists). Other variants of random forests differ in how each individual tree is constructed, how the original dataset is subsampled for training and testing, and how predictions are combined to form a consensus prediction. It has been shown that the standard random forests approach can be biased in regard to which of the z predictor variables it selects at any split, favoring covariates with many possible splits or missing values. The alternative method of *conditional inference forests* (Hothorn et al. 2006, Strobl et al. 2007) overcomes this bias.

In practice, random forests can be used to model typical response types seen in ecology. Many examples exist, including models of the probability of outbreak of spruce beetle in forests of north Colorado (Hart et al. 2014), predicting the conservation status of data deficient species (Bland et al. 2015), and modeling vegetation condition (Kocev et al. 2009). Presence-absence or multiclass observations are modeled with classification trees—so, for instance, the binary response "caught or not caught" for the

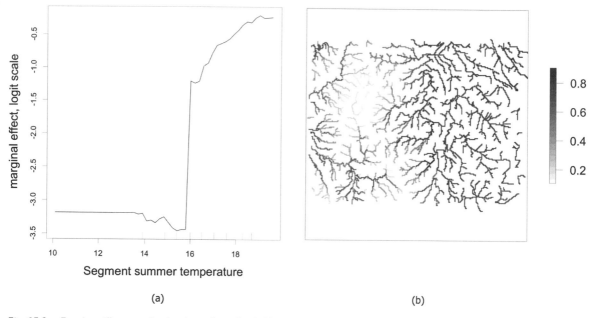

(a)   (b)

Fig. 15.8. For *Anguilla australis*, the short-finned eel, (a) partial dependence plot for the most important variable in the model, showing a preference for warmer river segments. Rug plot along the bottom shows the distribution of sites across the variable, in deciles. (b) Predicted distribution (probability of catch) in some of New Zealand's rivers.

short-finned eel was modeled here with classification trees. The majority vote for each class is predicted (with ties split randomly), along with an estimate of the probability of membership of each class, drawn from the proportions of votes in each class for the full ensemble (Cutler et al. 2007). Regression trees are used for numeric responses. This is straightforward if the dependent variable follows the typical distributional requirements for a model based on least-squares measures of performance—e.g., such that the residuals are normally distributed and have equal variance (i.e., are homoscedastic). Count data usually do not fit these requirements. For these, some transform the response (e.g., using a square-root transformation); others—particularly if there are only a few unique counts in the dataset—use each as a class and treat it as a classification problem. However, it would seem more sensible in the latter case to make use of the ordering of the categories and use the capacity of conditional inference forests to model ordinal data (Hothorn et al. 2006). These methods are actively under development, and so if you want

to model Poisson-distributed data, I suggest searching for new implementations.

Boosted Regression Trees

Boosting is a very different approach to making ensembles. First developed for classification, it has been extended to regression, and—as for many effective machine learning algorithms—numerous variants and implementations exist. A key feature of boosting is that it is sequential or stagewise: what is done in one step builds on what has been done previously, so a model (the base learner) fitted at any one step is influenced by the performance of those built previously. This is different from bagging, which creates a new, independent bootstrap sample at each step, and fits a base learner that is completely ignorant of what happens in other iterations to that sample. Both methods aim to create a set of base learners (models) that are complementary rather than similar. I have already described how random forests does this by using random subsets of the predic tors at each split. Boosting achieves it by focusing, at each step, on observations predicted poorly by previous models. Many implementations of boosting focus on using weights over the observations, with weights changing at each step to focus on (upweight) those observations predicted poorly so far. Often when machine learning practitioners are talking about "boosted trees," they are describing boosted classification trees, using weights. See Witten et al. (2016) for an accessible introduction to typical boosting algorithms.

From here on, I focus on a particular type of boosting—namely, *stochastic gradient boosting*—which is powerful for regression methods, and describe its application to regression trees. Even though the focus is on regression trees, outcomes that are classes can also be modeled with this approach in the same way that regression models can be used for binary or multinomial outcomes. Important insights for using boosting for regression came from Friedman et al. (2000), who interpreted boosting within a likelihood framework, making possible its extension to a broad variety of error distributions (e.g., leading to logistic or Poisson regression). In this framework, boosting can be viewed as a form of functional gradient descent where the aim is to minimize a loss function such as deviance by adding, at each step, a tree that best steps down the gradient of the loss function. The gradient descent is controlled to be slow, because slow learning tends to perform well. Gradient boosting for a continuous normally distributed response follows these general steps (based on James et al. [2013] and Schonlau [2005]):

1. For all i observations in the training data, set initial "model", $\hat{f}(x)$, to zero and compute the residuals r_i.
2. For all following regression trees $b = 1$ to B repeat until a stopping criterion is reached:
   a. Fit a regression tree $\hat{f}_b$ with a fixed number (usually small) of nodes to the residuals;
   b. Update the current model by adding the new tree, also shrinking its contribution:

$$\hat{f}(x) \leftarrow \hat{f}(x) + \lambda \hat{f}_b(x);$$

   c. Update the residuals:
   $$r_i \leftarrow r_i - \lambda \hat{f}_b(x_i).$$
3. Return the final boosted model:
$$\hat{f}(x) = \sum_{b=1}^{B} \lambda \hat{f}_b(x).$$

In other words, any current tree is being built on the residuals of the model (the collection of trees) fitted to date. By fitting to residuals, every new tree is focusing on observations not well predicted so far. This means that the boosting procedure is not only dealing with the variance of individual trees by forming an ensemble, but is also addressing bias by paying attention to poorly modeled observations. The steps above can be generalized beyond continuous normally distributed responses by replacing the zero in step 1 and residuals in all steps by functions of y values and deviance residuals appropriate to the distribution of interest. In the widely used *gbm* R package for fitted gradient boosted trees, eleven options are available at the time of writing, including Gaussian, Bernoulli (= logistic regression), multinomial, Poisson, Cox proportional hazards, and quantile regression.

In keeping with many machine learning methods, boosted regression trees (BRTs) employ several mechanisms to control overfitting and reduce bias and variance. The first two are already mentioned: use shallow trees; shrink the contribution of each tree. *Shallow trees* mean we are dealing with simple models at each step. Hastie et al. (2009: section 10.11) explain details, but in essence the choice of allowed tree complexity is meant to represent the true interaction order in the data. Whilst that is usually unknown, it is expected that low-order interactions are most common, and indeed in some data only main effects are important. An advantage of BRTs over random forests is that BRTs are designed to use shallow trees, and that makes it easy to fit an additive model. Specifically, if only main effects are anticipated, tree "stumps" (2 node, 1 split) can be used in a BRT and the model is simply additive. *Shrinkage* reduces the contribution of each tree to the model, slowing the learning rate of the boosting procedure. Intuitively, slow learning means small steps are taken rather than large ones, meaning that any missteps can more easily be rectified. This sort of shrinkage penalty is a form of regularization and is closely related to L1 penalties such as in lasso regression (Hastie et al. 2009). Finally, one more mechanism for controlling the fit is to use *subsampling* at each step (Friedman 2002), effectively shaking up the data. This makes the model stochastic, so each run is slightly different. Before each new tree is fitted, a random subsample of the residuals is selected without replacement, and this subset is used to fit the next tree. In practice, typical values for these parameters in the model are: tree complexity between 1 and 8 splits, learning rate of 0.001 to 0.05, subsamples of 50 to 75%. Elith et al. (2008) provide extensive discussion of these for ecologists, explaining how to choose appropriate values for any given dataset and giving worked code and examples now updated and available in the R package *dismo*. BRTs are more sensitive than random forests to choice of parameters, so it is important (but not difficult) to learn to tune the model well.

A final consideration is how many trees (how many iterations) to allow in the final model. The number should be optimized with respect to performance on hold-out data (e.g., using *k*-fold cross-validation), and varies with the learning rate. Learning slowly increases the optimal number of trees. While slower learning generally improves predictive performance, it also increases computational time and size of the fitted model object. Elith et al. (2008) recommend fitting more than 1,000 trees, because this implies a learning rate slow enough to temper the variation introduced by the stochasticity of the model.

As for random forests, these models can be understood through variance importance estimates, partial dependence plots (Fig 15.9), and all the usual measures of predictive performance. Buston and Elith (2011) and Elith et al. (2008) show BRT models for the clownfish and short-finned eel data pre-

sented above, demonstrating their utility in providing insights for supervised learning problems. A response function can be built up across the trees, and interactions can be visualized such as for the eel data (Fig. 15.9). Boosted regression trees have been extensively used in ecology for problems as diverse as modeling the zoonotic niche of the Ebola virus (Pigott et al. 2014), explaining species richness across biogeographic scales (Mouchet et al. 2015), and finding the environmental correlates of wildfire (Parisien and Moritz 2009).

Concluding Comments on Pros and Cons

Random forests and boosted regression trees, as examples of ensembles of trees, inherit the advantages of decision trees (ability to fit interactions automatically, handling continuous and categorical predictors, no need for transforming environmental covariates) and overcome the disadvantages, including high variance. Many software packages for random forests and BRTs allow handling of missing data using surrogate variables, but not all—this depends on the underlying tree algorithm that is used. Both random forests and BRTs make model fitting a relatively straightforward task, because the model does not have to be prespecified but can be learned from the data. Faced with a large dataset with potentially many predictors, this is a substantial benefit over classical regression methods (see Chapter 2 of this volume), where model specification can be quite challenging. The extension of BRTs to such a broad range of response types within a coherent regression framework makes the approach very appealing to ecologists. The ease of tuning random forests is similarly appealing. These models have demonstrated strong performance in predicting outcomes, often out-performing classical regression methods (e.g., Elith et al. 2006).

So far I have not discussed confidence intervals, p-values, or parameter estimates. The first can be constructed using resampling approaches, and grad-

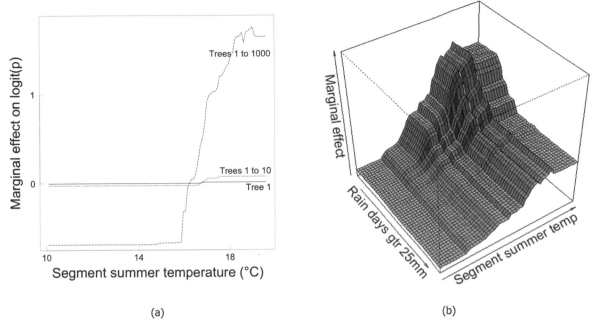

(a)   (b)

Fig. 15.9.   (a) Partial dependence plots from a boosted regression tree model fitted to the data for for *Anguilla australis*, the short-finned eel (using a learning rate of 0.005, tree complexity of 5 splits and 1800 trees in the full ensemble). Three plots are shown for the response to segment summer temperature, for the first tree, for trees 1 to 10, and for trees 1 to 1,000. This illustrates how a response is built up over multiple trees, the splits in each tree forming a small step in the function (for tree 1, hard to see at 16.7 °C) and each tree's contribution shrunk to having a small effect on the response. (b) Visualization of an important interaction in the model, showing the pronounced suitability of environments combining warm temperatures with low frequency of high-intensity rain events in the upstream catchment.

ually more coded examples are becoming available. In many senses these can address the lack of p-values (i.e., explicit estimates of parameter uncertainty) since p-values are quantifying uncertainty around fitted relationships. These models do not provide the parameter estimates available from classical regression models (Chapter 2). The true degrees of freedom of the model is also unknown, so any measure (such as AIC; see Chapter 4, this volume) requiring it cannot be estimated. These issues are usually not problematic (except in the sense that they call for a new way of thinking) since the alternative approaches of estimating training and test error deal with the underlying issue of tuning the model and .estimating predictive performance. More limiting is the fact that hierarchical (structured) data that would usually be fitted in a model such as a generalized lin-

ear mixed model is harder to handle well in these methods. Workarounds such as structuring the cross-validation used to learn the model (e.g., Buston and Elith 2011) are possible but not as satisfying as explicitly modeling the structure.

These methods do fit complex models and cannot be made smooth. Whether this is mostly aesthetically unappealing to ecologists or more deeply troublesome for analysis depends both on attitudes and on the purpose of analysis, as explored in detail in Merow et al. (2014). Sometimes models that reflect trends in data are useful, as can a tendency toward fitting the data well. At other times a smoother model might be much more reliable. In my view these models won't be the right tool in all instances, but they are an extremely useful part of the toolbox of an ecologist and are well worth exploring.

Acknowledgments

The author thanks Peter Buston, Gurutzeta Guillera-Arroita and Rebecca Hutchinson for very helpful comments on the draft and John Leathwick and Trevor Hastie for imparting their understanding of statistical learning methods through long collaborations. Research for this work was supported by the Australian Research Council's Centre of Excellence for Environmental Decisions (CE11001000104).

LITERATURE CITED

Bland, L. M., B. Collen, C. D. L. Orme, and J. Bielby. 2015. Predicting the conservation status of data-deficient species. Conservation Biology 29:250–259.

Boulesteix, A.-L., S. Janitza, J. Kruppa, and I. R. König. 2012. Overview of random forest methodology and practical guidance with emphasis on computational biology and bioinformatics. Wiley Interdisciplinary Reviews: Data Mining and Knowledge Discovery 2:493–507.

Breiman, L. 1996. Bagging predictors. Machine Learning 24:123–140.

Breiman, L. 2001a. Random forests. Machine Learning 45:5–32.

Breiman, L. 2001b. Statistical modeling: The two cultures. Statistical Science 16:199–215.

Breiman, L., J. H. Friedman, R. A. Olshen, and C. J. Stone. 1984. Classification and regression trees. Wadsworth International Group, Belmont, CA.

Buston, P. 2004. Does the presence of non-breeders enhance the fitness of breeders? An experimental analysis in the clown anemonefish *Amphiprion percula*. Behavioral Ecology and Sociobiology 57:23–31.

Buston, P. M. 2002. Group structure of the clown anemonefish *Amphiprion percula*. PhD Dissertation. Cornell University, Ithaca, NY.

Buston, P. M., and J. Elith. 2011. Determinants of reproductive success in dominant pairs of clownfish: A boosted regression tree analysis. Journal of Animal Ecology 80:528–538.

Cutler, D. R., T. C. Edwards, K. H. Beard, A. Cutler, and K. T. Hess. 2007. Random forests for classification in ecology. Ecology 88:2783–2792.

De'ath, G., and K. E. Fabricius. 2000. Classification and regression trees: A powerful yet simple technique for ecological data analysis. Ecology 81: 3178–3192.

Dieterich, T. G. 2000. Ensemble methods in machine learning. In Multiple classifier systems. MCS 2000.

Lecture Notes in Computer Science, vol 1857, pp. 1–15. Springer, Berlin.

Efron, B., and R. J. Tibshirani. 1993. An Introduction to the bootstrap, Chapman and Hall, London.

Elith, J., C. H. Graham, R. P. Anderson, M. Dudík, S. Ferrier, A. Guisan, R. J. Hijmans, F. Huettmann, J. R Leathwick, A. Lehmann, J. Li, L. G. Lohmann, B. A. Loiselle, G. Manion, C. Moritz, M. Nakamura, Y. Nakazawa, J. M. Overton, A. T. Peterson, S. J. Phillips, K. S. Richardson, R. Scachetti-Pereira, R. E. Schapire, J. Soberón, S. Williams, M. S. Wisz, and N. E. Zimmermann. 2006. Novel methods improve prediction of species' distributions from occurrence data. Ecography 29:29–151.

Elith, J., J. E. Leatherwick, and T. Hastie 2008. A working guide to boosted regression trees. Journal of Animal Ecology 77:802–813.

Friedman, J. H. 2002. Stochastic gradient boosting. Computational Statistics and Data Analysis 38:367–378.

Friedman, J. H., T. Hastie, and R. Tibshirani. 2000. Additive logistic regression: A statistical view of boosting. Annals of Statistics 28:337–407.

Guillera-Arroita, G., J. J. Lahoz-Monfort, J. Elith, A. Gordon, H. Kujala, P. E. Lentini, M. A. McCarthy, R. Tingley & B. A. Wintle (2015) Is my species distribution model fit for purpose? Matching data and models to applications. Global Ecology and Biogeography, 24, 276–292.

Hart, S. J., T. T. Veblen, K. S. Eisenhart, D. Jarvis, and D. Kulakowski. 2014. Drought induces spruce beetle (*Dendroctonus rufipennis*) outbreaks across northwestern Colorado. Ecology 95:930–939.

Hastie, T., R. Tibshirani, and J. H. Friedman. 2009. The elements of statistical learning: Data mining, inference, and prediction, 2nd ed. Springer-Verlag, New York.

Hothorn, T., K. Hornik, and A. Zeileis. 2006. Unbiased recursive partitioning: A conditional inference framework. Journal of Computational and Graphical Statistics 15:651–674.

James, G., D. Witten, T. Hastie, and R. Tibshirani. 2013. An introduction to statistical learning with applications in R. Springer, New York.

Kocev, D., S. Džeroski, M. D. White, G. R. Newell, and P. Griffioen. 2009. Using single- and multi-target regression trees and ensembles to model a compound index of vegetation condition. Ecological Modelling 220:1159–1168.

Merow, C., M. J. Smith, T. C. Edwards Jr, A. Guisan, S. M. McMahon, S. Normand, W. Thuiller, R. Wüest, N. E. Zimmermann, and J. Elith. 2014. What do we gain from simplicity versus complexity in species distribution models? Ecography 37:1267–1281.

Mouchet, M., C. Levers, L. Zupan, T. Kuemmerle, C. Plutzar, K. Erb, S. Lavorel, W. Thuiller, and H. Haberl. 2015. Testing the effectiveness of environmental variables to explain European terrestrial vertebrate species richness across biogeographical scales. Plos ONE 10:e0131924.

Parisien, M. A., and M. A. Moritz. 2009 Environmental controls on the distribution of wildfire at multiple spatial scales. Ecological Monographs 79:127–154.

Pigott, D. M., N. Golding, A. Mylne, Z. Huang, A. J. Henry, D. J. Weiss, O. J. Brady, M. U. G. Kraemer, D. L. Smith, C. L. Moyes, S. Bhatt, P. W. Gething, P. W. Horby, I. I. Bogoch, J. S. Brownstein, S. R. Mekaru, A. J. Tatem, K. Khan, and S. I. Hay. 2014. Mapping the zoonotic niche of Ebola virus disease in Africa. eLife, 3:e04395.

Roberts, D.W., V. Bahn, S. Ciuti, M. S. Boyce, J. Elith, G. Guillera-Arroita, S. Hauenstein, J. J. Lahoz-Monfort, B. Schroder, W. Thuiller, D. Warton, B. A. Wintle, F.

Hartig, and C. F. Dormann. 2017. Cross-validation strategies for data with temporal, spatial, hierarchical, or phylogenetic structure. Ecography 40:913–929.

Schonlau, M. 2005. Boosted regression (boosting): An introductory tutorial and a Stata plugin. Stata Journal 5:330–354.

Strobl, C., A. -L. Boulesteix, A. Zeileis, and T. Hothorn. 2007. Bias in random forest variable importance measures: Illustrations, sources and a solution. BMC Bioinformatics 8:25.

Valavi, R., J. Elith, J. J. Lahoz-Monfort, & G. Guillera-Arroita. 2019. blockCV: An r package for generating spatially or environmentally separated folds for k-fold cross-validation of species distribution models. Methods in Ecology and Evolution 10:225–232.

Witten, I. H., E. Frank, M. A. Hall, and C. J. Pal. 2016. Data Mining: Practical machine learning tools and techniques. Morgan Kaufmann. Cambridge, MA, USA.

16 — Causal Modeling and the Role of Expert Knowledge

Bruce G. Marcot

> Nature possesses stable causal mechanisms that, on a detailed level of descriptions, are deterministic functional relationships between variables, some of which are unobservable.
>
> —Pearl (2000:43)

> Time, space, and causality are only metaphors of knowledge, with which we explain things to ourselves.
>
> —Friedrich Nietzsche (quoted in Braezeale 1990)

This chapter addresses the role of expert knowledge in constructing models of wildlife-habitat and stressor relationships and compares objectives and results of guided model creation with machine-learning and statistical model construction for various modeling objectives. A critical look is given to defining expertise and how expert knowledge and experience can be codified and verified. I then discuss how models can be structured from machine learning, from expert knowledge, and from a synthesis of both approaches to ensure credibility and validity of expert knowledge–based models. Next, I address pitfalls and uncertainties in the use of expert knowledge, and the kinds of constructs best used to represent knowledge and expert understanding, including mind mapping, influence diagrams, and Bayesian networks.

Causality as a Concept and a Modeling Construct

What Is Causality in an Ecological Model?

What constitutes causality in an ecological model, and how do we know it when we see it? This seems a trivial question, but trivial it is not. Ecological models are generally constructed from three major sources: directly from empirical data, represented by mathematical or theoretical constructs, or interpreted from expert knowledge and experience. Empirical-based models are typically constructed from a variety of statistical frameworks. Mathematical or theoretical-based models are borne of known or hypothesized analytic relationships. Expert-based models are derived from practical experience and personal expertise.

In all three cases, demonstrating and verifying causality is more challenging than it may appear. For one, empirical-based statistical models do not, and cannot, identify causal relationships between some ecological outcome and affector covariates; statistical models are based essentially on correlations, including even some statistical approaches purported to reveal causality, such as structural equation modeling. Mathematical and theoretical models, like expert-based models, are generally constructed with the assumption of causality, but, again, the definitive evidence of cause still hides in the shadows.

So what is causality in ecological modeling, and how can it be identified, demonstrated, constructed, and verified? When some condition C can be seen to induce or affect some effect E, from a statistical perspective a true causal relationship can be asserted only when all other alternative explanations can be ruled out. This is the intent of clinical trials in medical experimentation, where condition C and its absence *not-C* are assigned randomly, with all other conditions held constant and accounted for, and with such a trial replicated many times over. Such experimental designs are, at best, very difficult to achieve in environmental laboratory conditions, and essentially impossible in natural field conditions with mixes of direct and indirect effects, time-lag effects, variable site histories, and other knots in the causal tapestry.

Take, for example, landscape ecology, where each landscape study area is a sample size of one; we assume away many complicating variables and focus on the assumed proximate causes, that is, the most immediate influence, while the ultimate influences can muddle analysis and result in misinterpretation of true causes. At the very least, we can ponder the nature of hidden and unstudied causes, represented in models as *latent variables*, which are those inferred from the mathematical relation among other observed variables (Rohr et al. 2010; Fig. 16.1). More confusing are *confounding variables* that are simply not measured, or in some cases are unmeasurable, but that nonetheless influence outcomes. Ignoring latent and confounding variables could result in assigning causality to the wrong factors, such as to some variable that correlates both with some outcome of interest and to some other, unobserved variable that is the true cause of the outcome.

Some approaches (e.g., Shipley 2013) purport to utilize statistical techniques to test causal relationships without latent variables and without prior knowledge of how a system might work. Such approaches might be useful for formulating initial hypotheses about causality, but again they cannot definitively determine causal structures in the absence of repeated randomized trials or time-series explorations of relationships in before-after conditions. In attempting to account for effects of latent variables, Guillemette and Larsen (2002) studied factors influencing the abundance, distribution, and behavior of wintering common eiders (*Somateria mollissima*) and removed the confounding variable of prey abundance from their model by randomizing its effect over the study area. Similarly, King et al. (2005) factored out their spatially autocorrelated confounding variables when modeling ecological indicators of watershed land cover. But such approaches serve more to hide away those hidden influences rather than explicitly incorporate them into the modeling structures.

Why Determine Causality?

So what is the importance of determining causality in ecological modeling? In some cases, it may not be a study objective if correlations suffice to provide some degree of descriptive power. However, if explanation is the objective, then causal modeling provides a trustworthy basis for forecasting, projecting, and predicting outcomes. Moreover, identifying causal relations can provide key information on management controllability of a system, such as the degree to which prohibitions on poaching might serve to conserve or restore an at-risk species that is also subject to other environmental stressors. However, Perdicoulis and Glasson (2012) found that environmental impact assessments, at least in the United Kingdom, typically do not explicitly identify causal relations.

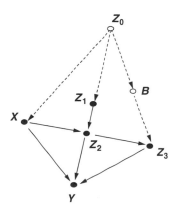

Fig. 16.1. Causal diagram of factors controlling a population of eelworms (Nematoda). Shown are direct effects among measured quantities (solid circles linked by solid arrows) and unmeasured quantities (open circles linked by dotted arrows). The unmeasured quantities represent latent variables. *Source: Pearl (1995).*

Depicting Causality in Ecological Models

Causality can be depicted in ecological models with a variety of constructs. One approach is to model a chain of influences—for example, as Markov processes, where a condition is influenced by other, or prior, conditions just one step away in the sequence. It is then the joint conditional influence of all steps that produce the result. Such a construct is useful if each step depicts a proximate causal relationship.

When conditions are influenced by unknown continuous functions of other factors or with error distributions, they can be depicted with interaction terms as is done in generalized additive models (GAMs) and generalized linear models (GLMs). For example, Hofmeister et al. (2017) used GAMs to explain the causal effects of the size and conditions of vegetation fragments on bird communities in central Europe. From those relationships, the authors surmised the types of timber harvesting that could be more or less detrimental to the birds. But still, their GAMs were based on correlations interpreted as direct causal influences, such as common bird species being most influenced by distance to the forest edge and size and vegetation of the forest fragments. Ando et al. (2017) used GLMs to determine that density of Jezo spruce (*Picea jezoensis* var. *hondoensis*) in Japan was adversely influenced by basal area of nearby ma-

ture Jezo spruce trees and by the amount of cover of the moss *Pleurozium schreberi*. However, as with the previously mentioned study, the GLMs were based on correlations of conditions; the authors inferred causality from the study results.

When conditions are influenced by more than one factor or when multiple factors combine in their causal influence, then some forms of network models can be useful. They can take the form of path regression models (e.g., Fig. 16.2), which denote partial correlations among variables (e.g., Chbouki et al. 2005), network theory models (Upadhyay et al. 2017), and Bayesian probability network models (Borsuk et al. 2006).

Ultimately, ecological modeling is the art of correctly interpreting correlations and simplifying multiple causal influences as cumulative effects. Time-dynamic simulations, as with agent-based simulations or individual movement models, can be useful to represent influences of potential causal factors in ecological systems.

Expert Knowledge as a Basis for Causal Modeling

Finally, expert-based models can be useful constructs for depicting and exploring potential causal structures in ecological systems. Developing models

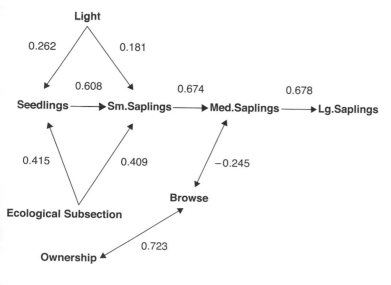

Figure 16.2. Example of a path regression model depicting the strength of causal relationships inferred from the partial correlations (path coefficients). Correlations are depicted on arrows pointing from affectors to response variables of the density of seedling and sapling eastern hemlocks (*Tsuga canadensis*) in conifer-hardwood forests of upper Midwestern United States. Source: Rooney et al. (2000).

from personal experience can be an enticing enterprise, but can also be fraught with many biases, as discussed further below.

Most fundamental to using expert knowledge as a basis for causal modeling is the need to identify what, or who, is an expert? The noun *expert* dates to the early fifteenth century and refers to "a person wise through experience"; the adjective *expert*, which dates to the late fourteenth century and "having had experience; skillful," comes from the Old French *espert*, meaning "experienced, practices, skilled" and from the Latin *expertus*, "tried, proved, known by experience."

Such denotations avoid reference to superficial or pedestrian understanding. Steels (1990) defined "expert knowledge" and "expertise" in terms of the degree to which inference can be made from one's understanding, the depth of that knowledge, and the degree to which such knowledge can be useful for problem-solving methods. Caley et al. (2014) developed a scoring system to rate the degree of expertise in taxonomy based on some 18 factors distilled into descriptions of the person's quality of work and total productivity. Other similar definitions of expertise and expert knowledge exist in the literature.

Here, the focus is on building credible and reliable ecological causal models from expert knowledge, and not from inexpert value judgment or personal opinion. Approaches to collecting and representing expert knowledge can range from single-expert interviews to highly structured expert panels (Ayyub 2001; Ayyub and Klir 2006; Cooke 1991). A rigorous approach to expert paneling is presented farther below.

Determining Causality in Ecological Systems
Causality and Study Designs

No model can tell causality; that is inferred by the researcher from the context of the system being modeled. One must proceed cautiously when interpreting correlation, especially spatial or temporal auto-

correlation, as causation. Well-designed experimental studies that implement management guidelines can go a long way to helping researchers infer—or at least hypothesize—causality and determining the degree to which management affects ecological systems in desired, or undesired, ways; this is the heart of active adaptive management (Gunderson 1999; Williams 2011). Study designs and approaches to adaptive management can run the gamut of some seven types (Marcot 1998)—literature review, expert judgment, demonstration, anecdote, retrospective study, nonexperimental study, and experimental study; the last of these provides the most definitive evidence from which to infer causality, although it is the most difficult to perform in situ.

Analytical Approaches to Determine Causality

In evaluating study results, a variety of analytical approaches can be useful for inferring causality. One such construct is structural equation modeling (SEM) with path analysis (Pearl 2011) and graph-theoretic representation (Grace et al. 2012), which can account for latent variables (Bollen 1989) and indirect effects (Clough 2012). SEM is more of a method for building causal relationships among variables, such as with construction of influence diagrams, than it is a specific analytic structure per se. An approach analogous to SEM is that of Bayesian networks. SEM and Bayesian network modeling share some traits but take different approaches (Table 16.1; Pearl 2000).

Another means of inferring causality is with *d*-separation (*d* is for *dependence* or *directional*), which is a procedure to help determine if two variables are independently conditional on a third variable (Clough 2012). Typically, *d*-separation is used in causal webs, influence diagrams, and probability networks. Also used are hidden Markov models (e.g., Etterson 2013) that can help reveal correlates and potential causal factors in state-path animal movement data. Sugihara et al. (2012) suggested using

Table 16.1. Congruence and isomorphisms between structural equation modeling (SEM) and Bayesian network (BN) modeling.

Structural equation modeling	Bayesian network modeling
EFA (exploratory factor analysis; Ullman 2006)	Induction of naive Bayes networks from data sets
CFA (confirmatory factor analysis) and network induction and updating (Ullman 2006)	Incorporation of case files to update probability tables (e.g., by use of the expectation maximization algorithm)
SEM diagramming	Influence diagramming to denote logical and causal relations among variables based on expert knowledge
Path regression modeling to identify degree of correlation and influence of covariates	Structure-induction algorithms to denote variable relations based on case data sets
Latent variables (unobserved, not directly observed; Ullman 2006)	Latent (hidden, summary) nodes
Counterfactual analysis	Influence analysis (sensu Marcot 2012)
Depiction of uncertainty: error terms	Depiction of uncertainty: posterior probability distributions
Confidence intervals	Credible intervals
Explanatory power of covariates: standardized regression coefficients or partial correlation coefficients in a path regression model	Explanatory power of covariates: sensitivity values of node

hindcasting to measure the degree to which historical records of some presumed causal precondition can reliably estimate some outcome effects, calling this approach "convergent cross mapping".

Still another approach is what is called power probabilistic contrast (PC) theory, which is used more in psychology (Buehner et al. 2003; Cheng 1997) but is generally applicable to inferring causality in any system. This method determines the power of a potential cause c normalized by the influence on the effect e (Collins and Shanks 2006). In general, a main contrast effect is calculated as $\Delta P(e|c) = P(e|c) - P(e|\sim c)$, or the difference between the probability of the effect, given the cause, minus the probability of the effect, not given the cause (the "not-cause"). High main-contrast values suggest a greater degree of causal linkage between c and e.

Using Expert Knowledge in Causal Modeling

Given how tricky—and misleading—it can be to definitively determine proximate, ultimate, and indirect causality in ecological systems, it is no surprise that many models are constructed from expert knowledge. This is no new approach, having been used in modeling and analysis of environmental systems for many years (e.g., O'Keefe et al. 1987). There is a potential dark side, however, to relying on expert knowledge in structuring ecological models, and it is related to numerous pitfalls and uncertainties.

Pitfalls and Uncertainties in Using Expert Knowledge

For one, expertise can be biased in various ways. One set of biases can be characterized in terms of a "psychology of uncertainty," which reflects the ways that people estimate probabilities, frequencies, or implications of events or situations. For example, Balph and Romesburg 1986 addressed the role of observer-expectancy bias as one aspect of systematic error in avian studies. In another example, people tend to more heavily weight more immediate events and costs over future events and costs, even if future events and costs might be far more dire. This is the "immediacy effect" (Gideon and Roelofsma 1995) and explains why we are less concerned with asteroid strikes than with potholes in our streets. It may also play out in emphasizing more proximate causes, such as current weather, over less immediate ultimate or more indirect causes, such as climate change, even if the latter tend to have greater control over the system in question.

Another potential bias in using expert knowledge to structure ecological models pertains to having incomplete experience or, worse, being unaware of having incomplete experience. This is "ignorance of ignorance," sometimes referred to as the "unknown unknowns." In a sense, then, what you don't know *can* hurt you, or at least bias the model. A variant of this bias is the Dunning-Kruger effect that states that people who lack the expertise to perform well are often unaware of this fact (Kruger and Dunning 1999).

A host of other potential biases can arise in group activities such as in expert paneling to derive collective knowledge for structuring ecological models; even group facilitators can hold bias and adversely influence outcomes (Table 16.2). For instance, the emotional state of an expert can taint how he or she recollects experience (Tambini et al. 2017). A rigor-

ous approach to holding expert panels is suggested below.

In summary, expertise can be biased, and expertise can be partial. And expertise is based on personal experience, meaning past or current conditions, not novel future conditions.

Modeling Frameworks for Structuring Expert Knowledge

So what are some useful tools for depicting and structuring expert knowledge? Some modeling frameworks that can help organize thinking are mind maps and cognitive maps, which, at their simplest, are diagrams of variables and their causal, correlational, or logical connections (Lee and Danileiko 2014). Think of a diagram of a food web, which is, in essence, a cognitive map of the trophic structure of an ecosystem (e.g., Fig. 16.3). When parameterized with bioenergetic flow rates, food webs can be useful cognitive maps for exploring the implications of species loss (Zhao et al. 2017) and associated trophic cascades of ecosystems (Canning and Death 2017).

Table 16.2. Potential biases when using expert knowledge to structure an ecological model. These biases pertain to eliciting knowledge from an expert or from expert panels.

Bias	Description
Expert bias: emotional	Unconsciously representing some causal effect to be more or less effective than it actually is, because of expert's mood or attitude toward the subject.
Expert bias: expectation, motivational	Providing an answer expecting that the recipient or user of the answer will misuse the information or will behave in a manner with which the expert does not agree.
Expert bias: lexicon uncertainty	Differing on definitions of key terms.
Expert bias: lack of knowledge parity	Differing in levels of understanding and knowledge of a key topic.
Group bias: anchoring	Adhering to information either recently encountered even if irrelevant.
Group bias: bandwagoning	Everyone on a panel going along with one answer or idea.
Group bias: domineering	Dominating the discussion by a single voice or personality or intimidating others to concede to his or her view.
Cognitive bias: plausibility	Giving a line of thought undue weight because it seems plausible and is thus deemed to be probable.
Facilitator bias: herding	Guiding the group to one idea and downplaying others.
Facilitator bias: charisma	Favoring views of the more charismatic or "big name" experts.
Facilitator bias: last opinion	Favoring the last expressed opinion; also called the "last speaker effect."

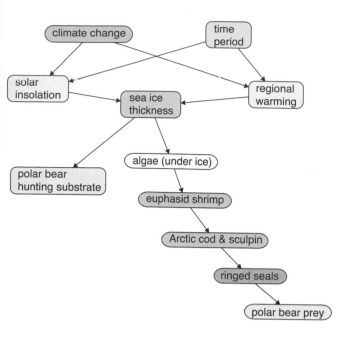

Figure 16.3.   Example of a cognitive map of a polar bear (*Ursus maritimus*) food web in the Arctic.

Vasslides and Jensen (2016) used fuzzy cognitive maps (depicting degrees or probabilities of connections among variables) to model an estuarine system and to compare four stakeholder groups' perceptions of social and ecological factors affecting the system. Similarly, Elsawah et al. (2015) used cognitive maps to help depict mental models of factors affecting viticulture irrigation in South Australia, and as a construct from which to build more quantitative agent-based models (see Chapter 10 of this volume). When parameterized with probabilities of states and interactions, cognitive maps become influence diagrams and the basis of probability-based models such as Bayesian networks.

How Reliable Is Expert Knowledge?

But how reliable is expert knowledge as a basis for constructing and using ecological models? Reliability and credibility are related to the degree to which any such model can be subjected to strict peer review, calibration, updating, and validation.

Ensuring Credibility and Validity of Expert-Based Causal Models

Insofar as possible, cross-validation using portions of independent data sets is the best measure, although with some expert-based models, as with testing stakeholder perceptions (Ozesmi and Ozesmi 2003), empirical case data are not really available or feasible. In such situations, one could compile a case database of the judgments of experts not used to construct the original model, and then test and adjust the model against that database, although this really amounts to calibrating a model against other experts, and not validating the model per se.

Crowd-Sourcing Expert Knowledge

One approach to developing models is using the "wisdom of the crowd" (Lyon and Pacuit 2013), as Eikelboom and Janssen (2017) did in involving stakeholders to address climate-change adaptation planning by using geodesign tools. Much due caution, however, is indicated when accepting opinions or judgments (Koriat 2012), even from multiple experts, without careful vetting and review.

This extends to use of, and cautions for, information derived from "citizen science" projects (Villaseñor et al. 2016). Kosmala et al. (2016) offered a set of criteria for helping ensure the veracity of citizen science monitoring data: iterative project development, volunteer training and testing, expert validation, replication across volunteers, and statistical modeling of systematic error.

Use of Expert Panels

Can the collective wisdom of multiple experts be wrong? (Spoiler alert: yes, but this can be corrected.) An approach to securing reliable input from multiple experts for developing causal ecological models is a rigorous use of expert panels. Following are steps for gathering knowledge from multiple experts in an organized process (after Marcot et al. 2012; also see Ayyub 2001; Burgman 2016; Krueger et al. 2017; and others).

STEP 1. CLEARLY STATE OBJECTIVES

This is self-evident, but it is surprising how often experts are consulted and models are constructed with vaguely stated goals and purposes. Remember, building a causal model for the aim of understanding and managing some ecological system is a method, not an objective. Objectives should be stated in terms of the specific condition to be evaluated and/or managed. For example, an objective could be to determine the viability outcome for some listed species under a suite of possible management activities.

The objectives should also clearly state if the purpose of securing expertise is to develop a single depiction representing all experts' collective knowledge, as with reaching consensus under the traditional Delphi paneling process (MacMillan and Marshall 2006), or if the objective is to depict the variation of knowledge among experts.

STEP 2. IDENTIFY PANELISTS AND PROVIDE PRE-MEETING MATERIALS

A key to holding a successful expert panel is to engage individuals who can work well with others in a panel setting, not try to dominate the panel, and who can "think beyond the data"—that is, who are comfortable extrapolating their experience beyond the boundaries of strict empirical studies. Depending on the project objectives, it may be useful to invite panelists who represent a spectrum of expertise and knowledge, such as from different geographic locations or ecosystems or ecological conditions. Also, it is useful to aim for an uneven number of panelists so that equally-numbered "teams" do not form on some issue. Panels of five or, at most, seven members seem to work best and still allow for independent contributions.

Pre-meeting materials are quite helpful for alerting invited panelists as to the specific objectives for the panel, the type and subject of their knowledge that is sought, how the panel will be held, and what will be done with their knowledge. Materials can include a few background papers or readings. For example, materials may include a status summary of the species of interest, including its biology and ecology, and threats to its viability, and also the set of management activities to be considered when evaluating the species' viability response.

The materials would also include a brief glossary of key terms—for instance, in our example here, definitions of "recovery," "viability," "extirpation," "threat," "stressor," and other terms related to the management alternatives to be considered. The materials should also define the terms used in an information or scoring sheet, such as levels of potential response in a viability rating scale, if the panelists will be asked to score outcomes. The overall aim here is to avoid "lexicon uncertainty," so that everyone uses the same definitions of the terms, and to reach "knowledge parity," so that everyone arriving at the panel has the same background understanding of concepts.

STEP 3. BEGIN PANEL REVIEW OF THE WORKSHEET AND TERMS

Here starts the actual panel meeting, best done in person with a panel facilitator. Typically, exper panels are convened to provide specific information or to rate or rank some alternative conditions, using a text worksheet or some scoring table. The facilitator would begin by reviewing the overall objectives for the panel, the background pre-meeting materials, the worksheet, and key terms. Again, the objective is to ensure that all panelists understand the purpose, context, and terminology the same way and the same degree.

STEP 4. PERFORM INITIAL (ROUND 1) SCORING

If the purpose of the panel is to provide scores of outcomes, such as levels of viability of an at-risk species under a suite of potential management actions, this step entails having the panelists independently and silently write down their scores, encouraged by the facilitator to do independent thinking. As an example, panelists might be asked to allocate 100 points among one or more of possible levels on a viability outcome scale, for each of a set of potential management actions, whereby spreading their points (to sum to 100) among >1 outcome level would represent their degree of uncertainty of outcomes. Panelists may also be asked to write down brief explanations of the type and strength of evidence that led them to denote which outcome would be most likely, and the key uncertainties that may have led them to spread their points among >1 outcome level.

STEP 5. ENGAGE IN STRUCTURED DISCLOSURE AND DISCUSSION

Next, the facilitator would have each panelist in turn reveal his or her scores and explain the evidence and uncertainties he or she considered. Each panelist would have equal time and opportunity for this explanation, uninterrupted. At the end of the disclosures, the panelists would then have an opportunity to ask questions of each other and offer their perspectives. The purpose of the discussion phase is for panelists to learn from each other, not for attempting to convince other panelists of the merits of one's scores and ideas. If the panelists were selected to represent a diversity of experience and knowledge,

then the contribution of every panelist has equal value.

Some expert panels may be held with an audience of others with specific subject knowledge that may be pertinent to the issue at hand, such as with managers or biologists who do not specifically serve on the panel. The facilitator would have had the audience remain silent to this point, but now could open the floor for any brief contributions or clarification questions that the audience members may wish to provide. Again, the aim here is not for audience members to convert panelists' thinking, but to inform for the purpose of mutual learning.

STEP 6. PERFORM SECOND (ROUND 2) SCORING

At this time, the facilitator has the panelists do another round of scoring in which they are free to retain their round 1 scores or update any scores based on what they may have learned from the structured disclosure and discussion from step 5. Again, scoring is to be done silently and independently, if the aim is to collect individual knowledge from each panelist.

STEP 7. ENGAGE IN STRUCTURED DISCLOSURE AND DISCUSSION

The facilitator then engages the panelists in another round of structured disclosure and discussion. As may be necessary, there may then be further rounds of scoring, disclosure, and discussion, but typically two rounds suffice to satisfy panelists' contributions.

STEP 8. REVIEW RESULTS

The facilitator can then quickly compile and review the final scores, and panelists may be given a last opportunity to clarify or explain their contributions. At this point the expert panel procedures are completed.

The overall roles of the panel facilitator are to ensure that panel discussions remain focused on the science; that the panelists adhere to the procedures of scoring, disclosure, and discussion; that the panel is held to the prescribed schedule for completion, that audience members follow such procedures; and that the panelists' information is duly recorded and presented. It may be useful to have a scribe present at an expert panel to record discussions and information not captured in the panelists' worksheets. It is also useful, in the final report of the panel outcome, to acknowledge the panelists' participation but to keep their individual contributions anonymous; this encourages the panelists to speak freely without worry of specific attribution for any statements or contributions that could be misinterpreted or taken out of context.

Following such a rigorous paneling procedure can help ensure, in an efficient and credible manner, that expert knowledge can be garnered to suggest outcomes, reduce areas of uncertainty, identify topics requiring further exploration or study, and to best represent collective knowledge and experience.

Using Expert Knowledge in Causal Modeling

In the end, expert knowledge, such as that gathered through an expert panel or another expert knowledge elicitation approach, can provide the basis for developing cognitive maps, mind maps, influence diagrams, and models representing causal influences in some ecological system. Beyond the initial mining of expert knowledge, it is then the use of peer review that can help ensure reliability, the use of validation to ensure robustness, and the updating of the knowledge base to ensure longevity and utility of the resulting models. The target is to use uncertainty as information to guide and temper management decisions in a risk analysis, risk management, and overall structured decision-making framework (Sloman 2009).

LITERATURE CITED

Ando, Y., Y. Fukasawa, and Y. Oishi. 2017. Interactive effects of wood decomposer fungal activities and bryophytes on spruce seedling regeneration on coarse woody debris. Ecological Research 32(2):173–182.

Ayyub, B. M. 2001. Elicitation of expert opinions for uncertainty and risks. CRC Press, Boca Raton, FL. 328 pp.

Ayyub, B. M., and G. J. Klir. 2006. Uncertainty modeling and analysis in engineering and the sciences. Chapman & Hall/CRC, Boca Raton, FL. 378 pp.

Balph, D. F., and C. Romesburg. 1986. The possible impact of observer bias on some avian research. Auk 103(4):831–832.

Bollen, K. A. 1989. Structural equations with latent variables. John Wiley & Sons, New York. 528 pp.

Borsuk, M. E., P. Reichert, A. Peter, E. Schager, and P. Burkhardt-Holm. 2006. Assessing the decline of brown trout (*Salmo trutta*) in Swiss rivers using a Bayesian probability network. Ecological Modelling 192:224–244.

Breazeale, D., ed. 1990. Philosophy & truth: Selections from Nietzsche's notebooks of the early 1870's. Humanities Press, London. 166 pp.

Buehner, M. J., P. W. Cheng, and D. Clifford. 2003. From covariation to causation: A test of the assumption of causal power. Journal of Experimental Psychology: Learning, Memory, and Cognition 29:1119–1140.

Burgman, M. A. 2016. Trusting judgements: How to get the best out of experts. Cambridge University Press, Cambridge, UK. 203 pp.

Caley, M. J., R. A. O'Leary, R. Fisher, S. Low-Choy, S. Johnson, and K. Mengersen. 2014. What is an expert? A systems perspective on expertise. Ecology and Evolution 4(3):231–242.

Canning, A. D., and R. G. Death. 2017. Trophic cascade direction and flow determine network flow stability. Ecological Modelling 355:18–23.

Chbouki, S., B. Shipley, and A. Bamouh. 2005. Path models for the abscission of reproductive structures in three contrasting cultivars of faba bean (*Vicia faba*). Canadian Journal of Botany 83(3):264–271.

Cheng, P. W. 1997. From covariation to causation: A causal power theory. Psychological Review 104(2):367–405.

Clough, Y. 2012. A generalized approach to modeling and estimating indirect effects in ecology. Ecology 93(8):1809–1815.

Collins, D. J., and D. R. Shanks. 2006. Short article conformity to the power PC theory of causal induction depends on the type of probe question. Quarterly Journal of Experimental Psychology 59(2):225–232.

Cooke, R. M. 1991. Experts in uncertainty: Opinion and subjective probability in science. Oxford University Press, New York.

Eikelboom, T., and R. Janssen. 2017. Collaborative use of geodesign tools to support decision-making on adaptation to climate change. Mitigation and Adaptation Strategies for Global Change 22(2):247–266.

Elsawah, S., J. H. A. Guillaume, T. Filatova, J. Rook, and A. Jakeman. 2015. A methodology for eliciting, representing, and analysing stakeholder knowledge for decision making on complex socio-ecological systems: From cognitive maps to agent-based models. Journal of Environmental Management 151:500–516.

Etterson, M. A. 2013. Hidden Markov models for estimating animal mortality from anthropogenic hazards. Ecological Applications 23(8):1915–1925.

Gideon, K. and P. Roelofsma. 1995. Immediacy and certainty in intertemporal choice. Organizational Behavior and Human Decision Processes 63(3):287–297.

Grace, J. B., D. R. Schoolmaster, G. R. Guntenspergen, A. M. Little, B. R. Mitchell, K. M. Miller, and E. W. Schweiger. 2012. Guidelines for a graph-theoretic implementation of structural equation modeling. Ecosphere 3(8): article 73. http://dx.doi.org/10.1890/ES12-00048.1.

Gunderson, L. 1999. Resilience, flexibility and adaptive management—Antidotes for spurious certitude? Conservation Ecology 3(1):7. http://www.consecol.org/vol3/iss1/art7.

Guillemette, M., and J. K. Larsen. 2002. Postdevelopment experiments to detect anthropogenic disturbances: The case of sea ducks and wind parks. Ecological Applications 12(3):868–877.

Hofmeister, J., J. Hošek, M. Brabec, and R. Kočvara. 2017. Spatial distribution of bird communities in small forest fragments in central Europe in relation to distance to the forest edge, fragment size and type of forest. Forest Ecology and Management 401:255–263.

King, R. S., M. E. Baker, D. F. Whigham, D. E. Weller, T. E. Jordan, P. F. Kazyak, and M. K. Hurd. 2005. Spatial considerations for linking watershed land cover to ecological indicators in streams. Ecological Applications 15(1):137–153.

Koriat, A. 2012. When are two heads better than one and why? Science 336(6079):360–362.

Kosmala, M., A. Wiggins, A. Swanson, and B. Simmons. 2016. Assessing data quality in citizen science. Frontiers in Ecology and Evolution 14(10):551–560.

Krueger, K. L., D. L. Bottom, W. G. Hood, G. E. Johnson, K. K. Jones, and R. M. Thom. 2017. An expert panel process to evaluate habitat restoration actions in the Columbia River estuary. Journal of Environmental Management 188:337–350.

Kruger, J., and D. Dunning. 1999. Unskilled and unaware of it: How difficulties in recognizing one's own incompetence lead to inflated self-assessments. Journal of Personality and Social Psychology 77(6):1121–1134.

Lee, M. D., and I. Danileiko. 2014. Using cognitive models to combine probability estimates. Judgment and Decision Making 9(3):259–273.

Lyon, A., and E. Pacuit. 2013. The wisdom of crowds: Methods of human judgement aggregation. In Hand-

book of Human Computation, edited by Michelucci, pp. 599–614. Springer, New York.

MacMillan, D. C., and K. Marshall. 2006. The Delphi process—An expert-based approach to ecological modelling in data-poor environments. Animal Conservation 9(1):11–20.

Marcot, B. G. 1998. Selecting appropriate statistical procedures and asking the right questions: A synthesis. In Statistical methods for adaptive management studies, edited by V. Sit and B. Taylor, pp. 129–142. B.C. Ministry of Forests Research Branch, Victoria, BC. www.for.gov.bc.ca/hfd/pubs/docs/lmh/lmh42.htm.

Marcot, B. G. 2012. Metrics for evaluating performance and uncertainty of Bayesian network models. Ecological Modelling 230:50–62.

Marcot, B. G., C. Allen, S. Morey, D. Shively, and R. White. 2012. An expert panel approach to assessing potential effects of bull trout reintroduction on federally listed salmonids in the Clackamas River, Oregon. North American Journal of Fisheries Management 32(3):450–465.

O'Keefe, J. H., D. B. Danilewitz, and J. A. Bradshaw. 1987. An "expert system" approach to the assessment of the conservation status of rivers. Biological Conservation 40:69–84.

Ozesmi, U., and S. Ozesmi. 2003. A participatory approach to ecosystem conservation: Fuzzy cognitive maps and stakeholder group analysis in Uluabat Lake, Turkey. Environmental Management 31(4):518–531.

Pearl, J. 1995. Causal diagrams for empirical research. Biometrika 82(4):669–710.

Pearl, J. 2000. Causality: Models, reasoning, and inference. Cambridge University Press, Cambridge, UK.

Pearl, J. 2011. The causal foundations of structural equation modeling. In Handbook of structural equation modeling, edited by R. H. Hoyle, pp. 68–91. Guilford Press, New York.

Perdicoulis, A., and J. Glasson. 2012. How clearly is causality communicated in EIA? Journal of Environmental Assessment Policy and Management 14(3): art. no. 1250020.

Rohr, R. P., H. Scherer, P. Kehrli, C. Mazza, and L.-F. Bersier. 2010. Modeling food webs: Exploring unexplained structure using latent traits. American Naturalist 176(2):170–177.

Rooney, T. P., R. J. McCormick, S. L. Solheim, and D. M. Waller. 2000. Regional variation in recruitment of hemlock seedlings and saplings in the Upper Great Lakes, USA. Ecological Applications 10(4):1119–1132.

Shipley, B. 2013. The AIC model selection method applied to path analytic models compared using a d-separation test. Ecology 94(3):560–564.

Sloman, S. 2009. Causal models: How people think about the world and its alternatives. Oxford University Press 224 pp.

Steels, L. 1990. Components of expertise. AI Magazine 11(2):28–49.

Sugihara, G., R. May, H. Ye, C.-H. Hsieh, E. Deyle, M. Fogarty, and S. Munch. 2012. Detecting causality in complex ecosystems. Science 338:496–500.

Tambini, A., U. Rimmele, E. A. Phelps, and L. Davachi. 2017. Emotional brain states carry over and enhance future memory formation. Nature Neuroscience 20:271–278.

Ullman, J. B. 2006. Structural equation modeling: Reviewing the basics and moving forward. Journal of Personality Assessment 87(1):35–50.

Upadhyay, S., A. Roy, M. Ramprakash, J. Idiculla, A. S. Kumar, and S. Bhattacharya. 2017. A network theoretic study of ecological connectivity in Western Himalayas. Ecological Modelling 359:246–257.

Vasslides, J. M., and O. P. Jensen. 2016. Fuzzy cognitive mapping in support of integrated ecosystem assessments: Developing a shared conceptual model among stakeholders. Journal of Environmental Management 166:348–356.

Villaseñor, E., L. Porter-Bolland, F. Escobar, M. R. Gauriguata, and P. Moreno-Casasola. 2016. Characteristics of participatory monitoring projects and their relationship to decision-making in biological resource management: A review. Biodiversity and Conservation 25(11):2001–2019.

Williams, B. K. 2011. Passive and active adaptive management: Approaches and an example. Journal of Environmental Management 92(5):1371–1378.

Zhao, L., H. Zhang, W. Tian, R. Li, and X. Xu. 2017. Viewing the effects of species loss in complex ecological networks. Mathematical Biosciences 285:55–60.

Andrew N. Tri,
Bruce G. Marcot, and
Leonard A. Brennan

17 — Summary and Synthesis
Looking to the Future

Furious activity is no substitute for analytical thought.
—Sir Alastair Pilkington (quoted in Kenward 1972:429)

It is axiomatic that there will always be a need for the analysis of quantitative data in wildlife science. Statistics and estimation of some sort or another will almost assuredly play a key role in many, but not all areas of data analysis as well. To many wildlife scientists these days, an a-statistical approach to data analysis is anathema, but nevertheless such an approach is possible. Can there *really* be wildlife science without statistics? Guthery et al. (2005) and Rader et al. (2011) provide two rare contemporary examples of peer-reviewed papers that eschewed any kind of P-value or AIC but still drew strong inferences from field data. We argue, however, that without quantification and analysis wildlife science will not progress. The insights of the human mind, rather than the actual statistics or statistical tests themselves, are the real drivers of scientific progress.

While it is the a priori scientific construct that drives how data are collected and ultimately should be analyzed, it is absolutely essential to understand and appreciate that natural history is the foundation of wildlife science (Herman 2002). In addition to a sound understanding of quantitative techniques, a deep appreciation of observational natural history is essential for wildlife science to progress. Without linkages with natural history, all is speculation, and lots of speculation can be plain wrong.

Analytical Guidance

It is our hope as editors that the readers who have made it through this book will find many of the analytical techniques less bewildering and less overwhelming than they seemed before reading this book. If so, we will have done our job. One of our goals in this chapter is to provide a roadmap (Fig. 17.1) that can guide the reader through the landscape of analytical techniques covered in this book.

The appropriate technique for analyzing a particular data set depends on a number of factors such as the type of hypothesis being considered, whether the study is experimental or observational, how the variance in the data set is structured between or among groupings, and so on. Sorting out these issues is best done at the beginning of a study, of course. However, there can also be a role for post-hoc analyses as well. For example, after collecting a string of 34 continuous months of bird foraging data, Brennan et al. (2000) realized that their data could be analyzed using time series analyses, which added considerable insights into the behavioral dynamics they had observed.

Wildlife scientists are studying increasingly complex systems and dynamics as we move farther into the twenty-first century. We live in an analytical age.

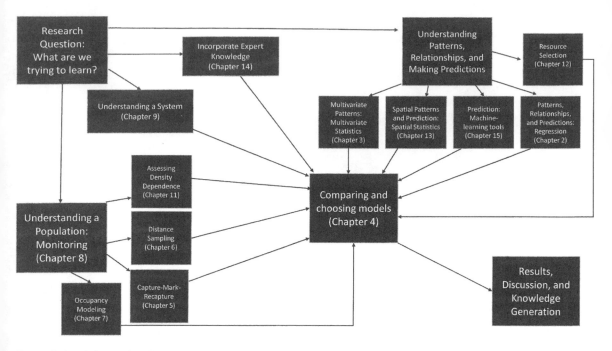

Figure 17.1.   Relationships between and among analytical techniques covered in this book.

Many contemporary problems require the holistic integration of data with both traditional and modern (and often sophisticated) statistical techniques. Analytical techniques, with their complicated assumptions and coding requirements, are often daunting to the new user. The discrepancy between the statistical training offered to graduate students and what is required in the professional world is widening (Touchon and McCoy 2016). Nearly one-third of ecology doctoral programs in the United States do not require a statistics course; of the programs that did require graduate-level statistics, only 6.5% required a course that covered advanced techniques or contemporary statistics (Touchon and McCoy 2016). It is of the utmost importance that early-career professionals have sufficient training to interpret, understand, implement, and critique the complex analysis that is becoming the norm in our field.

The Future

This book could be subtitled "Tools for the Twenty-First Century." We imagine that 50 to 100 years from now, most frequentist analyses—with the exception of regression—will probably be a footnote in the history of wildlife science. There will likely be a renaissance or golden age of long-term data sets and resultant analyses that provide strong inference concerning a myriad of phenomena. We will have a far more rigorous scientific basis for conservation of renewable natural resources than we do today. Techniques such as machine learning (Chapter 15) and causal modeling (Chapter 16) will be common, rather than rare, in a wildlife science graduate student's analytical tool box. Geographic information systems will be frequently used to create automated sampling survey designs, as foretold by Strindberg et al. (2004). "Big data" will allow researchers to find patterns never conceived before, but with caveats about their limitations (Hastie et al. 2015). All electronic devices will be wireless, and nearly all of them will be linked to satellites. There is more than a good chance that we will have a 5-gram radio-transmitter with a GPS-satellite uplink and 6-month battery that can be easily carried by a 150-g bird without impeding flight. Unmanned aerial vehicles will replace fixed-wing and helicopter-based aerial surveys.

New analytical techniques are devised all the time. It may seem hopeless to keep up, but readers should remain undaunted. Beneath the advanced statistical techniques exist the basic fundamentals of science and ecology. The scientific method, good study design, and a critical understanding of natural history are more important than ever for progress in wildlife science.

Sometimes the drive to use the newest and novel statistical method can titillate researchers, causing them to utilize complex analyses when simpler ones will suffice. Hurlbert (1984:208) cautioned that "because an obsessive preoccupation with quantification sometimes coincides, in a reviewer or editor, with a blindness . . . , it is often easier to get a paper published if one uses erroneous statistics than if one uses no statistical analysis at all"; this concern was about pseudo-replication, but the same can be said about running an overly complex analysis that is unnecessary. We sometimes overlook the elegant simplicity of traditional statistics or classical graphics in favor of a complex, sexy analysis with a lot of statistical jargon. Why use an axe when a scalpel will do? Oftentimes, the best analysis is the simplest one that is sufficient to test your research hypothesis and is widely understood by a wide audience.

It is important that we don't abandon a traditional technique in favor of a contemporary one, simply because it is new or novel. Just because a technique is "old" does not mean it is a poor technique. Consider logistic regression: this technique was introduced over a half century ago and is still as useful today as it was then. Conversely, new techniques aren't always a panacea either. Ecological data are messy and rarely conform to the traditional assumptions of frequentist statistics. New methods may require fewer assumptions but are harder to fit and are more data-hungry. New techniques can appropri-

ately handle messy data, but they are not a solution for poor study design, insufficient sample sizes, or unrefined hypotheses.

Volume II?

When we were organizing this book, space limitations resulted in our making hard choices about what topics to include and what topics to leave out. If this book finds a niche and is successful, we hope a companion volume can be developed. Chapters in such a volume could include, in no particular order, analysis of landscape metrics, ecological niche or species distribution modeling, null models, logistic regression, neural networks, time series analyses, cluster analysis, analysis of foraging behavior, population viability analysis, and of course, climate change impacts on wildlife populations and the habitats that support them.

LITERATURE CITED

Brennan, L. A., M. L. Morrison and D. L. Dahlsten. 2000. Comparative foraging dynamics of chestnut-backed and mountain chickadees in the western Sierra Nevada. Northwestern Naturalist 1:129–147.

Guthery, F. S., A. R. Rybak, S. D. Fuhlendorf, T. L. Hiller, S. G. Smith, W. H. Puckett Jr., and R. A. Baker. 2005. Aspects of the thermal ecology of bobwhites in north Texas. Wildlife Monographs 159.

Hastie, T., R. Tibshirani, and M. Wainwright. 2015. Statistical learning with sparsity: the Lasso and generalizations. Monographs on statistics and applied probability 143. CRC Press, Chapman and Hall, 367 pp.

Herman, S. G. 2002. Natural history and wildlife biology: Time for a reunion. Journal of Wildlife Management 66:933–946.

Hurlbert S. H. 1984. Pseudoreplication and the design of ecological field experiments. Ecological Monographs 54:187–211.

Kenward, M. 1972. Reflections on glass research. New Scientist 55:426–429.

Rader, M. J., L. A. Brennan, F. Hernandez, and N. J. Silvy. 2011. Simulating northern bobwhite population responses to nest predation, habitat and weather. Journal of Wildlife Management. 75:582–587.

Strindberg, S., S. T. Buckland, and L. Thomas. 2004. Design of distance sampling surveys and geographic information systems. In Advanced distance sampling: Estimating abundance of biological populations, edited by S. T. Buckland, D. R. Anderson, K. P. Burnham, J. L. Laake, D. L. Borchers, and L. Thomas, pp. 190–228. Oxford University Press, Oxford.

Touchon, J. C., and M. W. McCoy. 2016. The mismatch between current statistical practice and doctoral training in ecology. Ecosphere 7:e01394.

Index